F

S

# CLASSICAL MECHANICS

### 5th Edition

# CLASSICAL MECHANICS

## 5th Edition

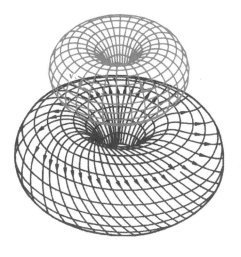

## Tom W.B. Kibble
## Frank H. Berkshire
Imperial College London

Imperial College Press

*Published by*

Imperial College Press
57 Shelton Street
Covent Garden
London WC2H 9HE

*Distributed by*

World Scientific Publishing Co. Pte. Ltd.
5 Toh Tuck Link, Singapore 596224
*USA office:* Suite 202, 1060 Main Street, River Edge, NJ 07661
*UK office:* 57 Shelton Street, Covent Garden, London WC2H 9HE

**Library of Congress Cataloging-in-Publication Data**
Kibble, T. W. B.
    Classical mechanics / Tom W. B. Kibble, Frank H. Berkshire, -- 5th ed.
    p. cm.
    Includes bibliographical references and index.
    ISBN 1860944248 -- ISBN 1860944353 (pbk).
    1. Mechanics, Analytic. I. Berkshire, F. H. (Frank H.). II. Title

    QA805 .K5 2004
    531'.01'515--dc 22                   2004044010

**British Library Cataloguing-in-Publication Data**
A catalogue record for this book is available from the British Library.

*To Anne and Rosie*

# Preface

This book, based on courses given to physics and applied mathematics students at Imperial College, deals with the mechanics of particles and rigid bodies. It is designed for students with some previous acquaintance with the elementary concepts of mechanics, but the book starts from first principles, and little detailed knowledge is assumed. An essential prerequisite is a reasonable familiarity with differential and integral calculus, including partial differentiation. However, no prior knowledge of differential equations should be needed. Vectors are used from the start in the text itself; the necessary definitions and mathematical results are collected in an appendix.

Classical mechanics is a very old subject. Its basic principles have been known since the time of Newton, when they were formulated in the *Principia*, and the mathematical structure reached its mature form with the works of Lagrange in the late eighteenth century and Hamilton in the nineteenth. Remarkably enough, within the last few decades the subject has once again become the focus of very active fundamental research. Some of the most modern mathematical tools have been brought to bear on the problem of understanding the qualitative features of dynamics, in particular the transition between regular and turbulent or chaotic behaviour. The fourth edition of the book was extended to include new chapters providing a brief introduction to this exciting work. In this fifth edition, the material is somewhat expanded, in particular to contrast continuous and discrete behaviours. We have also taken the opportunity to revise the earlier chapters, giving more emphasis to specific examples worked out in more detail.

Many of the most fascinating recent discoveries about the nature of the Earth and its surroundings — particularly since the launching of artificial satellites — are direct applications of classical mechanics. Several of these are discussed in the following chapters, or in problems.

For physicists, however, the real importance of classical mechanics lies not so much in the vast range of its applications as in its role as the base on which the whole pyramid of modern physics has been erected. This book, therefore, emphasizes those aspects of the subject which are of importance in quantum mechanics and relativity — particularly the conservation laws, which in one form or another play a central role in all physical theories.

For applied mathematicians, the methods of classical mechanics have evolved into a much broader theory of dynamical systems with many applications well outside of physics, for example to biological systems.

The first five chapters are primarily concerned with the mechanics of a single particle, and Chapter 6, which could be omitted without substantially affecting the remaining chapters, deals with potential theory. Systems of particles are discussed in Chapters 7 and 8, and rigid bodies in Chapter 9. The powerful methods of Lagrange are introduced at an early stage, and in simple contexts, and developed more fully in Chapters 10 and 11. Chapter 12 contains a discussion of Hamiltonian mechanics, emphasizing the relationship between symmetries and conservation laws — a subject directly relevant to the most modern developments in physics. It also provides the basis for the later treatment of order and chaos in Hamiltonian systems in Chapter 14. This follows the introduction to the geometrical description of continuous dynamical systems in Chapter 13, which includes a discussion of various non-mechanics applications. Appendices from the fourth edition on (A) Vectors, (B) Conics, (C) Phase-plane analysis near critical points are supplemented here by a new appendix (D) Discrete dynamical systems — maps.

The writing of the first edition of this book owed much to the advice and encouragement of the late Professor P.T. Matthews. Several readers have also helped by pointing out errors, particularly in the answers to problems.

We are grateful to the following for permission to reproduce copyright material: Springer–Verlag and Professor Oscar E. Lanford III for Fig. 13.20; Cambridge University Press for Fig. 13.22 and, together with Professors G.L. Baker and J.P. Gollub, for Fig. D.5; Institute of Physics Publishing Limited and Professor M.V. Berry for Figs. 14.11, 14.12 and 14.13; the Royal Society and Professors W.P. Reinhardt and I. Dana for the figure in the answer to Appendix D, Problem 13.

T.W.B. Kibble
F.H. Berkshire
*Imperial College London*

# Contents

# Useful Constants and Units

**Physical constants**

| | |
|---|---|
| speed of light | $c = 2.99792458 \times 10^8 \, \mathrm{m\,s^{-1}}$ |
| gravitational constant | $G = 6.673 \times 10^{-11} \, \mathrm{N\,m^2\,kg^{-2}}$ |
| Planck's constant$/2\pi$ | $\hbar = 1.05457 \times 10^{-34} \, \mathrm{J\,s}$ |
| mass of hydrogen atom | $m_\mathrm{H} = 1.67352 \times 10^{-27} \, \mathrm{kg}$ |
| mass of electron | $m_\mathrm{e} = 9.10938 \times 10^{-31} \, \mathrm{kg}$ |
| charge of proton | $e = 1.60218 \times 10^{-19} \, \mathrm{C}$ |
| permittivity of vacuum | $\epsilon_0 = 8.85419 \times 10^{-12} \, \mathrm{F\,m^{-1}}$ |
| permeability of vacuum | $\mu_0 = 4\pi \times 10^{-7} \, \mathrm{H\,m^{-1}}$ |

**Defined standard values**

| | |
|---|---|
| standard gravitational acceleration | $g_\mathrm{n} = 9.80665 \, \mathrm{m\,s^{-2}}$ |
| normal atmospheric pressure | $1\,\mathrm{atm} = 1.01325 \times 10^5 \, \mathrm{Pa}$ |
| | $= 1.01325 \, \mathrm{bar}$ |

**Properties of Earth**

| | |
|---|---|
| mass | $M = 5.974 \times 10^{24} \, \mathrm{kg}$ |
| | $GM = 3.9860 \times 10^{14} \, \mathrm{m^3\,s^{-2}}$ |
| radius (polar) | $R_\mathrm{p} = 6356.8 \, \mathrm{km}$ |
| (equatorial) | $R_\mathrm{e} = 6378.1 \, \mathrm{km}$ |
| (mean)$[= (R_\mathrm{e}^2 R_\mathrm{p})^{1/3}]$ | $R = 6371.0 \, \mathrm{km}$ |
| semi-major axis of orbit | $a = 1.49598 \times 10^8 \, \mathrm{km}$ |
| eccentricity of orbit | $e = 0.016722$ |
| orbital period (sidereal year) | $\tau = 3.15575 \times 10^7 \, \mathrm{s}$ |
| mean orbital velocity | $v = 29.785 \, \mathrm{km\,s^{-1}}$ |
| surface escape velocity (mean) | $v_\mathrm{e} = 11.18 \, \mathrm{km\,s^{-1}}$ |
| rotational angular velocity | $\omega = 7.2921 \times 10^{-5} \, \mathrm{s^{-1}}$ |

## Properties of Sun and Moon

mass of Sun                                    $M_S = 1.989 \times 10^{30}\,\mathrm{kg}$
                                               $= 3.329\,5 \times 10^5\,M$
                                               $GM_S = 1.327\,12 \times 10^{20}\,\mathrm{m^3\,s^{-2}}$
mass of Moon                                   $M_M = 7.348 \times 10^{22}\,\mathrm{kg}$
                                               $= 0.012\,300\,M$
semi-major axis of lunar orbit                 $a_M = 3.8440 \times 10^5\,\mathrm{km}$
lunar orbital period (sidereal month)          $\tau_M = 2.3606 \times 10^6\,\mathrm{s}$

## SI units

*kilogramme* (kg): mass of the international standard kilogramme kept at Sèvres in France.

*second* (s): 9 192 631 770 oscillation periods of the hyperfine transition between the levels $F = 4, m_F = 0$ and $F = 3, m_F = 0$ in the ground state of $^{133}$Cs.

*metre* (m): distance travelled by light in vacuum in $(1/299\,792\,458)\,\mathrm{s}$.

*ampere* (A): defined so that the force per unit length between two infinitely long parallel wires of negligible cross section 1 m apart in vacuum, each carrying a current of 1 A is $2 \times 10^{-7}\,\mathrm{N\,m^{-1}}$ (or, equivalently, so that the constant $\mu_0$ has the precise value $4\pi \times 10^{-7}\,\mathrm{N\,A^{-2}}$).

(The *kelvin*, *candela* and *mole* are not used in this book.)

## Subsidiary units

| | | | | |
|---|---|---|---|---|
| *newton* | $1\,\mathrm{N} = 1\,\mathrm{kg\,m\,s^{-2}}$ | *pascal* | $1\,\mathrm{Pa} = 1\,\mathrm{N\,m^{-2}}$ |
| *joule* | $1\,\mathrm{J} = 1\,\mathrm{N\,m}$ | *ton(ne)* | $1\,\mathrm{t} = 10^3\,\mathrm{kg}$ |
| *watt* | $1\,\mathrm{W} = 1\,\mathrm{J\,s^{-1}}$ | *bar* | $1\,\mathrm{bar} = 10^5\,\mathrm{Pa}$ |
| *coulomb* | $1\,\mathrm{C} = 1\,\mathrm{A\,s}$ | *hertz* | $1\,\mathrm{Hz} = 1\,\mathrm{s^{-1}}$ |

## British and American units

| | | | |
|---|---|---|---|
| *foot* | $1\,\mathrm{ft} = 0.3048\,\mathrm{m}$ | *pound* | $1\,\mathrm{lb} = 0.452\,592\,37\,\mathrm{kg}$ |

## Prefixes denoting multiples and submultiples

| | | | |
|---|---|---|---|
| $10^3$ | kilo (k) | $10^{-3}$ | milli (m) |
| $10^6$ | mega (M) | $10^{-6}$ | micro ($\mu$) |
| $10^9$ | giga (G) | $10^{-9}$ | nano (n) |
| $10^{12}$ | tera (T) | $10^{-12}$ | pico (p) |
| $10^{15}$ | peta (P) | $10^{-15}$ | femto (f) |
| $10^{18}$ | exa (E) | $10^{-18}$ | atto (a) |

# List of Symbols

The following list is not intended to be exhaustive. It includes symbols of frequent occurrence or special importance. The figures refer to the section in which the symbol is defined.

| | | |
|---|---|---|
| $A, A_\alpha$ | complex amplitude | 2.3, 11.3 |
| $\boldsymbol{A}$ | vector potential | A.7 |
| $\boldsymbol{a}, a$ | acceleration | 1.2 |
| $a$ | amplitude of oscillation | 2.2 |
| $a$ | semi-major axis of orbit | 4.4, B.1 |
| $a$ | equatorial radius | 6.5 |
| $\boldsymbol{B}$ | magnetic field | 5.2, A.7 |
| $b$ | semi-minor axis of orbit | 4.4, B.1 |
| $b$ | impact parameter | 4.3, 4.5 |
| $c$ | polar radius | 6.5 |
| $c$ | propagation velocity | 10.6 |
| $\boldsymbol{c}$ | control parameters | 13.1 |
| $\boldsymbol{d}$ | dipole moment | 6.2, 6.4 |
| $E$ | total energy | 2.1, 3.1, 8.5 |
| $\boldsymbol{E}$ | electric field | 6.1, A.7 |
| e | base of natural logarithms | |
| $e$ | (minus) electron charge | 4.7 |
| $e$ | eccentricity of orbit | 4.4, B.2 |
| $\boldsymbol{e}_1, \boldsymbol{e}_2, \boldsymbol{e}_3$ | principal axes | 9.4, A.10 |
| $\boldsymbol{F}, \boldsymbol{F}_i, \boldsymbol{F}_{ij}$ | force | 1.2, 8.1 |
| $F_\alpha$ | generalized force | 3.7, 10.2 |
| $\boldsymbol{F}$ | phase-space velocity | 13.1 |
| $F_i$ | constants of the motion | 14.1 |
| $f$ | particle flux | 4.5 |

| | | |
|---|---|---|
| $G$ | gravitational constant   . | 1.2 |
| $\boldsymbol{G}$ | moment of force | 3.3 |
| $G$ | generator of transformation | 12.7 |
| $g, \boldsymbol{g}$ | gravitational acceleration | 5.3, 6.1 |
| $\boldsymbol{g}^*$ | observed gravitational acceleration | 5.3 |
| $H$ | Hamiltonian function | 12.1 |
| $I$ | action integral | 3.7 |
| $I, I_{xx}$ | moment of inertia | 9.2, 9.3 |
| $I_{xy}$ | product of inertia | 9.3 |
| $I_i, I_i^*$ | principal moment of inertia | 9.4, 9.5 |
| $I_i$ | action variables | 14.3 |
| i | $\sqrt{-1}$ | |
| $i$ | particle index | 1.2, 8.1 |
| $\boldsymbol{i}, \boldsymbol{j}, \boldsymbol{k}$ | unit vectors | A.1, 1.1 |
| $\boldsymbol{J}, \boldsymbol{J}^*$ | angular momentum | 3.3, 7.1, 8.3 |
| $k, k_{\alpha\beta}$ | oscillator constant | 2.2, 11.2 |
| $k$ | inverse square law constant | 4.3 |
| $L$ | Lagrangian function | 3.7, 10.2 |
| $l$ | semi-latus rectum of orbit | 4.3, 4.4, B.2 |
| $\ln$ | natural logarithm | |
| $M$ | total mass | 7.1, 8.1 |
| $M$ | Jacobian matrix | 14.1 |
| $\boldsymbol{M}$ | invariant torus | 14.1 |
| $m, m_i$ | mass | 1.2, 6.1 |
| $N$ | number of particles | 1.2, 8.1 |
| $n$ | number of degrees of freedom | 10.1 |
| $\boldsymbol{P}$ | total momentum | 7.1, 8.1 |
| $\boldsymbol{p}, \boldsymbol{p}_i, \boldsymbol{p}^*$ | momentum | 1.2, 7.2 |
| $p_i, p_\alpha$ | generalized momentum | 3.7, 12.1 |
| $p$ | exponential rate coefficient | 2.2 |
| $Q$ | quality factor | 2.5 |
| $Q$ | quadrupole moment | 6.2, 6.4 |
| $q, q_i$ | electric charge | 1.2, 6.1 |
| $q_i$ | curvilinear co-ordinate | 3.7, A.8 |
| $q_\alpha$ | generalized co-ordinate | 10.1 |
| $R$ | radius of Earth | 4.3 |
| $\boldsymbol{R}$ | position of centre of mass | 7.1, 8.1 |
| $r$ | radial co-ordinate | 3.5 |

| | | |
|---|---|---|
| $r, r_i$ | position vector | 1.1, A.1 |
| $r, r_{ij}$ | relative position | 1.2, 7.1 |
| $r_i^*$ | position in CM frame | 7.2, 8.3 |
| $T, T^*$ | kinetic energy | 2.1, 3.1, 7.2, 8.5 |
| $T, T_\epsilon$ | twist map | D.3 |
| $t$ | time | 1.1 |
| $U$ | effective potential energy function | 4.2, 12.4 |
| $V$ | potential energy | 2.1, 3.1, 8.5, 13.3 |
| $v, v_i$ | velocity | 1.2 |
| $dW, \delta W$ | work done in small displacement | 2.4, 10.2 |
| $X, Y, Z$ | co-ordinates of centre of mass $R$ | 10.1 |
| $X, Y$ | fixed points of maps | D.1 |
| $x, y, z$ | Cartesian co-ordinates | A.1, 1.1 |
| $x, x_i, x, y$ | phase-space co-ordinates | 13.1, 13.2 |
| $x_n, y_n$ | iterates of maps | D.1 |
| $\alpha$ | rotation number | D.3 |
| $\gamma$ | damping constant | 2.5 |
| $\delta$ | small variation | 3.6 |
| $\delta$ | Feigenbaum number | D.1 |
| $\epsilon$ | oblateness | 6.5 |
| $\epsilon_0$ | permittivity of free space | 1.2, A.7 |
| $\theta$ | polar angle, Euler angle | 3.5, 9.9 |
| $\Theta, \theta, \theta^*$ | scattering angle | 4.4, 7.3 |
| $\lambda, \lambda_i$ | Lyapunov exponent | 13.7, D.1, D.2 |
| $\lambda_i$ | eigenvalues | C.1 |
| $\mu$ | reduced mass | 7.1 |
| $\mu_0$ | permeability of free space | A.7 |
| $\xi, \xi, \eta$ | linearized phase-space co-ordinates | 13.2, C.1 |
| $\pi$ | circumference/diameter ratio of circle | |
| $\rho$ | cylindrical polar co-ordinate | 3.5 |
| $\rho$ | mass or charge density | 6.3, A.7 |
| $\sigma, d\sigma$ | cross-section | 4.5 |
| $\tau$ | period | 2.2, 4.4 |
| $\Phi$ | gravitational potential | 6.1 |
| $\phi$ | electrostatic potential | 6.1, A.7 |
| $\varphi$ | azimuth angle, Euler angle | 3.5, 9.9 |
| $\phi_i$ | angle variables | 14.3 |
| $\psi$ | Euler angle | 9.9 |

| | | |
|---|---|---|
| $d\Omega$ | solid angle | 4.5 |
| $\boldsymbol{\omega}, \boldsymbol{\Omega}$ | angular velocity | 5.1, 9.2, 9.7 |
| $\omega$ | angular frequency | 2.2, 11.3 |
| $\omega_i$ | natural angular frequencies | 14.1 |
| $\boldsymbol{\nabla}$ | vector differential operator | A.5 |
| $\boldsymbol{a} \cdot \boldsymbol{b}$ | scalar product of $\boldsymbol{a}$ and $\boldsymbol{b}$ | A.2 |
| $\boldsymbol{a} \wedge \boldsymbol{b}$ | vector product of $\boldsymbol{a}$ and $\boldsymbol{b}$ | A.3 |

# Chapter 1

# Introduction

Classical mechanics is one of the most familiar of scientific theories. Its basic concepts — mass, acceleration, force, and so on — have become very much a part of our everyday modes of thought. So we may easily regard their physical meaning as more obvious than it really is. For this reason, a large part of this introductory chapter will be devoted to a critical examination of the fundamental concepts and principles of mechanics.

Every scientific theory starts from a set of hypotheses, which are suggested by our observations, but represent an idealization of them. The theory is then tested by checking the predictions deduced from these hypotheses against experiment. When persistent discrepancies are found, we try to modify the hypotheses to restore the agreement with observation. If many such tests are made and no serious disagreement emerges, then the hypotheses gradually acquire the status of 'laws of nature'. When results that apparently contradict well-established laws appear, as they often do, we tend to look for other possible explanations — for simplifying assumptions we have made that may be wrong, or neglected effects that may be significant.

It must be remembered however that, no matter how impressive the evidence may be, we can never claim for these laws a universal validity. We may only be confident that they provide a good description of that class of phenomena for which their predictions have been adequately tested. One of the earliest examples is provided by Euclid's axioms. On any ordinary scale, they are unquestionably valid, but we are not entitled to assume that they should necessarily apply on either a cosmological or a sub-microscopic scale. Indeed, they have been modified in Einstein's theory of gravitation ('general relativity').

The laws of classical mechanics are no exception. Since they were first formulated by Galileo and by Newton in his *Principia*, their range of known validity has been enormously extended, but in two directions they have been found to be inadequate. For the description of the small-scale phenomena of atomic and nuclear physics, classical mechanics has been superseded by quantum mechanics, and for phenomena involving speeds approaching that of light, by relativity.

This is not to say that classical mechanics has lost its value. Indeed both quantum mechanics and the special and general theories of relativity are extensions of classical mechanics in the sense that they reproduce its results in appropriate limiting cases. Thus the fact that these theories have been confirmed actually reinforces our belief in the correctness of classical mechanics within its own vast range of validity. Indeed, it is a remarkably successful theory, which provides a coherent and satisfying account of phenomena as diverse as the planetary orbits, the tides and the motion of a gyroscope. Moreover, even outside this range, many of the results of classical mechanics still apply. In particular, the conservation laws of energy, momentum and angular momentum are, so far as we yet know, of universal validity.

## 1.1   Space and Time

The most fundamental assumptions of physics are probably those concerned with the concepts of space and time. We assume that space and time are continuous, that it is meaningful to say that an event occurred at a specific point in space and a specific instant of time, and that there are universal standards of length and time (in the sense that observers in different places and at different times can make meaningful comparisons of their measurements). These assumptions are common to the whole of physics, and, though all are being challenged, there is as yet no compelling evidence that we have reached the limits of their range of validity.

In 'classical' physics, we assume further that there is a universal time scale (in the sense that two observers who have synchronized their clocks will always agree about the time of any event), that the geometry of space is Euclidean, and that there is no limit in principle to the accuracy with which we can measure all positions and velocities. These assumptions have been somewhat modified in quantum mechanics and relativity. Here, however, we shall take them for granted, and concentrate our attention on the more specific assumptions of classical mechanics.

## The relativity principle

In Aristotle's conception of the universe, the fact that heavy bodies fall downwards was explained by supposing that each element (earth, air, fire, water) has its own appointed sphere, to which it tends to return unless forcibly prevented from so doing. The element *earth*, in particular, tends to get as close as it can to the centre of the Universe, and therefore forms a sphere about this point. In this kind of explanation, the central point plays a special, distinguished role, and position in space has an absolute meaning.

In Newtonian mechanics, on the other hand, bodies fall downward because they are attracted towards the *Earth*, rather than towards some fixed point in space. Thus position has a meaning only relative to the Earth, or to some other body. In just the same way, velocity has only a relative significance. Given two bodies moving with constant relative velocity, it is impossible in principle to decide which of them is at rest, and which moving. This statement, which is of fundamental importance, is the *principle of relativity*.

Acceleration, however, still retains an absolute meaning, since it is experimentally possible to distinguish between motion with uniform velocity (*i.e.*, constant in magnitude and direction) and accelerated motion. If we are sitting inside an aircraft, we can easily detect its acceleration, but we cannot measure its velocity — though by looking out we can estimate its velocity *relative* to objects outside. (In Einstein's theory of general relativity, even acceleration becomes a relative concept, at least on a small scale. This is made possible by the fact that, to an observer in a confined region of space, the effects of being accelerated and of being in a gravitational field are indistinguishable.)

If two unaccelerated observers perform the same experiment, they must arrive at the same results. It makes no difference whether it is performed on the ground or in a smoothly travelling vehicle. However, an accelerated observer who performs the experiment may well get a different answer. The relativity principle asserts that all unaccelerated observers are equivalent; it says nothing about accelerated observers.

## Inertial frames

It is useful at this point to introduce the concept of a frame of reference. To specify positions and time, each observer may choose a zero of the time scale, an origin in space, and a set of three Cartesian co-ordinate axes. We shall refer to these collectively as a *frame of reference*. The position and time of any event may then be specified with respect to this frame by the three Cartesian co-ordinates $x, y, z$ and the time $t$. If we are located on a solid body, such as the Earth, we may, for example, choose some point of the body as the origin, and take the axes to be rigidly fixed to it (though, as we discuss later, this frame is not quite unaccelerated).

In view of the relativity principle, the frames of reference used by different unaccelerated observers are completely equivalent. The laws of physics expressed in terms of our $x, y, z, t$ must be identical with those of someone else's $x', y', z', t'$. They are not, however, identical with the laws expressed in terms of the co-ordinates used by an accelerated observer. The frames used by unaccelerated observers are called *inertial* frames.

We have not yet said how we can tell whether a given observer is unaccelerated. We need a criterion to distinguish inertial frames from the others. Formally, an inertial frame may be defined to be one with respect to which any isolated body, far removed from all other matter, would move with uniform velocity. This is of course an idealized definition, since in practice we can never get infinitely far away from other matter. For all practical purposes, however, an inertial frame is one whose orientation is fixed relative to the 'fixed' stars, and in which the Sun (or more precisely the centre of mass of the solar system) moves with uniform velocity. It is an essential assumption of classical mechanics that such frames exist. Indeed, this assumption (together with a definition of inertial frames) is the real physical content of *Newton's first law* (a body acted on by no forces moves with uniform velocity in a straight line).

It is generally convenient to use only inertial frames, but there is no necessity to do so. Sometimes it proves convenient to use a non-inertial (in particular, rotating) frame, in which the laws of mechanics take on a more complicated form. For example, we shall discuss in Chapter 5 the use a frame rigidly fixed to the rotating Earth.

### *Vectors*

It is often convenient to use a notation which does not refer explicitly to a particular set of co-ordinate axes. Instead of using Cartesian co-ordinates $x, y, z$ we may specify the position of a point $P$ with respect to a given origin $O$ by the length and direction of the line $OP$. A quantity which is specified by a magnitude and a direction is called a *vector*; in this case the *position vector* $\boldsymbol{r}$ of $P$ with respect to $O$. Many other physical quantities are also vectors: examples are velocity and force. They are to be distinguished from *scalars* — like mass and energy — which are completely specified by a magnitude alone.

We assume here that readers are familiar with the ideas of vector algebra; if not, they will find a discussion which includes all the results we shall need in Appendix A.

Throughout this book, vectors will be denoted by boldface letters (like $\boldsymbol{a}$). The magnitude of the vector will be denoted by the corresponding letter in ordinary italic type ($a$), or by the use of vertical bars ($|\boldsymbol{a}|$). The scalar and vector products of two vectors $\boldsymbol{a}$ and $\boldsymbol{b}$ will be written $\boldsymbol{a} \cdot \boldsymbol{b}$ and $\boldsymbol{a} \wedge \boldsymbol{b}$ respectively. We shall use $\hat{\boldsymbol{r}}$ to denote the unit vector in the direction of $\boldsymbol{r}$, $\hat{\boldsymbol{r}} = \boldsymbol{r}/r$. The unit vectors along the $x$-, $y$-, $z$-axes will be denoted by $\boldsymbol{i}$, $\boldsymbol{j}$, $\boldsymbol{k}$, so that

$$\boldsymbol{r} = x\boldsymbol{i} + y\boldsymbol{j} + z\boldsymbol{k}.$$

We shall use the vector notation in formulating the basic laws of mechanics, both because of the mathematical simplicity thereby attained, and because the physical ideas behind the mathematical formalism are often much clearer in terms of vectors.

## 1.2 Newton's Laws

Classical mechanics describes how physical objects move, how their positions change with time. Its basic laws may be applied to objects of any size (above the atomic level) and of any shape and internal structure, and, in classical hydrodynamics, to fluids too. It is not immediately obvious, however, what is meant by the 'position' of a large object of complex shape. Only in the idealized case of point particles (which do not exist in nature) does this concept have an intuitively obvious meaning. We shall therefore consider first only small bodies which can be effectively located at a point, so that the position of each, at time $t$, can be specified by a position vector $\boldsymbol{r}(t)$.

When we come to deal with large extended bodies, in Chapter 8, we shall make the additional assumption that any such body may be divided up into a large number of very small bodies, each of which may be treated as a point particle. (We shall also need to make some assumptions about the nature of the internal forces between these particles.) We shall then find that if we are interested in the overall motion of even a very large object, such as a planet, we may often legitimately treat it as a point particle located at the *centre of mass* of the body. The laws themselves prescribe the meaning of the 'position' of an extended body.

We shall begin by simply stating Newton's laws, and defer to the following section a discussion of the physical significance of the concepts involved, particularly those of *mass* and *force*.

Let us consider an isolated system comprising $N$ bodies, which we label by an index $i = 1, 2, \ldots, N$. By saying that the system is *isolated*, we mean that all other bodies are sufficiently remote to have a negligible influence on it. Each of the $N$ bodies is assumed to be small enough to be treated as a point particle. The position of the $i$th body with respect to a given inertial frame will be denoted by $\boldsymbol{r}_i(t)$. Its velocity and acceleration are

$$\boldsymbol{v}_i(t) = \dot{\boldsymbol{r}}_i(t),$$
$$\boldsymbol{a}_i(t) = \dot{\boldsymbol{v}}_i(t) = \ddot{\boldsymbol{r}}_i(t),$$

where the dots denote differentiation with respect to the time $t$. For example

$$\dot{\boldsymbol{r}} \equiv \frac{\mathrm{d}\boldsymbol{r}}{\mathrm{d}t}.$$

Each body is characterized by a scalar constant, its *mass* $m_i$. Its momentum $\boldsymbol{p}_i$ is defined to be mass × velocity:

$$\boldsymbol{p}_i = m_i \boldsymbol{v}_i.$$

The equation of motion, which specifies how the body will move is *Newton's second law* (mass × acceleration = force):

$$\dot{\boldsymbol{p}}_i = m_i \boldsymbol{a}_i = \boldsymbol{F}_i, \tag{1.1}$$

where $\boldsymbol{F}_i$ is the total force acting on the body. This force is composed of a sum of forces due to each of the other bodies in the system. If we denote the force *on* the $i$th body *due to* the $j$th body by $\boldsymbol{F}_{ij}$, then

$$\boldsymbol{F}_i = \boldsymbol{F}_{i1} + \boldsymbol{F}_{i2} + \cdots + \boldsymbol{F}_{iN} = \sum_{j=1}^{N} \boldsymbol{F}_{ij}, \tag{1.2}$$

where of course $\boldsymbol{F}_{ii} = \boldsymbol{0}$, since there is no force on the $i$th body due to itself. Note that since the sum on the right side of (1.2) is a vector sum, this equation incorporates the 'parallelogram law' of composition of forces.

The two-body forces $\boldsymbol{F}_{ij}$ must satisfy *Newton's third law*, which asserts that 'action' and 'reaction' are equal and opposite,

$$\boldsymbol{F}_{ji} = -\boldsymbol{F}_{ij}. \tag{1.3}$$

Moreover, $\boldsymbol{F}_{ij}$ is a function of the positions and velocities (and internal structure) of the $i$th and $j$th bodies, but is unaffected by the presence of the other bodies. (It can be argued that this is an unnecessarily restrictive assumption. It would be perfectly possible to include also, say, three-body forces, which depend on the positions and velocities of three particles simultaneously. However, within the realm of validity of classical mechanics, no such forces are known, and their inclusion would be an inessential complication.) Because of the relativity principle, the force can in fact depend only on the *relative* position

$$\boldsymbol{r}_{ij} = \boldsymbol{r}_i - \boldsymbol{r}_j$$

(see Fig. 1.1), and the *relative* velocity

$$\boldsymbol{v}_{ij} = \boldsymbol{v}_i - \boldsymbol{v}_j.$$

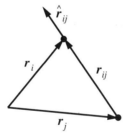

Fig. 1.1

If the forces are known, as functions of the positions and velocities, then from (1.1) we can predict the future motion of the bodies. Given their initial positions and velocities, we can solve these equations (analytically or numerically) to find their positions at a later time.

There is here an implicit assumption of perfect knowledge and infinite precision of calculation. It is now recognized (see Chapters 13, 14) that this assumption is, in general, false, leading to a loss of predictability. However, for the time being, we shall assume that our solution can be effected.

All that then remains is to specify the precise laws by which the two-body forces are to be determined. The most important class of forces consists of the *central, conservative* forces, which depend only on the relative positions of the two bodies, and have the form

$$\boldsymbol{F}_{ij} = \hat{\boldsymbol{r}}_{ij} f(r_{ij}), \tag{1.4}$$

where, as usual, $\hat{\boldsymbol{r}}_{ij}$ is the unit vector in the direction of $\boldsymbol{r}_{ij}$ and $f(r_{ij})$ is a scalar function of the relative distance $r_{ij}$. If $f(r_{ij})$ is positive the force $\boldsymbol{F}_{ij}$ is a *repulsive* force directed outwards along the line joining the bodies; if $f(r_{ij})$ is negative, it is an *attractive* force, directed inwards.

According to *Newton's law of universal gravitation*, there is a force of this type between *every* pair of bodies, proportional in magnitude to the product of their masses. It is given by (1.4) with

$$f(r_{ij}) = -\frac{Gm_im_j}{r_{ij}^2}, \tag{1.5}$$

where $G$ is Newton's gravitational constant, whose value is

$$G = 6.673 \times 10^{-11}\,\mathrm{N\,m^2\,kg^{-2}}.$$

Since the masses are always positive, this force is always attractive.

In addition, if the bodies are electrically charged, there is an electrostatic Coulomb force given by

$$f(r_{ij}) = \frac{q_iq_j}{4\pi\epsilon_0 r_{ij}^2}, \tag{1.6}$$

where $q_i$ and $q_j$ are the electric charges, and $\epsilon_0$ is another constant,

$$\epsilon_0 = 8.854\,19 \times 10^{-12}\,\mathrm{F\,m^{-1}}.$$

Note that the analogue of Newton's constant $G$ is

$$1/4\pi\epsilon_0 = 8.987\,55 \times 10^9\,\mathrm{N\,m^2\,C^{-2}}.$$

Electric charges may be of either sign, and therefore the electrostatic force may be repulsive or attractive according to the relative sign of $q_i$ and $q_j$.

Note the enormous difference in the orders of magnitude of the constants $G$ and $1/4\pi\epsilon_0$ when expressed in SI units. This serves to illustrate the fact that gravitational forces are really exceptionally weak. They appear significant to us only because we happen to live close to a body of very large mass. Correspondingly large charges never appear, because positive and negative charges largely cancel out, leaving macroscopic bodies with a net charge close to zero.

In bodies with structure, central, conservative forces between their constituent parts can evidently give rise to forces which are still conservative (*i.e.*, which are independent of velocity, and satisfy some further conditions that need not worry us here — see §3.1 and §A.6), but no longer central (*i.e.*, not directed along the line joining the bodies). This can happen, for example, if there is a distribution of electric charge within each body.

They can also give rise, in a less obvious way, to non-conservative, velocity-dependent forces, as we shall see in Chapter 2. Many resistive and frictional forces can be understood as macroscopic effects of forces which are really conservative on a small scale. The chief distinguishing feature of conservative forces is the existence of a quantity which is *conserved, i.e.*, whose total value never changes, namely the *energy* of the system. Frictional forces have the effect of transferring some of this energy from the large-scale motion of the bodies to small-scale movements in their interior, and therefore appear non-conservative on a large scale.

In a sense, therefore, we may regard central, conservative forces as the norm. It would be wrong to conclude, however, that we can explain everything in terms of them. In the first place, the concepts of classical mechanics cannot be applied to the really small-scale structure of matter. For that, we need quantum mechanics. More serious is the existence of *electromagnetic* forces, which are of great importance even in the realm of classical physics, but which cannot readily be accommodated in the framework of classical mechanics. The force between two charges in relative motion is neither central nor conservative, and does not even satisfy Newton's third law (1.3). This is a consequence of the finite velocity of propagation of electromagnetic waves. The force on one charge depends not only on the instantaneous position of the other, but on its past history. The effect of a disturbance of one charge is not felt immediately by the other, but after an interval of time sufficient for a light signal to propagate from one to the other. This particular difficulty may be resolved by introducing the concept of the electromagnetic *field*. Then we may suppose that one charge does not act directly on the other, but on the field in its immediate vicinity; this in turn affects the field further out, and so on. By supposing that the field itself can carry energy and momentum, we can reinstate the conservation laws, which are among the most important consequences of Newton's laws.

However, this does not completely remove the difficulty, for there is still an apparent contradiction between this classical electromagnetic theory and the principle of relativity discussed in §1.1. This arises from the fact that if the speed of light is constant with respect to one inertial frame — as it should be according to electromagnetic theory — then the usual rules for combining velocities would lead to the conclusion that it is not constant with respect to a relatively moving frame, in contradiction with the statement that all inertial frames are equivalent. This paradox can only be resolved by the introduction of Einstein's theory of relativity (*i.e.*, 'special' relativity). Classical electromagnetic theory and classical mechanics *can* be incorporated into a single self-consistent theory, but only by ignoring the relativity principle and sticking to one 'preferred' inertial frame.

## 1.3   The Concepts of Mass and Force

It is an important general principle of physics (though not universally applied!) that no quantity should be introduced into the theory which cannot, at least in principle, be measured. Now, Newton's laws involve not only the concepts of velocity and acceleration, which can be measured by measuring distances and times, but also the new concepts of mass and force. To give the laws a physical meaning we have, therefore, to show that these are measurable quantities. This is not quite as trivial as it might seem, because any experiment designed to measure these quantities must necessarily involve Newton's laws themselves in its interpretation. Thus the operational definitions of mass and force — the prescriptions of how they may be measured — which are required to make the laws physically significant, are actually contained in the laws themselves. This is by no means an unusual or logically objectionable situation, but it may clarify the status of these concepts to reformulate the laws in such a way as to isolate their definitional element.

Let us consider first the measurement of mass. Since the units of mass are arbitrary, we have to specify a way of comparing the masses of two given bodies. It is important to realize that we are discussing here the *inertial* mass, which appears in Newton's second law, (1.1) and not the *gravitational* mass, which appears in (1.5). The two are of course proportional, but this *equivalence principle* is a physical law derived from experimental observation (in particular from Galileo's observations of falling bodies, from which he deduced that in a vacuum all bodies would fall equally fast) rather than an *a priori* assumption. To verify the law, we must be able to measure each kind of mass separately. This rules out, for example, the use of a balance, which compares gravitational masses.

Clearly, we can compare the inertial masses of two bodies by subjecting them to equal forces and comparing their accelerations, but this does not help unless we have some way of knowing that the forces *are* equal. However there is one case in which we *do* know this, because of Newton's third law. If we isolate the two bodies from all other matter, and compare their mutually induced accelerations, then according to (1.1) and (1.3),

$$m_1\boldsymbol{a}_1 = -m_2\boldsymbol{a}_2, \tag{1.7}$$

so that the accelerations are oppositely directed, and inversely proportional to the masses. If we allow two small bodies to collide, then during the collision the effects of more remote bodies are generally negligible in comparison with their effect on each other, and we may treat them approximately as an isolated system. (Such collisions will be discussed in detail in Chapters 2 and 7.) The mass ratio can then be determined from measurements of their velocities before and after the collision, by using (1.7) or its immediate consequence, the law of *conservation of momentum*,

$$m_1\boldsymbol{v}_1 + m_2\boldsymbol{v}_2 = \text{constant.} \tag{1.8}$$

If we wish to separate the definition of mass from the physical content of equation (1.7), we may adopt as a fundamental axiom the following:

> In an isolated two-body system, the accelerations always satisfy the relation $\boldsymbol{a}_1 = -k_{21}\boldsymbol{a}_2$, where the scalar $k_{21}$ is, for two given bodies, a constant independent of their positions, velocities and internal states.

If we choose the first body to be a standard body, and conventionally assign it unit mass (say $m_1 = 1$ kg), then we may *define* the mass of the second to be $k_{21}$ in units of this standard mass (here $m_2 = k_{21}$ kg).

Note that for consistency, we must have $k_{12} = 1/k_{21}$. We must also assume of course that if we compare the masses of three bodies in this way, we obtain consistent results:

> For any three bodies, the constants $k_{ij}$ satisfy $k_{31} = k_{32}k_{21}$.

It then follows that for *any* two bodies, $k_{32}$ is the mass ratio: $k_{32} = m_3/m_2$.

To complete the list of fundamental axioms, we need one which deals with systems containing more than two bodies, analogous to the law of composition of forces, (1.2). This may be stated as follows:

The acceleration induced in one body by another is some definite function of their positions, velocities and internal structure, and is unaffected by the presence of other bodies. In a many-body system, the acceleration of any given body is equal to the sum of the accelerations induced in it by each of the other bodies individually.

These laws, which appear in a rather unfamiliar form, are actually completely equivalent to Newton's laws, as stated in the previous section. In view of the apparently fundamental role played by the concept of force in Newtonian mechanics, it is remarkable that we have been able to reformulate the basic laws without mentioning this concept. It can of course be introduced, by defining it through Newton's second law, (1.1). The utility of this definition arises from the fact that forces satisfy Newton's third law, (1.3), while accelerations satisfy only the more complicated law, (1.7). Since the mutually induced accelerations of two given bodies are always proportional, they are essentially determined by a single function, and it is useful to introduce the more symmetric concept of force, for which this becomes obvious.

It is interesting to note, finally, that one consequence of our basic laws is the additive nature of mass. Let us take a three-body system. Then, returning to the notation of the previous section, the equations of motion for the three bodies are

$$
\begin{aligned}
m_1 \boldsymbol{a}_1 &= \boldsymbol{F}_{12} + \boldsymbol{F}_{13}, \\
m_2 \boldsymbol{a}_2 &= \boldsymbol{F}_{21} + \boldsymbol{F}_{23}, \\
m_3 \boldsymbol{a}_3 &= \boldsymbol{F}_{31} + \boldsymbol{F}_{32}.
\end{aligned}
\tag{1.9}
$$

If we add these equations, then, in view of (1.3), the terms on the right cancel in pairs, and we are left with

$$
m_1 \boldsymbol{a}_1 + m_2 \boldsymbol{a}_2 + m_3 \boldsymbol{a}_3 = \boldsymbol{0},
\tag{1.10}
$$

which is the generalization of (1.7). Now, if we suppose that the force between the second and third is such that they are rigidly bound together to form a composite body, their accelerations must be equal: $\boldsymbol{a}_2 = \boldsymbol{a}_3$. In that case, we get

$$
m_1 \boldsymbol{a}_1 = -(m_2 + m_3) \boldsymbol{a}_2,
$$

which shows that the mass of the composite body is just $m_{23} = m_2 + m_3$.

## 1.4   External Forces

To find the motion of the various bodies in any dynamical system, we have to solve two closely interrelated problems. First, given the positions and velocities at any one instant of time, we have to determine the forces acting on each body. Second, given the forces acting, we have to compute the new positions and velocities after a short interval of time has elapsed. In a general case, these two problems are inextricably bound up with each other, and must be solved simultaneously. If, however, we are concerned with the motions of a small body, or group of small bodies, then we can often neglect its effect on other bodies, and in that case the two problems can be separated.

For example, in discussing the motion of an artificial satellite, we can clearly ignore its effect on the Earth. Since the motion of the Earth is already known, we can calculate the force on the satellite as a function of its position and (if atmospheric resistance is included) its velocity. Then, taking the force as known, we can solve separately the problem of its motion. In the latter problem, we are really concerned with the satellite alone. The Earth enters simply as a known external influence.

In many cases, therefore, it is useful to concentrate our attention on a small part of a dynamical system, and to represent the effect of everything outside this by external forces, which we suppose to be known in advance, as functions of position, velocity and time. This is the kind of problem with which we shall be mainly concerned in the next few chapters. Typically, we shall consider the motion of a particle under a known external force. In Chapter 6, we consider, for the gravitational and electrostatic cases, the complementary problem of determining the force from a knowledge of the positions of other bodies. Later, in Chapter 7, we return to the more complex type of problem in which the system of immediate interest cannot be taken to be merely a single particle.

## 1.5   Summary

To some extent, the selection of a group of basic concepts, in terms of which others are to be defined, is a matter of choice. We have chosen to regard position and time (relative to some frame of reference) as basic. From this point of view, Newton's laws must be regarded as containing definitions in addition to physical laws. The first law contains the definition of an inertial frame, together with the physical assertion that such frames exist, while the second and third laws contain definitions of mass and force. These laws, supplemented by the laws of force, such as the law of universal gravitation, provide the equations from which we can determine the motion of any dynamical system.

## Problems

*Note.* Here and in later chapters, starred problems are somewhat harder.

1. An object $A$ moving with velocity $v$ collides with a stationary object $B$. After the collision, $A$ is moving with velocity $\frac{1}{2}v$ and $B$ with velocity $\frac{3}{2}v$. Find the ratio of their masses. If, instead of bouncing apart, the two bodies stuck together after the collision, with what velocity would they then move?

2. The two components of a double star are observed to move in circles of radii $r_1$ and $r_2$. What is the ratio of their masses? (*Hint*: Write down their accelerations in terms of the angular velocity of rotation, $\omega$.)

3. Consider a system of three particles, each of mass $m$, whose motion is described by (1.9). If particles 2 and 3, even though not rigidly bound together, are regarded as forming a composite body of mass $2m$ located at the mid-point $r = \frac{1}{2}(r_2+r_3)$, find the equations describing the motion of the two-body system comprising particle 1 and the composite body $(2+3)$. What is the force on the composite body due to particle 1? Show that the equations agree with (1.7). When the masses are *un*equal, what is the correct definition of the position of the composite $(2+3)$ that will make (1.7) still hold?

4. Find the distance $r$ between two protons at which the electrostatic repulsion between them will equal the gravitational attraction of the Earth on one of them. (Proton charge $= 1.6 \times 10^{-19}$ C, proton mass $= 1.7 \times 10^{-27}$ kg.)

5. Consider a transformation to a relatively uniformly moving frame of reference, where each position vector $r_i$ is replaced by $r_i' = r_i - vt$. (Here $v$ is a constant, the relative velocity of the two frames.) How does a relative position vector $r_{ij}$ transform? How do momenta and forces transform? Show explicitly that if equations (1.1) to (1.4) hold in the original frame, then they also hold in the new one.

6. A body of mass 50 kg is suspended by two light, inextensible cables of

lengths 15 m and 20 m from rigid supports placed 25 m apart on the same level. Find the tensions in the cables. (Note that by convention 'light' means 'of negligible mass'. Take $g = 10\,\mathrm{m\,s^{-2}}$. This and the following two problems are applications of vector addition.)

7. *An aircraft is to fly to a destination 800 km due north of its starting point. Its airspeed is $800\,\mathrm{km\,h^{-1}}$. The wind is from the east at a speed of $30\,\mathrm{m\,s^{-1}}$. On what compass heading should the pilot fly? How long will the flight take? If the wind speed increases to $50\,\mathrm{m\,s^{-1}}$, and the wind backs to the north-east, but no allowance is made for this change, how far from its destination will the aircraft be at its expected arrival time, and in what direction?

8. *The two front legs of a tripod are each 1.4 m long, with feet 0.8 m apart. The third leg is 1.5 m long, and its foot is 1.5 m directly behind the midpoint of the line joining the other two. Find the height of the tripod, and the vectors representing the positions of its three feet relative to the top. (*Hint*: Choose a convenient origin and axes and write down the lengths of the legs in terms of the position vector of the top.) Given that the tripod carries a weight of mass 2 kg, find the forces in the legs, assuming they are purely compressional (*i.e.*, along the direction of the leg) and that the legs themselves have negligible weight. (Take $g = 10\,\mathrm{m\,s^{-2}}$.)

9. *Discuss the possibility of using force rather than mass as the basic quantity, taking for example a standard weight (at a given latitude) as the unit of force. How should one then define and measure the mass of a body?

10. The first estimate of Newton's constant was made by the astronomer Nevil Maskelyne in 1774 by measuring the angle between the directions of the apparent plumb-line vertical on opposite sides of the Scottish mountain Schiehallion (height 1081 m, chosen for its regular conical shape). Find a rough estimate of the angle through which a plumb line is deviated by the gravitational attraction of the mountain, by modelling the mountain as a sphere of radius 500 m and density $2.7 \times 10^3\,\mathrm{kg\,m^{-3}}$, and assuming that its gravitational effect is the same as though the total mass were concentrated at its centre. (This latter assumption will be justified, for a spherical object, in Chapter 6.)

# Chapter 2

# Linear Motion

In this chapter we discuss the motion of a body which is free to move only in one dimension. The problems discussed are chosen to illustrate the concepts and techniques which will be of use in the more general case of three-dimensional motion.

## 2.1 Conservative Forces; Conservation of Energy

We consider first a particle moving along a line, under a force which is given as a function of its position, $F(x)$. Then the equation of motion (1.1) is

$$m\ddot{x} = F(x). \tag{2.1}$$

Since this equation is of second order in the time derivatives, we shall have to integrate twice to find $x$ as a function of $t$. Thus the solution will contain two arbitrary constants that may be fixed by specifying the initial values of $x$ and $\dot{x}$.

When the force depends only on $x$, we can always find a 'first integral', a function of $x$ and $\dot{x}$ whose value is constant in time. Let us consider the *kinetic energy*,

$$T = \tfrac{1}{2}m\dot{x}^2. \tag{2.2}$$

Differentiating, we find for its rate of change

$$\dot{T} = m\dot{x}\ddot{x} = F(x)\dot{x}, \tag{2.3}$$

using (2.1). Integrating with respect to time, we find

$$T = \int F(x)\dot{x}\,\mathrm{d}t = \int F(x)\,\mathrm{d}x. \tag{2.4}$$

If we define the *potential energy*,

$$V(x) = -\int_{x_0}^{x} F(x')\,\mathrm{d}x',$$ (2.5)

we can write (2.4) in the form

$$T + V = E = \text{constant.}$$ (2.6)

Here $x_0$ is an arbitrary constant, corresponding to the arbitrary additive constant of integration in (2.4). There is a corresponding arbitrariness up to an additive constant in the *total energy E*.

Note that (2.5) can be inverted to give the force in terms of the potential energy:

$$F(x) = -\frac{\mathrm{d}V}{\mathrm{d}x} \equiv -V'(x).$$ (2.7)

The equation (2.6) is the law of *conservation of energy*. A force of this type, depending only on $x$, is called a *conservative* force. A great deal of information about the motion can be obtained from this conservation law, even without integrating again to find $x$ explicitly as a function of $t$. If the initial position and velocity are given, we can calculate the value of the constant $E$. Then (2.6) in the form

$$\tfrac{1}{2}m\dot{x}^2 = E - V(x)$$ (2.8)

gives the velocity of the particle (except for sign) when it is at any given position $x$. Since the kinetic energy is obviously positive, the motion is confined to the region where

$$V(x) \le E.$$

An example may help to explain this.

*Example:* **The simple pendulum**

A simple pendulum comprises a bob of mass $m$ supported by a light rigid rod of length $l$. (We choose a rod rather than a string so that we may consider motion *above* the point of support. By 'light' we mean 'of negligible mass'.) The bob starts with velocity $v$ from the equilibrium position. What kinds of motion are possible for different values of $v$?

The distance moved by the bob is $x = l\theta$, where $\theta$ is the angular displacement from the downward vertical. Corresponding to the restoring force $F = -mg\sin\theta = -mg\sin(x/l)$, the potential energy function is

$$V = mgl[1 - \cos(x/l)] = mgl(1 - \cos\theta). \tag{2.9}$$

It is plotted in Fig. 2.1. (Note that the points $\theta = \pi$ and $\theta = -\pi$ may be identified.)

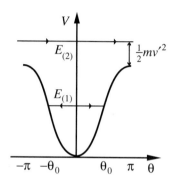

Fig. 2.1

Initially $\theta = 0$, and the bob is moving with velocity $v$. Since we have chosen the arbitrary constant in $V$ so that $V(0) = 0$, the total energy is $E = \frac{1}{2}mv^2$. Now if $E < 2mgl$ ($E_{(1)}$ in the figure), the motion will be confined between the two angles $\pm\theta_0$ where $V(\theta_0) = E$, i.e.

$$1 - \cos\theta_0 = v^2/2gl.$$

These are the points where the kinetic energy vanishes, so that the pendulum bob is instantaneously at rest. The motion is an oscillation of amplitude $\theta_0$.

On the other hand, if the bob is pushed so hard that $E > 2mgl$ ($E_{(2)}$ in the figure), then the kinetic energy will never vanish. When the bob reaches the upward vertical, it still has positive energy, namely

$$\tfrac{1}{2}mv'^2 = E - 2mgl = \tfrac{1}{2}mv^2 - 2mgl.$$

In this case, the motion is a continuous rotation rather than an oscillation.

## 2.2    Motion near Equilibrium; the Harmonic Oscillator

A particle can be in equilibrium only if the total force acting on it is zero. For a conservative force this means, by (2.7), that the potential energy curve is horizontal at the position of the particle. Let us now consider the motion of a particle near a position of equilibrium. Without loss of generality, we may choose the equilibrium point to be the origin, $x = 0$, and choose the arbitrary constant in $V$ so that $V(0) = 0$. If we are interested only in small displacements, we may expand $V(x)$ in a Maclaurin–Taylor series,

$$V(x) = V(0) + xV'(0) + \tfrac{1}{2}x^2V''(0) + \cdots ,$$

where the primes denote differentiation with respect to $x$. Since we have chosen $V(0) = 0$, the constant term is absent, and since the equilibrium condition is $V'(0) = 0$, the linear term is absent also. Thus near $x = 0$ we can write, approximately,

$$V(x) = \tfrac{1}{2}kx^2, \qquad k = V''(0). \tag{2.10}$$

(We assume for simplicity that $V''(0)$ does not also vanish.)

Because motion near almost any point of equilibrium is described approximately by this potential energy function, it is remarkably ubiquitous; it plays as important a role in quantum mechanics as in classical mechanics. It will therefore be useful to analyze it in some detail. The potential energy curve here is a parabola (see Fig. 2.2). Thus for $k > 0$ and any energy $E > 0$ (Fig. 2.2(a)), there are two points where $V(x) = E$, namely

$$x = \pm a, \qquad a = \sqrt{2E/k}. \tag{2.11}$$

The motion is an oscillation between these two points.

On the other hand, if $k < 0$, the curve is an inverted parabola (see Fig. 2.2(b)). In this case, two kinds of motion are possible. If $E < 0$ the particle may approach to some minimum distance, where it comes momentarily to rest before reversing direction ($E_{(2)}$ in figure). But if $E > 0$ ($E_{(1)}$ in figure), it has enough energy to surmount the barrier and will never come to rest: it will continue moving in the same direction for ever, with decreasing speed as it approaches $x = 0$ and then increasing speed.

The force corresponding to the potential energy function (2.10) is, by (2.7),

$$F(x) = -kx. \tag{2.12}$$

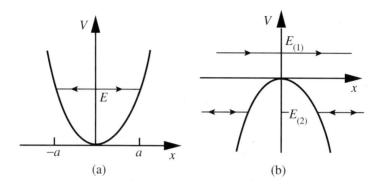

Fig. 2.2

It is an attractive or repulsive force, according as $k > 0$ or $k < 0$, proportional to displacement from the point of equilibrium.

The equation of motion may be written

$$m\ddot{x} + kx = 0. \tag{2.13}$$

This equation is very easy to solve. We could proceed directly from the energy conservation equation in the form (2.8) and integrate again to obtain

$$\int \left( \frac{2E}{m} - \frac{k}{m}x^2 \right)^{-1/2} \mathrm{d}x = \int \mathrm{d}t. \tag{2.14}$$

However, since we shall encounter a number of similar equations later, it will be useful to discuss an alternative method of solution that can be adapted later to other examples.

### Solution of the harmonic oscillator equation

Equation (2.13) is a *linear* differential equation; that is, one involving only linear terms in $x$ and its derivatives. Such equations have the important property that their solutions satisfy the *superposition principle*: if $x_1(t)$ and $x_2(t)$ are solutions, then so is any linear combination

$$x(t) = a_1 x_1(t) + a_2 x_2(t), \tag{2.15}$$

where $a_1$ and $a_2$ are constants; for, clearly,

$$m\ddot{x} + kx = a_1(m\ddot{x}_1 + kx_1) + a_2(m\ddot{x}_2 + kx_2) = 0.$$

Moreover, if $x_1$ and $x_2$ are *linearly independent* solutions (that is, unless one is simply a constant multiple of the other), then (2.15) is actually the general solution. For, since (2.13) is of second order, we could obtain its solution by integrating twice, and the general solution must therefore contain just two arbitrary constants of integration. So to find the general solution, all we have to do is to find any two independent solutions $x_1(t)$ and $x_2(t)$.

Let us consider first the case where $k < 0$, so that $V(x)$ has a maximum at $x = 0$. Then (2.13) can be written

$$\ddot{x} - p^2 x = 0, \qquad p = \sqrt{-k/m}. \tag{2.16}$$

It is easy to verify that this equation is satisfied by the functions $x = e^{pt}$ and $x = e^{-pt}$. Thus the general solution is

$$x = \tfrac{1}{2}Ae^{pt} + \tfrac{1}{2}Be^{-pt}. \tag{2.17}$$

(We introduce the factor of $\tfrac{1}{2}$ for later convenience; it is a matter of convention whether we call the arbitrary constants $A$ and $B$ or $\tfrac{1}{2}A$ and $\tfrac{1}{2}B$.) Clearly, a small displacement will in general lead to an exponential increase of $x$ with time, which continues until the approximation involved in (2.10) ceases to be valid. Thus the equilibrium is unstable, as we should expect when $V$ has a maximum.

We now turn to the case where $k > 0$ and $V$ has a minimum at $x = 0$. Then the potential energy function (2.10) is that of a *simple harmonic oscillator*. The equation of motion (2.13) becomes

$$\ddot{x} + \omega^2 x = 0, \qquad \omega = \sqrt{k/m}. \tag{2.18}$$

It is again very easy to check that the functions $x = \cos \omega t$ and $x = \sin \omega t$ are solutions, and the general solution is therefore

$$x = c \cos \omega t + d \sin \omega t. \tag{2.19}$$

The arbitrary constants $c$ and $d$ are to be determined by the initial conditions. If at $t = 0$ the particle is at $x_0$ with velocity $v_0$, then we easily find

$$c = x_0, \qquad d = v_0/\omega.$$

An alternative form of (2.19) is

$$x = a \cos(\omega t - \theta), \tag{2.20}$$

where the constants $a, \theta$ are related to $c, d$ by

$$c = a \cos \theta, \qquad d = a \sin \theta.$$

The constant $a$ is called the *amplitude*. It is identical to the constant introduced in (2.11), and defines the limits between which the particle oscillates, $x = \pm a$. The motion is a periodic oscillation, of *period* $\tau$ given by

$$\tau = 2\pi/\omega. \qquad (2.21)$$

(See Fig. 2.3.) The *frequency* $f$ is the number of oscillations per unit time, namely

$$f = \frac{1}{\tau} = \frac{\omega}{2\pi}. \qquad (2.22)$$

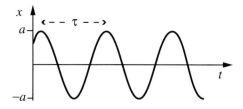

Fig. 2.3

Notice that this discussion applies to the motion of a particle near a stable equilibrium point of *any* potential energy function. For sufficiently small displacements, any system of this kind behaves like a simple harmonic oscillator. In particular, the period or frequency of small oscillations may always be found from the second derivative of the potential energy function at the equilibrium position.

### *Example:* **Charged particle between two fixed charges**

A particle of charge $q$ is constrained to move along a line between two other equal charges $q$, fixed at $x = \pm a$. What is the period of small oscillations?

Here, the force (in the region $|x| < a$) is

$$F = \frac{q^2}{4\pi\epsilon_0(a+x)^2} - \frac{q^2}{4\pi\epsilon_0(a-x)^2},$$

corresponding to the potential energy function

$$V = \frac{q^2}{4\pi\epsilon_0}\left(\frac{1}{a+x} + \frac{1}{a-x}\right) = \frac{q^2}{4\pi\epsilon_0}\left(\frac{2a}{a^2-x^2}\right).$$

Clearly, the position of equilibrium, where $V' = 0$, is $x = 0$, and we easily find

$$k = V''(0) = \frac{q^2 a}{\pi\epsilon_0}\frac{a^2+3x^2}{(a^2-x^2)^3}\bigg|_{x=0} = \frac{q^2}{\pi\epsilon_0 a^3}.$$

Hence, $\omega^2 = q^2/\pi\epsilon_0 a^3 m$, and $\tau = (2\pi/q)\sqrt{\pi\epsilon_0 a^3 m}$.

## 2.3   Complex Representation

It is often very convenient, in discussing periodic phenomena, to use complex numbers. In particular, this allows us to treat both cases, $k > 0$ and $k < 0$, of the previous section together. If we allow $p$ to be complex, then the solution in both cases is given by (2.16) and (2.17). When $k > 0$, $p$ becomes purely imaginary, $p = i\omega$, and the solution (2.17) is

$$x = \tfrac{1}{2}Ae^{i\omega t} + \tfrac{1}{2}Be^{-i\omega t}. \tag{2.23}$$

Of course, $x$ must be a real number, so that $A$ and $B$ must be complex conjugates. If we write

$$A = c - id, \qquad B = c + id,$$

and use the relation

$$e^{i\omega t} = \cos\omega t + i\sin\omega t,$$

we recover the solution (2.19). Similarly, if we use the polar form of a complex number,

$$A = ae^{-i\theta}, \qquad B = ae^{i\theta},$$

we obtain the solution in the form (2.20).

It is also useful to note that if $z = x + iy$ is a complex solution of (2.13),

$$m\ddot{z} + kz = (m\ddot{x} + kx) + i(m\ddot{y} + ky) = 0,$$

then its real and imaginary parts, $x = \mathrm{Re}(z)$ and $y = \mathrm{Im}(z)$, must separately each satisfy the equation. This method of obtaining solutions will be useful later.

If we plot the complex solution,

$$z = x + iy = Ae^{i\omega t},$$

in the $xy$-plane, we see that $z$ moves in a circle around the origin with angular velocity $\omega$ (see Fig. 2.4). For this reason, $\omega$ is usually called the

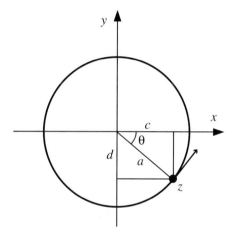

Fig. 2.4

*angular frequency.* The constant $A$ is a complex amplitude, whose absolute value (the real amplitude $a$) is the radius of the circle, and whose phase $\theta$ defines the initial direction of the vector from the origin to $z$. The real part of the solution, $x$, may be regarded as the projection of this circular motion on the real axis.

## 2.4    The Law of Conservation of Energy

This law was originally derived from Newton's laws for the case where the force is a function only of $x$. It has, however, a much wider application.

By introducing additional forms of energy (heat, chemical, electromagnetic and so on), it has been extended far beyond the field of mechanics, to the point where it is now recognized as one of the most fundamental of all physical laws. The existence of such a law, and of the laws of conservation of momentum and angular momentum, is in fact closely related to the relativity principle, as we shall see later in Chapter 12.

Many non-conservative (or *dissipative*) forces may be regarded as macro-scopic effects of forces which are really conservative on a small scale. For example, when a particle penetrates a retarding medium, such as the at-mosphere, it experiences a force which is velocity-dependent, and therefore non-conservative. However, if we look at the situation on a sub-microscopic scale, we see that what happens is that the particle makes a series of col-lisions with the molecules of the medium. In each collision, energy is con-served, and some of the kinetic energy of the incoming particle is transferred to the molecules with which it collides. (Such collisions will be discussed in detail in Chapter 7.) By means of further collisions, this energy is gradually distributed among the surrounding molecules. The net result is to retard the incoming particle, and to increase the average energy of the molecules in the medium. This increased energy appears macroscopically as heat, and results in a rise in the temperature of the medium.

For an arbitrary force $F$, the rate of change of the kinetic energy $T$ is given by (2.3): $\dot{T} = F\dot{x}$. Thus the increase in kinetic energy in a time interval $\mathrm{d}t$, during which the particle moves a distance $\mathrm{d}x$, is

$$\mathrm{d}T = \mathrm{d}W, \tag{2.24}$$

where

$$\mathrm{d}W = F\mathrm{d}x \tag{2.25}$$

is called the *work* done by the force $F$ in the infinitesimal displacement $\mathrm{d}x$. (Thus for a conservative force, the potential energy $V(x)$ is equal to minus the work done by the force in the displacement from the fixed point $x_0$ to $x$.) The work is therefore a measure of the amount of energy converted to kinetic energy from other forms, and the rate $P$ of doing work (the *power*) is defined as $F\dot{x}$. In a real mechanical system, there is usually some loss of mechanical (kinetic or potential) energy to heat or other forms. Correspondingly, there will be dissipative forces acting on the system. (Of course, it is also possible to convert heat to mechanical energy, as in an internal combustion engine.)

## 2.5    The Damped Oscillator

We saw in §2.2 that a particle near a position of stable equilibrium under a conservative force may always be treated approximately as a simple harmonic oscillator. If there is energy loss, we must include in the equation of motion a force depending on the velocity. So long as we are concerned only with small displacements from the equilibrium position, we may treat both $x$ and $\dot{x}$ as small quantities, and neglect $x^2$, $x\dot{x}$ and $\dot{x}^2$ in comparison. Thus we are led to consider the *damped harmonic oscillator* for which

$$F = -kx - \lambda\dot{x},$$

where $\lambda$ is another constant. The equation of motion now becomes

$$m\ddot{x} + \lambda\dot{x} + kx = 0. \tag{2.26}$$

Equations of this form turn up in many branches of physics. For example, the oscillations of an electrical circuit containing an inductance $L$, resistance $R$ and capacitance $C$ in series are described by the equation

$$L\ddot{q} + R\dot{q} + \frac{1}{C}q = 0,$$

in which the variable $q$ represents the charge on one plate of the condenser.

The equation (2.26) may be solved as before by looking for solutions of the form $x = e^{pt}$. Substituting in (2.26), we obtain for $p$ the equation

$$mp^2 + \lambda p + k = 0,$$

whose roots are

$$p = -\gamma \pm \sqrt{\gamma^2 - \omega_0^2}, \tag{2.27}$$

where

$$\gamma = \frac{\lambda}{2m}, \quad \text{and} \quad \omega_0 = \sqrt{\frac{k}{m}}. \tag{2.28}$$

Note that $\omega_0$ is the frequency of the undamped oscillator, as in (2.18). The rate at which work is done by the force $-\lambda\dot{x}$ is $-\lambda\dot{x}^2$. If $\lambda$ were negative, the particle would therefore be gaining, rather than losing, energy. So we shall assume that $\lambda$ is positive.

## Large damping

If $\lambda$ is so large that $\gamma > \omega_0$, then both roots for $p$ are real and negative:

$$p = -\gamma_\pm, \qquad \gamma_\pm = \gamma \pm \sqrt{\gamma^2 - \omega_0^2}.$$

The general solution is then

$$x = \tfrac{1}{2}Ae^{-\gamma_+ t} + \tfrac{1}{2}Be^{-\gamma_- t}, \tag{2.29}$$

where $A$ and $B$ are arbitrary constants. Hence the displacement tends exponentially to zero. For large times, the dominant term is that containing in the exponent the smaller coefficient, $\gamma_-$. Thus the characteristic time in which the displacement $x$ is reduced by a factor $1/e$ is of order $1/\gamma_-$.

## Small damping

Now let us consider the case where $\lambda$ is small, so that $\gamma < \omega_0$. Then the two roots for $p$ are complex conjugates,

$$p = -\gamma \pm i\omega, \qquad \omega = \sqrt{\omega_0^2 - \gamma^2}. \tag{2.30}$$

The general solution may be written in the alternative forms

$$\begin{aligned} x &= \tfrac{1}{2}Ae^{i\omega t - \gamma t} + \tfrac{1}{2}Be^{-i\omega t - \gamma t} \\ &= \mathrm{Re}(Ae^{i\omega t - \gamma t}) \\ &= ae^{-\gamma t}\cos(\omega t - \theta), \end{aligned} \tag{2.31}$$

where

$$A = ae^{-i\theta}, \qquad B = ae^{i\theta}.$$

From the last form of (2.31) we see that this solution represents an oscillation with exponentially decreasing amplitude $ae^{-\gamma t}$, and angular frequency $\omega$. (See Fig. 2.5.) Note that $\omega$ is always less than the frequency $\omega_0$ of the undamped oscillator. The time in which the amplitude is reduced by a factor $1/e$, called the *relaxation time* of the oscillator, is

$$1/\gamma = 2m/\lambda.$$

It is often convenient to introduce the *quality factor*, or simply '$Q$', of the oscillator, defined as the dimensionless number

$$Q = \frac{m\omega_0}{\lambda} = \frac{\omega_0}{2\gamma}. \tag{2.32}$$

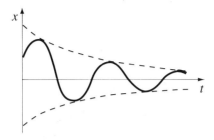

Fig. 2.5

If the damping is small then $Q$ is large. In a single oscillation period, the amplitude is reduced by the factor $\mathrm{e}^{-2\pi\gamma/\omega}$, or approximately $\mathrm{e}^{-\pi/Q}$. The number of periods in a relaxation time is roughly $Q/\pi$.

### Critical damping

The limiting case, $\gamma = \omega_0$, is the case of *critical* damping, in which $\omega = 0$ and the two roots for $p$ coincide. Then the solution (2.29) or (2.31) involves only one arbitrary constant, $A + B$, and cannot be the general solution. We need to find a second independent solution in this case. In fact, it is easy to verify by direct substitution that the equation (2.26) in this special case is also satisfied by the function $te^{-pt}$. Thus the general solution is

$$x = (a + bt)\mathrm{e}^{-\gamma t}. \tag{2.33}$$

Critical damping is often the ideal. For example, in a measuring instrument, we want to damp out the oscillations of the pointer about its correct position as quickly as possible, but too much damping would lead to a very slow response. Let us assume that $k$ is fixed and the amount of damping varied. When the damping is less than critical ($\gamma < \omega_0$), the characteristic time of response is the relaxation time $1/\gamma$, which of course *decreases* as $\gamma$ is increased. However, when $\gamma > \omega_0$, the characteristic time is $1/\gamma_-$, as we noted above. It is not hard to verify that as $\gamma$ increases, $\gamma_-$ decreases, so that the response time $1/\gamma_-$ *increases* (see Fig. 2.6). Thus the shortest possible response time is obtained by choosing $\gamma = \omega_0$, that is for critical damping.

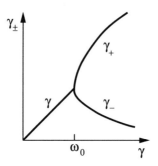

Fig. 2.6

## 2.6   Oscillator under Simple Periodic Force

In an isolated system the forces are functions of position and velocity, but
not explicitly of the time. However, we are often interested in the response
of an oscillatory system to an applied external force, which is given as a
function of the time, $F(t)$. Then we have to deal with the equation

$$m\ddot{x} + \lambda\dot{x} + kx = F(t). \tag{2.34}$$

Now, if $x_1(t)$ is any solution of this equation, and $x_0(t)$ is a solution of
the corresponding *homogeneous* equation (2.26) for the unforced oscillator,
then, as one can easily check by direct substitution, $x_1(t) + x_0(t)$ will be
another solution of (2.34). Hence we only need to find *one* particular solution of this equation. Its general solution is then obtained by adding to
this particular solution the general solution of the homogeneous equation
(2.26). (This will be the *general* solution because it contains two arbitrary
constants.)

   Let us consider first the case where the applied force is periodic in time,
with the simple form

$$F(t) = F_1 \cos\omega_1 t, \tag{2.35}$$

where $F_1$ and $\omega_1$ are real constants. It is convenient to write this in the
form

$$F(t) = \mathrm{Re}(F_1 e^{i\omega_1 t}),$$

and to solve first the equation with a complex force,

$$m\ddot{z} + \lambda\dot{z} + kz = F_1 e^{i\omega_1 t}. \tag{2.36}$$

Then the real part $x$ of this solution will be a solution of the equation (2.34) with the force (2.35).

We now look for a particular solution of (2.36) which is periodic in time with the same period as the applied force,

$$z = A_1 e^{i\omega_1 t} = a_1 e^{i(\omega_1 t - \theta_1)},$$

where $A_1 = a_1 e^{-i\theta_1}$ is a complex constant. Substituting in (2.36), we obtain an equation for $A_1$:

$$(-m\omega_1^2 + i\lambda\omega_1 + k)A_1 = F_1. \tag{2.37}$$

Dividing by $me^{-i\theta_1}$ and rearranging the terms gives

$$(\omega_0^2 + 2i\gamma\omega_1 - \omega_1^2)a_1 = \frac{F_1}{m}e^{i\theta_1},$$

where $\gamma$ and $\omega_0$ are defined as before by (2.28). Hence, equating real and imaginary parts,

$$(\omega_0^2 - \omega_1^2)a_1 = (F_1/m)\cos\theta_1,$$
$$2\gamma\omega_1 a_1 = (F_1/m)\sin\theta_1.$$

The amplitude is found by squaring these equations and adding, to give

$$a_1 = \frac{F_1/m}{\sqrt{(\omega_0^2 - \omega_1^2)^2 + 4\gamma^2\omega_1^2}}. \tag{2.38}$$

Dividing one equation by the other yields the phase:

$$\tan\theta_1 = \frac{2\gamma\omega_1}{\omega_0^2 - \omega_1^2}. \tag{2.39}$$

If $F_1 > 0$, the correct choice between the two solutions for $\theta_1$ differing by $\pi$ is the one lying in the range $0 < \theta_1 < \pi$.

We have found a particular solution of (2.36). Its real part, the corresponding particular solution of our original equation (2.34), is simply

$$x = a_1 \cos(\omega_1 t - \theta_1). \tag{2.40}$$

The general solution, as we discussed above, is obtained by adding to this particular solution the general solution of the corresponding homogeneous equation, namely (2.29) or (2.31). In the case where the damping is less than critical ($\gamma < \omega_0$), we obtain

$$x = a_1 \cos(\omega_1 t - \theta_1) + a e^{-\gamma t} \cos(\omega t - \theta). \tag{2.41}$$

Here, of course, $a_1$ and $\theta_1$ are given by (2.38) and (2.39), but $a$ and $\theta$ are arbitrary constants to be fixed by the initial conditions.

The second term in the general solution (2.41), which represents a free oscillation, dies away exponentially with time. It is therefore called the *transient*. After a long time, the displacement $x$ will be given by the first term of (2.41), that is, by (2.40), alone. Thus, no matter what initial conditions we choose, the oscillations are ultimately governed solely by the external force. Note that their period is the period of the applied force, not the period of the unforced oscillator.

### Resonance

The amplitude $a_1$ and phase $\theta_1$ of the forced oscillations are strongly dependent on the angular frequencies $\omega_0$ and $\omega_1$. In particular, if the damping is small, the amplitude can become very large when the frequencies are almost equal. If we fix $\gamma$ and the forcing frequency $\omega_1$, and vary the oscillator frequency $\omega_0$, the amplitude is a maximum when $\omega_0 = \omega_1$. In this case, we say that the system is in *resonance*. At resonance, the amplitude is

$$a_1 = \frac{F_1}{2m\gamma\omega_1} = \frac{F_1}{\lambda\omega_1}, \tag{2.42}$$

which can be very large if the damping constant $\lambda$ is small. If, on the other hand, we fix the parameters $\gamma$ and $\omega_0$ of the oscillator, and vary the forcing frequency $\omega_1$, the maximum amplitude actually occurs for a frequency slightly lower than $\omega_0$, namely

$$\omega_1 = \sqrt{\omega_0^2 - 2\gamma^2}.$$

However, if $\gamma$ is small, this does not differ much from $\omega_0$. Note that the natural frequency $\omega$ of the oscillator lies between this resonant frequency and the frequency $\omega_0$ of the undamped oscillator.

Near resonance, the dependence of the amplitude $a_1$ on the forcing frequency $\omega_1$ is of the form illustrated in Fig. 2.7(a). The *width* of the resonance, that is the range of frequencies over which the amplitude is large,

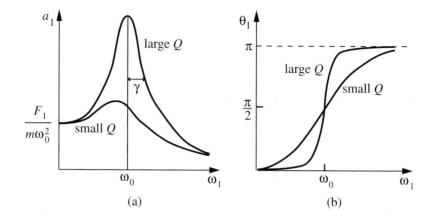

Fig. 2.7

is determined by $\gamma$. For, the amplitude is reduced to $1/\sqrt{2}$ of its peak value when the two terms in the denominator of (2.38) become roughly equal, and for small $\gamma$ this occurs when $\omega_1 = \omega_0 \pm \gamma$. Therefore, $\gamma$ is called the *half-width* of the resonance. Notice the inverse relationship between the width and peak amplitude: the narrower the resonance, the higher is its peak.

The quality factor $Q = \omega_0/2\gamma$ provides a quantitative measure of the sharpness of the resonance peak. Indeed, the ratio of the amplitude at resonance, given by (2.42), to the amplitude at $\omega_1 = 0$ (which describes the response to a time-independent force) is precisely $Q$.

The phenomenon of resonance occurs with any oscillatory system, and is of great practical importance. Since quite small forces can set up large oscillations if the frequencies are in resonance, great care must be taken in the design of any mechanical structure to avoid this possibility. This led to problems for example with the original design of the Millennium Bridge in London.

The constant $\theta_1$ specifies the phase relation between the applied force and the induced oscillations. If the force is slowly oscillating, $\omega_1$ is small, and $\theta_1 \approx 0$, so that the induced oscillations are in phase with the force. In that case, the amplitude (2.38) is

$$a_1 \approx \frac{F_1}{m\omega_0^2} = \frac{F_1}{k}.$$

Thus the position $x$ at any time $t$, given by (2.40), is approximately the equilibrium position under the force $F_1 \cos \omega_1 t - kx$. At resonance, the phase shift is $\theta_1 = \frac{1}{2}\pi$, and the induced oscillations lag behind by a quarter period. (The variation of $\theta_1$ with $\omega_1$ is shown in Fig. 2.7(b).) For very rapidly oscillating forces, $\theta_1 \approx \pi$, and the oscillations are almost exactly out of phase. In this limiting case, $a_1 \approx F_1/m\omega_1^2$ and the oscillations correspond to those of a free particle under the applied oscillatory force. Note that the value of the damping constant is important only in the region near the resonance.

## 2.7   General Periodic Force

The solution we have obtained can immediately be generalized to the case where the applied force is a sum of periodic terms,

$$F(t) = F_1 e^{i\omega_1 t} + F_2 e^{i\omega_2 t} + \cdots = \sum_r F_r e^{i\omega_r t}. \qquad (2.43)$$

We can always ensure that this force is real by including with each term its complex conjugate, so it is not really necessary to write Re in front of the sum. (In particular, a simple periodic force may be written as a sum of two terms proportional to $e^{i\omega t}$ and $e^{-i\omega t}$.) Because of the linearity of (2.34), the corresponding solution is

$$x = \sum_r A_r e^{i\omega_r t} + \text{transient}, \qquad (2.44)$$

where each $A_r$ is related to the corresponding $F_r$ by (2.37). This may easily be verified by direct substitution in (2.34).

   In particular, let us take

$$F(t) = \sum_{n=-\infty}^{\infty} F_n e^{in\omega t}, \qquad (2.45)$$

where $n$ runs over all integers, positive, negative and zero. Here $\omega$ represents a fundamental angular frequency and the frequencies $n\omega$ are its harmonics. Each term in the sum is unchanged by the replacement $t \to t + 2\pi/\omega$, so $F(t)$ is clearly periodic in the sense that

$$F(t + \tau) = F(t), \quad \text{where} \quad \tau = 2\pi/\omega. \qquad (2.46)$$

In fact, *any* periodic force with period $\tau$ can be expressed in the form (2.45), which is called a *Fourier series*. If we are given $F(t)$ it is easy to find the corresponding coefficients $F_n$. For one easily sees that

$$\frac{1}{\tau} \int_0^\tau e^{i(n-m)t} dt = \delta_{nm}, \tag{2.47}$$

where $\delta_{nm}$ is Kronecker's delta symbol defined by

$$\delta_{nm} = \begin{cases} 1 & \text{if } n = m, \\ 0 & \text{if } n \neq m. \end{cases} \tag{2.48}$$

Thus if we multiply (2.45) by $e^{-im\omega t}$ and integrate over a period, the only term in the sum which survives is the one with $n = m$. Hence we obtain

$$F_m = \frac{1}{\tau} \int_0^\tau F(t) e^{-im\omega t} dt. \tag{2.49}$$

A simple example involves a force with 'square-wave' form.

### *Example:* **Square-wave force**

Find the position after a long time of an oscillator subjected to a force with the square-wave form shown in Fig. 2.8, and given by

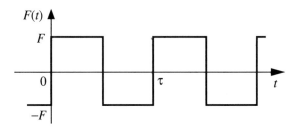

Fig. 2.8

$$F(t) = \begin{cases} F, & r\tau < t \leq (r + \tfrac{1}{2})\tau, \\ -F, & (r - \tfrac{1}{2})\tau < t \leq r\tau, \end{cases}$$

for $r = 0, \pm 1, \pm 2, \ldots$.

Since $\omega = 2\pi/\tau$, we find

$$F_n = \frac{F}{\tau} \int_0^{\tau/2} e^{-2\pi int/\tau} dt - \frac{F}{\tau} \int_{\tau/2}^{\tau} e^{-2\pi int/\tau} dt$$

$$= \frac{F}{i\pi n} [1 - (-1)^n].$$

Note that $F_n = 0$ for all even values of $n$, because the expression in square brackets vanishes. This happens because in addition to repeating after the period $\tau$, $F(t)$ repeats with opposite sign after half this period: $F(t + \frac{1}{2}\tau) = -F(t)$. For odd values of $n$, we find from (2.37) that the amplitudes $A_n$ are

$$A_n = \frac{2F}{i\pi nm(\omega_0^2 - n^2\omega^2 + 2in\gamma\omega)} \qquad (n \text{ odd}).$$

The position of the oscillator is then given by

$$x = \sum_{r=-\infty}^{\infty} A_{2r+1} e^{i(2r+1)\omega t} + \text{transient}.$$

After a long time, the transient can be ignored, and the position is given by the sum alone.

Note that if the damping is small and $\omega_0$ is close to one of the values $n\omega$, then one pair of amplitudes, $A_n$ and $A_{-n}$ will be much larger than the rest, and the oscillation will be almost simple harmonic motion at this frequency. In general, for large $n$, the amplitudes $A_n$ decrease rapidly in magnitude as $n$ increases, so that we can obtain a good approximation to $x$ by keeping only a few terms of the series.

In this way, one can obtain the solution to (2.34) for an arbitrary periodic force. As a matter of fact, this kind of treatment can be extended even further, by replacing the sum in (2.45) by an integral over all possible angular frequencies $\omega$. Any force whatever, subject to some very general mathematical requirements, can be written in this form. (This is the *Fourier integral theorem.*) However, we shall not pursue this method, because we shall find in the next section an alternative method of solving the problem, which is often simpler to use.

## 2.8 Impulsive Forces; the Green's Function Method

There are many physical situations, for example collisions, in which very large forces act for very short times. Let us consider a force $F$ acting on a particle during the time interval $\Delta t$. The resulting change of momentum of the particles is

$$\Delta p = p(t + \Delta t) - p(t) = \int_{t}^{t+\Delta t} F \, dt. \tag{2.50}$$

The quantity on the right is called the *impulse $I$* delivered to the particle. It is natural to consider an idealized situation in which the time interval $\Delta t$ tends to zero, and the whole of the impulse $I$ is delivered to the particle instantaneously. In other words, we let $\Delta t \to 0$ and $F \to \infty$ in such a way that the impulse $I$ remains finite. At the instant when the impulse is delivered the momentum of the particle changes discontinuously. Although in reality the force must always remain finite, this is a good description of, for example, the effect of a sudden blow.

In the next section, we shall discuss some simple collision problems. Here, we shall consider the effect of an impulsive force on an oscillator. Let us suppose first that the oscillator is at rest at its equilibrium position $x = 0$, and that at time $t = 0$ it experiences a blow of impulse $I$. Immediately after the blow, its position is still $x = 0$ (since the velocity remains finite, the position does not change discontinuously), while its velocity is $v_0 = I/m$. Inserting these initial conditions in the general solution (2.31) for the oscillatory case $\gamma < \omega_0$, we find that the position is given by

$$x = \begin{cases} 0, & t < 0, \\ \dfrac{I}{m\omega} e^{-\gamma t} \sin \omega t, & t > 0. \end{cases} \tag{2.51}$$

This solution is illustrated in Fig. 2.9.

Just as in the case of periodic forces, we can immediately generalize this result to a sum of impulsive forces. If the oscillator is subjected to a series of blows, of impulse $I_r$ at time $t_r$, its position may be found by adding together the corresponding set of solutions of the form (2.51). We obtain in this way

$$x(t) = \sum_{r} G(t - t_r) I_r + \text{transient}, \tag{2.52}$$

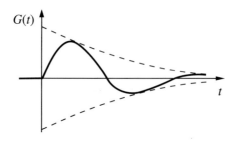

Fig. 2.9

where

$$G(t - t') = \begin{cases} 0, & t < t', \\ \dfrac{1}{m\omega} e^{-\gamma(t-t')} \sin\omega(t - t'), & t > t'. \end{cases} \quad (2.53)$$

The function $G(t - t')$ is called the *Green's function* of the oscillator. It represents the response to a blow of unit impulse delivered at the time $t'$.

We can now use this result to obtain a solution to (2.34) for an arbitrary force $F(t)$. The method is, in a sense, complementary to Fourier's. Instead of expressing $F(t)$ as a sum of periodic forces, we express it as a sum of impulsive forces. If we divide the time scale into small intervals $\Delta t$, then the impulse delivered to the particle between $t$ and $t + \Delta t$ will be approximately $F(t)\Delta t$. Moreover, if $\Delta t$ is short enough, it will be a good approximation to suppose that the whole of this impulse is delivered instantaneously at the time $t$. Thus the solution should be given approximately by (2.52), in which the interval between $t_r$ and $t_{r+1}$ is $\Delta t$ and $I_r = F(t_r)\Delta t$. The approximation will improve as $\Delta t$ is reduced, and become exact in the limit $\Delta t \to 0$. In this limit, the sum in (2.52) goes over into an integral, and we finally obtain

$$x(t) = \int_{t_0}^{t} G(t - t')F(t')\, dt' + \text{transient}. \quad (2.54)$$

The lower limit $t_0$ is the initial time at which the initial conditions determining the arbitrary constants in the transient term are to be imposed. The upper limit may be taken to be $t$ because $G(t - t')$ vanishes for $t' > t$. (In other words, blows subsequent to $t$ do not affect the position at $t$.)

The solution (2.54) is very useful in practice, because it is an explicit solution requiring the evaluation of only one integral. It is particularly well adapted to numerical solution of the problem when $F(t)$ is known numerically.

### *Example:* **Step-function force**

An oscillator is initially at rest and is subject to a step-function force,

$$F(t) = \begin{cases} 0, & t < 0, \\ F, & t > 0, \end{cases}$$

with $F$ constant. What is the position at time $t > 0$?

To perform the integration in (2.54), it is convenient to write the sine function as the imaginary part of an exponential:

$$\begin{aligned}
x(t) &= \int_0^t G(t - t')F \, \mathrm{d}t' = \frac{F}{m\omega} \int_0^t \mathrm{Im} \left[ e^{(-\gamma + i\omega)(t - t')} \right] \mathrm{d}t' \\
&= \frac{F}{m\omega} \mathrm{Im} \left[ \frac{1 - e^{-\gamma t + i\omega t}}{\gamma - i\omega} \right] \\
&= \frac{F}{m\omega(\gamma^2 + \omega^2)} \mathrm{Im} \left[ (\gamma + i\omega) \left( 1 - e^{-\gamma t + i\omega t} \right) \right] \\
&= \frac{F}{m\omega\omega_0^2} [\omega(1 - e^{-\gamma t} \cos \omega t) - \gamma e^{-\gamma t} \sin \omega t],
\end{aligned}$$

using $\gamma^2 + \omega^2 = \omega_0^2$. Note that for large $t$, this tends as it must to the displacement for a constant force $F$, namely $x = F/m\omega_0^2$.

## 2.9 Collision Problems

So far, we have been discussing the motion of a single particle under a known external force. We now turn to the problem of an isolated system of two bodies moving in one dimension under a mutual force $F$ which is given in terms of their positions and velocities. The equations of motion are then

$$m_1\ddot{x}_1 = F, \qquad m_2\ddot{x}_2 = -F. \tag{2.55}$$

As we have already seen in Chapter 1, the sum of these two equations leads to the law of *conservation of momentum*,

$$p_1 + p_2 = P = \text{constant}. \tag{2.56}$$

According to the relativity principle, $F$ must be a function only of the *relative* distance and relative velocity,

$$x = x_1 - x_2 \qquad \text{and} \qquad \dot{x} = \dot{x}_1 - \dot{x}_2.$$

When it is a function only of $x$ the force is conservative, and we can introduce a potential energy function $V(x)$ as before. The law of conservation of energy then takes the form

$$T + V = E = \text{constant}, \qquad\qquad (2.57)$$

with

$$T = \tfrac{1}{2}m_1\dot{x}_1^2 + \tfrac{1}{2}m_2\dot{x}_2^2.$$

This may easily be verified by differentiating (2.57) and using (2.55) and the relation between $V$ and $F$, which is still (2.7).

We are particularly interested in collision problems, in which the force is generally small except when the bodies are very close together. The potential energy function is then a constant (say zero) for large values of $x$, and rises very sharply for small values. (See Fig. 2.10(a).) An ideal *impulsive* conservative force corresponds to a potential energy function with

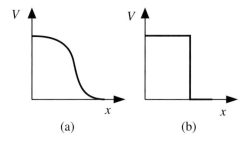

(a)                    (b)

Fig. 2.10

a discontinuity or step (Fig. 2.10(b)). So long as the initial kinetic energy is less than the height of the step, the bodies will bounce off one another. (If the kinetic energy exceeds this value, one body will pass through the other. For completely impenetrable bodies, the potential step would have to be infinitely high.)

From the law of conservation of energy, we know that the final value of the kinetic energy, when the bodies are again far apart, is the same as the initial value well before the collision. If we denote the initial velocities by $u_1, u_2$ and the final velocities by $v_1, v_2$, then (see Fig. 2.11)

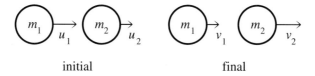

initial                    final

Fig. 2.11

$$\tfrac{1}{2}m_1v_1^2 + \tfrac{1}{2}m_2v_2^2 = \tfrac{1}{2}m_1u_1^2 + \tfrac{1}{2}m_2u_2^2. \tag{2.58}$$

Collisions of this type, in which there is no loss of kinetic energy, are known as *elastic* collisions. They are typical of very hard bodies, like billiard balls.

We saw earlier that a particle which bounces back from a potential barrier emerges with velocity equal (except for sign) to its initial velocity. We can easily derive a similar result for the case of two-body collisions. To do this we write the energy and momentum conservation equations in the forms

$$m_1v_1^2 - m_1u_1^2 = m_2u_2^2 - m_2v_2^2,$$
$$m_1v_1 - m_1u_1 = m_2u_2 - m_2v_2.$$

We can then divide the first equation by the second, to obtain

$$v_1 + u_1 = u_2 + v_2,$$

or, equivalently,

$$v_2 - v_1 = u_1 - u_2. \tag{2.59}$$

This shows that the *relative* velocity is just reversed by the collision.

Equation (2.59) and the momentum conservation equation (2.56) may be solved for the final velocities in terms of the initial ones. In particular, if the second body is initially at rest ($u_2 = 0$), the result is

$$v_1 = \frac{m_1 - m_2}{m_1 + m_2}u_1, \qquad v_2 = \frac{2m_1}{m_1 + m_2}u_1. \tag{2.60}$$

Note that if the masses are equal, then the first body is brought to rest by the collision, and its velocity transferred to the second body. If $m_1 > m_2$, the first body continues in the same direction, with reduced velocity, whereas if $m_1 < m_2$, it rebounds in the opposite direction. In the limit where $m_2$ is much larger than $m_1$, we obtain $v_1 = -u_1$, which agrees with the previous result for a particle rebounding from a fixed potential barrier.

## Inelastic collisions

In practice there is generally some loss of energy in a collision, for instance in the form of heat. In that case, the relative velocity will be reduced in magnitude by the collision. We define the *coefficient of restitution e* in a particular collision by

$$v_2 - v_1 = e(u_1 - u_2). \tag{2.61}$$

The usefulness of this quantity derives from the experimental fact that, for any two given bodies, $e$ is approximately a constant for a wide range of velocities.

The final velocities may again be found from (2.61) and the momentum conservation equation (2.56). When $u_2 = 0$, we get

$$v_1 = \frac{m_1 - em_2}{m_1 + m_2}u_1, \qquad v_2 = \frac{(1+e)m_1}{m_1 + m_2}u_1. \tag{2.62}$$

For $e = 1$ these reduce to the relations (2.60) for the case of an elastic collision. At the other extreme, we have the case of very soft bodies that stick together on impact. Then $e = 0$, and the collision is called *perfectly inelastic* (or, more accurately perhaps, *perfectly plastic*).

The energy loss in an inelastic collision is easily evaluated. The initial kinetic energy is $T = \frac{1}{2}m_1u_1^2$, while the final kinetic energy is $T' = \frac{1}{2}m_1v_1^2 + \frac{1}{2}m_2v_2^2$. Substituting for the final velocities from (2.62) we find, after a simple calculation, that the fractional loss of kinetic energy is

$$\frac{T - T'}{T} = \frac{(1 - e^2)m_2}{m_1 + m_2}. \tag{2.63}$$

This shows, incidentally, that $e$ must always be less than one, unless some energy is released, as for example by an explosion.

## 2.10   Summary

The conservation laws for energy and momentum are among the most important consequences of Newton's laws. They are valid quite generally for an isolated system, as we shall see in later chapters. For a system which is only part of a larger system, and subject to external forces, there may be some transfer of energy and momentum from the system to its surroundings (or *vice versa*). For a single particle moving on a straight line, the total (kinetic plus potential) energy is conserved if the force is a function only of $x$. If the force is velocity-dependent, however, there will be some loss of energy to the surroundings.

The importance of the harmonic oscillator, treated in detail in this chapter, lies in the fact that any system with one degree of freedom behaves like a harmonic oscillator when sufficiently near to a point of stable equilibrium. The methods we have discussed for solving the oscillator equation can therefore be applied to a great variety of problems. In particular, the phenomenon of resonance can occur in any system subjected to periodic forces. It occurs when there is a natural frequency of oscillation close to the forcing frequency, and leads, if the damping is small, to oscillations of very large amplitude.

---

## Problems

1. A harmonic oscillator of angular frequency $2\,\mathrm{s}^{-1}$ is initially at $x = -3\,\mathrm{m}$, with $\dot{x} = 8\,\mathrm{m\,s}^{-1}$. Write the solution in each of the three forms (2.19), (2.20) and (2.23). Find the first time at which $x = 0$, and the first at which $\dot{x} = 0$. (A sketch of the solution may help.)

2. A weight of mass $m$ is hung from the end of a spring which provides a restoring force equal to $k$ times its extension. The weight is released from rest with the spring unextended. Find its position as a function of time, assuming negligible damping.

3. When a mass is suspended from a spring, the equilibrium length is increased by 50 mm. The mass is then given a blow which starts it moving vertically at $200\,\mathrm{mm\,s}^{-1}$. Find the period and amplitude of the resulting oscillations, assuming negligible damping.

4. A pendulum of period 2 s is released from rest at an inclination of 5° to the (downward) vertical. What is its angular velocity when it reaches the vertical? When it first returns to its starting point, it is given an impulsive blow *towards* the vertical that increases the amplitude of swing to 10°. Find its subsequent angular position as a function of time.

5. Write down the potential energy function corresponding to the force $-GMm/x^2$ experienced by a particle of mass $m$ at a distance $x$ from a

planet of mass $M$ ($\gg m$). The particle is released from rest at a distance $a$ from the centre of the planet, whose radius is $R$, and falls under gravity. What is its speed when it strikes the surface? Evaluate this speed for a particle falling to the Earth from $a = 2R$. (Use $R = 6400$ km, and $GM/R^2 = g = 10\,\mathrm{m\,s^{-2}}$.)

6. A particle of mass $m$ moves under a force $F = -cx^3$, where $c$ is a positive constant. Find the potential energy function. If the particle starts from rest at $x = -a$, what is its velocity when it reaches $x = 0$? Where in the subsequent motion does it instantaneously come to rest?

7. A particle of mass $m$ moves (in the region $x > 0$) under a force $F = -kx + c/x$, where $k$ and $c$ are positive constants. Find the corresponding potential energy function. Determine the position of equilibrium, and the frequency of small oscillations about it.

8. A particle of mass $m$ has the potential energy function $V(x) = mk|x|$, where $k$ is a positive constant. What is the force when $x > 0$, and when $x < 0$? Sketch the function $V$ and describe the motion. If the particle starts from rest at $x = -a$, find the time it takes to reach $x = a$.

9. A particle of mass $m$ moves under a conservative force with potential energy function given by

$$V(x) = \begin{cases} \frac{1}{2}k(a^2 - x^2) & \text{for } |x| < a, \\ 0 & \text{for } |x| \geq a, \end{cases}$$

where $a$ and $k$ are constants, and $a > 0$. What is the force on the particle? Sketch the function $V$, for both cases $k > 0$ and $k < 0$, and describe the possible types of motion.

10. The particle of Problem 9 with $k = -m\omega^2 < 0$ is initially in the region $x < -a$, moving to the right with velocity $v$. When it emerges into the region $x > a$, will it do so earlier or later than if it were moving freely under no force? Find an expression for the time difference. (To do the required integral, use a substitution of the form $x = \text{constant} \times \sin\theta$.)

11. The potential energy function of a particle of mass $m$ is $V = -\frac{1}{2}c(x^2 - a^2)^2$, where $c$ and $a$ are positive constants. Sketch this function, and describe the possible types of motion in the three cases (a) $E > 0$, (b) $E < -\frac{1}{2}ca^4$, and (c) $-\frac{1}{2}ca^4 < E < 0$.

12. *The potential energy function of a particle of mass $m$ is $V = cx/(x^2 + a^2)$, where $c$ and $a$ are positive constants. Sketch $V$ as a function of $x$. Find the position of stable equilibrium, and the period of small oscillations about it. Given that the particle starts from this point

with velocity $v$, find the ranges of values of $v$ for which it (a) oscillates, (b) escapes to $-\infty$, and (c) escapes to $+\infty$.

13. *A particle falling under gravity is subject to a retarding force proportional to its velocity. Find its position as a function of time, if it starts from rest, and show that it will eventually reach a terminal velocity. [The equation of motion can be integrated once to give, with a suitable choice of origin and definition of $\gamma$ (differing from (2.28) by a factor of 2), $\dot{z} + \gamma z = -gt$. To integrate again, use an integrating factor, *i.e.* a function $f(t)$ such that when the equation is multiplied by $f(t)$ the left-hand side becomes an exact derivative, in fact the derivative of $zf$. The final stage requires an integration by parts.]

14. The terminal speed of the particle in Problem 13 is $50\,\mathrm{m\,s^{-1}}$. Find the time it takes to reach a speed of $40\,\mathrm{m\,s^{-1}}$, and the distance it has fallen in that time. (Take $g = 10\,\mathrm{m\,s^{-2}}$.)

15. *A particle moves vertically under gravity and a retarding force proportional to the *square* of its velocity. (This is more appropriate than a linear relation for larger particles — see Problem 8.10.) If $v$ is its upward or downward speed, show that $\dot{v} = \mp g - kv^2$, respectively, where $k$ is a constant. If the particle is moving upwards, show that its position at time $t$ is given by $z = z_0 + (1/k)\ln\cos[\sqrt{gk}\,(t_0 - t)]$, where $z_0$ and $t_0$ are integration constants. If its initial velocity at $t = 0$ is $u$, find the time at which it comes instantaneously to rest, and its height then. [Note that ln always denotes the natural logarithm: $\ln x \equiv \log_e x$. You may find the identity $\ln\cos x = -\frac{1}{2}\ln(1 + \tan^2 x)$ useful.]

16. *Show that if the particle of the previous question falls from rest its speed after a time $t$ is given by $v = \sqrt{g/k}\,\tanh(\sqrt{gk}\,t)$. What is its limiting speed? How long does it take to hit the ground if dropped from height $h$?

17. The pendulum described in §2.1 is released from rest at an angle $\theta_0$ to the downward vertical. Find its angular velocity as a function of $\theta$, and the period of small oscillations about the position of stable equilibrium. Write down the solution for $\theta$ as a function of time, assuming that $\theta_0$ is small.

18. *For the pendulum described in §2.1, find the equation of motion for small displacements from the position of *unstable* equilibrium, $\theta = \pi$. Show that if it is released from rest at a small angle to the upward vertical, then the time taken for the angular displacement to increase by a factor of 10 will be approximately $\sqrt{l/g}\ln 20$. Evaluate this time for

a pendulum of period 2 s, and find the angular velocity of the pendulum when it reaches the downward vertical.

19. *The particle of Problem 11 starts from rest at $x = -a$, and is given a small push to start it moving to the right. What is its velocity when it reaches the point $x$? Given that $t = 0$ is the instant when it reaches $x = 0$, find its position as a function of time. (Assume that the push has negligible effect on its energy.)

20. *A particle of mass $m$ moves in the region $x > 0$ under the force $F = -m\omega^2(x - a^4/x^3)$, where $\omega$ and $a$ are constants. Sketch the potential energy function. Find the position of equilibrium, and the period of small oscillations about it. The particle starts from this point with velocity $v$. Find the limiting values of $x$ in the subsequent motion. Show that the period of oscillation is independent of $v$. (To do the integration, transform to the variable $y = x^2$, then add a constant to 'complete the square', and finally use a trigonometric substitution.)

21. Repeat the calculation of Problem 2 assuming that the system is critically damped. Given that the final position of equilibrium is 0.4 m below the point of release, find how close to the equilibrium position the particle is after 1 s.

22. *A pendulum whose period in a vacuum is 1 s is placed in a resistive medium. Its amplitude on each swing is observed to be half that on the previous swing. What is its new period? Given that the pendulum bob is of mass 0.1 kg, and is subjected to a periodic force of amplitude 0.02 N and period 1 s, find the angular amplitude of the resulting forced oscillation. Compare your answer with the angular deviation that would be induced by a constant force of this magnitude in a pendulum at rest.

23. Write down the solution to the oscillator equation for the case $\omega_0 > \gamma$ if the oscillator starts from $x = 0$ with velocity $v$. Show that, as $\omega_0$ is reduced to the critical value $\gamma$, the solution tends to the corresponding solution for the critically damped oscillator.

24. *Solve the problem of an oscillator under a simple periodic force (turned on at $t = 0$) by the Green's function method, and verify that it reproduces the solution of §2.6. [Assume that the damping is less than critical. To do the integral, write $\sin \omega t$ as $(e^{i\omega t} - e^{-i\omega t})/2i$.]

25. *For an oscillator under a periodic force $F(t) = F_1 \cos \omega_1 t$, calculate the power, the rate at which the force does work. (Neglect the transient.) Show that the average power is $P = m\gamma\omega_1^2 a_1^2$, and hence verify that it is equal to the average rate at which energy is dissipated against the resistive force. Show that the power $P$ is a maximum, as a function

of $\omega_1$, at $\omega_1 = \omega_0$, and find the values of $\omega_1$ for which it has half its maximum value.

26. *Find the average value $\bar{E}$ of the total energy of an oscillator under a periodic force. If $W$ is the work done against friction in one period, show that when $\omega_1 = \omega_0$ the ratio $W/\bar{E}$ is related to the quality factor $Q$ by $W/\bar{E} = 2\pi/Q$.

27. Three perfectly elastic bodies of masses $5\,\mathrm{kg}$, $1\,\mathrm{kg}$, $5\,\mathrm{kg}$ are arranged in that order on a straight line, and are free to move along it. Initially, the middle one is moving with velocity $27\,\mathrm{m\,s^{-1}}$, and the others are at rest. Find how many collisions take place in the subsequent motion, and verify that the final value of the kinetic energy is equal to the initial value.

28. A ball is dropped from height $h$ and bounces. The coefficient of restitution at each bounce is $e$. Find the velocity immediately after the first bounce, and immediately after the $n$th bounce. Show that the ball finally comes to rest after a time

$$\frac{1+e}{1-e}\sqrt{\frac{2h}{g}}.$$

29. *An oscillator with free period $\tau$ is critically damped and subjected to a force with the 'saw-tooth' form

$$F(t) = c(t - n\tau) \qquad \text{for} \qquad (n - \tfrac{1}{2})\tau < t \le (n + \tfrac{1}{2})\tau,$$

for each integer $n$ ($c$ is a constant). Find the ratios of the amplitudes $a_n$ of oscillation at the angular frequencies $2\pi n/\tau$.

30. *A particle moving under a conservative force oscillates between $x_1$ and $x_2$. Show that the period of oscillation is

$$\tau = 2 \int_{x_1}^{x_2} \sqrt{\frac{m}{2[V(x_2) - V(x)]}}\,\mathrm{d}x.$$

In particular, if $V = \tfrac{1}{2}m\omega_0^2(x^2 - bx^4)$, show that the period for oscillations of amplitude $a$ is

$$\tau = \frac{2}{\omega_0} \int_{-a}^{a} \frac{\mathrm{d}x}{\sqrt{a^2 - x^2}\sqrt{1 - b(a^2 + x^2)}}.$$

Using the binomial theorem to expand in powers of $b$, and the substitution $x = a\sin\theta$, show that for small amplitude the period is approximately $\tau \approx 2\pi(1 + \tfrac{3}{4}ba^2)/\omega_0$.

31. Use the result of the preceding question to obtain an estimate of the correction factor to the period of a simple pendulum when its angular amplitude is $30°$. (Write $\cos\theta \approx 1 - \frac{1}{2}\theta^2 + \frac{1}{24}\theta^4$.)

32. *Find the Green's function of an oscillator in the case $\gamma > \omega_0$. Use it to solve the problem of an oscillator that is initially in equilibrium, and is subjected from $t = 0$ to a force increasing linearly with time, $F = ct$.

# Chapter 3

# Energy and Angular Momentum

In this chapter, we generalize the discussion of Chapter 2 to the case of motion in two or three dimensions. Throughout this chapter, we shall be concerned with the problem of a particle moving under a known external force $\boldsymbol{F}$.

## 3.1 Energy; Conservative Forces

The *kinetic energy* of a particle of mass $m$ free to move in three dimensions is defined to be

$$T = \tfrac{1}{2}m\dot{r}^2 = \tfrac{1}{2}m(\dot{x}^2 + \dot{y}^2 + \dot{z}^2). \tag{3.1}$$

The rate of change of the kinetic energy is, therefore,

$$\dot{T} = m(\dot{x}\ddot{x} + \dot{y}\ddot{y} + \dot{z}\ddot{z}) = m\dot{\boldsymbol{r}} \cdot \ddot{\boldsymbol{r}} = \dot{\boldsymbol{r}} \cdot \boldsymbol{F}, \tag{3.2}$$

by the equation of motion (1.1). The change in kinetic energy in a time interval $\mathrm{d}t$ during which the particle moves a (vector) distance $\mathrm{d}\boldsymbol{r}$ is then

$$\mathrm{d}T = \mathrm{d}W, \tag{3.3}$$

with

$$\mathrm{d}W = \boldsymbol{F} \cdot \mathrm{d}\boldsymbol{r} = F_x\mathrm{d}x + F_y\mathrm{d}y + F_z\mathrm{d}z. \tag{3.4}$$

This is the three-dimensional expression for the work done by the force $\boldsymbol{F}$ in the displacement $\mathrm{d}\boldsymbol{r}$. Note that it is equal to the distance travelled, $|\mathrm{d}\boldsymbol{r}|$, multiplied by the component of $\boldsymbol{F}$ in the direction of the displacement.

One might think at first sight that a conservative force in three dimensions should be defined to be a force $\boldsymbol{F}(\boldsymbol{r})$ depending only on the position $\boldsymbol{r}$ of the particle. However, this is not sufficient to ensure the existence of a law of conservation of energy, which is the essential feature of a conservative force. We require that

$$T + V = E = \text{constant}, \tag{3.5}$$

where $T$ is given by (3.1), and the *potential energy* $V$ is a function of position, $V(\boldsymbol{r})$. Now, the rate of change of the potential energy is

$$\dot{V}(\boldsymbol{r}) = \frac{\partial V}{\partial x}\dot{x} + \frac{\partial V}{\partial y}\dot{y} + \frac{\partial V}{\partial z}\dot{z},$$

or, in terms of the *gradient* of $V$,

$$\boldsymbol{\nabla}V = \boldsymbol{i}\frac{\partial V}{\partial x} + \boldsymbol{j}\frac{\partial V}{\partial y} + \boldsymbol{k}\frac{\partial V}{\partial z},$$

(see (A.21)), by

$$\dot{V} = \dot{\boldsymbol{r}} \cdot \boldsymbol{\nabla}V. \tag{3.6}$$

Thus, differentiating (3.5), and using (3.2) and (3.6) for $\dot{T}$ and $\dot{V}$, we obtain

$$\dot{\boldsymbol{r}} \cdot (\boldsymbol{F} + \boldsymbol{\nabla}V) = 0.$$

Since this must hold for any value of the velocity $\dot{\boldsymbol{r}}$ of the particle, the quantity in parentheses must vanish, *i.e.*

$$\boldsymbol{F}(\boldsymbol{r}) = -\boldsymbol{\nabla}V(\boldsymbol{r}). \tag{3.7}$$

This is the analogue for the three-dimensional case of (2.7). In terms of components, it reads

$$F_x = -\frac{\partial V}{\partial x}, \qquad F_y = -\frac{\partial V}{\partial y}, \qquad F_z = -\frac{\partial V}{\partial z}. \tag{3.8}$$

For example, suppose that $V$ has the form

$$V = \tfrac{1}{2}kr^2 = \tfrac{1}{2}k(x^2 + y^2 + z^2), \tag{3.9}$$

which describes a three-dimensional harmonic oscillator (discussed in more detail in Chapter 4). Then (3.8) yields

$$\boldsymbol{F} = (-kx, -ky, -kz) = -k\boldsymbol{r}. \tag{3.10}$$

It is not hard to find the necessary and sufficient condition on the force $\boldsymbol{F}(\boldsymbol{r})$ to ensure the existence of a potential energy function $V(\boldsymbol{r})$ satisfying (3.7). Any vector function $\boldsymbol{F}(\boldsymbol{r})$ of the form (3.7) obeys the relation

$$\boldsymbol{\nabla} \wedge \boldsymbol{F} = \boldsymbol{0}, \tag{3.11}$$

that is, its *curl* vanishes (see (A.26)). For instance, the $z$ component of (3.11) is

$$\frac{\partial F_y}{\partial x} - \frac{\partial F_x}{\partial y} = 0,$$

and this is true because of (3.8) and the symmetry of the second partial derivative:

$$\frac{\partial^2 V}{\partial x \, \partial y} = \frac{\partial^2 V}{\partial y \, \partial x}.$$

Thus (3.11) is a necessary condition for the force $\boldsymbol{F}(\boldsymbol{r})$ to be conservative. It can be shown (see Appendix A) that it is also a sufficient condition. More specifically, if the force $\boldsymbol{F}(\boldsymbol{r})$ satisfies this condition, then the work done by the force in a displacement from $\boldsymbol{r}_0$ to $\boldsymbol{r}$ is independent of the path chosen between these points. Therefore, we can define the potential energy $V(\boldsymbol{r})$ in terms of the force by a relation analogous to (2.5).

## 3.2 Projectiles

As a first example of motion in two dimensions let us consider a projectile moving under gravity.

The equation of motion, assuming that atmospheric friction is negligible, is

$$m\ddot{\boldsymbol{r}} = m\boldsymbol{g}, \tag{3.12}$$

where $\boldsymbol{g}$ is the *acceleration due to gravity*, a vector of constant magnitude $g = 9.81 \,\mathrm{m\,s^{-2}}$ directed in the downward vertical direction. (We ignore the small effects due to the Earth's rotation, which will be discussed in Chapter

5, as well as the variation of $g$ with height — see Chapter 6.) If we choose the $z$-axis vertically upwards, then $\boldsymbol{g} = -g\boldsymbol{k}$, so that (3.12) takes the form

$$\ddot{x} = 0, \qquad \ddot{y} = 0, \qquad \ddot{z} = -g. \qquad (3.13)$$

If we choose the $x$- and $y$-axes so that initially the particle is moving in the $xz$-plane, then it will do so always. Thus the motion is effectively two-dimensional. The solutions of (3.13) are

$$x = a + bt, \qquad z = c + dt - \tfrac{1}{2}gt^2,$$

where $a, b, c, d$ are constants to be fixed by the initial conditions.

### *Example:* **Projectile range**

A projectile is launched from the surface of the Earth (assumed flat) with velocity $v$ at an angle $\alpha$ to the horizontal. What is its range?

We may choose the point of launch as the origin, and the time of launch as $t = 0$. At that instant,

$$\dot{x} = v \cos \alpha, \qquad \dot{z} = v \sin \alpha.$$

Thus the solution with the required initial conditions is

$$x = vt \cos \alpha, \qquad z = vt \sin \alpha - \tfrac{1}{2}gt^2. \qquad (3.14)$$

(The trajectory in the $xz$-plane is obtained by eliminating $t$ between these equations, and is easily seen to be a parabola. However, the explicit equation is not needed here.) The projectile will hit the ground when $z = 0$ again, *i.e.*, at the time $t = 2(v \sin \alpha)/g$. Substituting into the first of Eqs. (3.14) yields for the range

$$x = 2(v^2 \cos \alpha \sin \alpha)/g. \qquad (3.15)$$

It is interesting to note that to maximize the range we should choose $\alpha = \pi/4$; the maximum value is $x_{\max} = v^2/g$.

Finally, let us consider the much less trivial problem of a projectile subject also to a resistive force, such as atmospheric drag. As in Chapter 2, we assume that the retarding force is proportional to the velocity, so that the equation of motion becomes

$$m\ddot{\boldsymbol{r}} = -\lambda\dot{\boldsymbol{r}} + m\boldsymbol{g}. \tag{3.16}$$

The actual dependence of atmospheric drag on velocity is definitely non-linear, but this equation nevertheless gives a reasonable qualitative picture of the motion.

Defining $\gamma = \lambda/m$, we may write (3.16) as

$$\ddot{x} = -\gamma\dot{x}, \qquad \ddot{z} = -\gamma\dot{z} - g.$$

These equations may be integrated by the use of integrating factors (see Chapter 2, Problem 13). If the initial velocity is $\boldsymbol{v} = (u, 0, w)$, the required solution is

$$x = \frac{u}{\gamma}(1 - \mathrm{e}^{-\gamma t}), \qquad z = \left(\frac{w}{\gamma} + \frac{g}{\gamma^2}\right)(1 - \mathrm{e}^{-\gamma t}) - \frac{gt}{\gamma}.$$

Eliminating $t$ between these two equations yields the equation of the trajectory,

$$z = \frac{\gamma w + g}{\gamma u}x + \frac{g}{\gamma^2}\ln\left(1 - \frac{\gamma x}{u}\right). \tag{3.17}$$

A distinctive feature of this equation is that, even if the launch point is far above the ground, there is a maximum value of $x$. The projectile can never reach beyond $x = u/\gamma$, and as it approaches that value, $z \to -\infty$. Thus the trajectory ends in a near-vertical drop, when all the forward momentum has been spent.

## 3.3   Moments; Angular Momentum

The *moment about the origin* of a force $\boldsymbol{F}$ acting on a particle at position $\boldsymbol{r}$ is defined to be the vector product

$$\boldsymbol{G} = \boldsymbol{r} \wedge \boldsymbol{F}. \tag{3.18}$$

The components of the vector $\boldsymbol{G}$ are the *moments about the x-, y- and z-axes*,

$$G_x = yF_z - zF_y, \qquad G_y = zF_x - xF_z, \qquad G_z = xF_y - yF_x.$$

The direction of the vector $\boldsymbol{G}$ is that of the normal to the plane of $\boldsymbol{r}$ and $\boldsymbol{F}$. It may be regarded as defining the axis about which the force tends to rotate the particle. The magnitude of $G$ is

$$G = rF \sin \theta = bF,$$

where $\theta$ is the angle between $r$ and $F$, and $b$ is the perpendicular distance
from the origin to the line of the force. (See Fig. 3.1.) Moments of forces

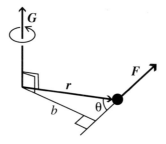

Fig. 3.1

play a particularly important role in the dynamics of rigid bodies, as we
shall see in Chapters 8 and 9.

Correspondingly, we define the vector *angular momentum* (sometimes
called *moment of momentum*) *about the origin* of a particle at position $r$,
and moving with momentum $p$, to be

$$J = r \wedge p = mr \wedge \dot{r}. \tag{3.19}$$

Its components, the *angular momenta about the $x$-, $y$- and $z$-axes*, are

$$J_x = m(y\dot{z} - z\dot{y}), \qquad J_y = m(z\dot{x} - x\dot{z}), \qquad J_z = m(x\dot{y} - y\dot{x}).$$

The momentum $p$ is often called *linear momentum* when it is important to
emphasize the distinction between $p$ and $J$.

The rate of change of angular momentum, obtained by differentiating
(3.19) is

$$\dot{J} = m\frac{\mathrm{d}}{\mathrm{d}t}(r \wedge \dot{r}) = m(\dot{r} \wedge \dot{r} + r \wedge \ddot{r}). \tag{3.20}$$

(Recall that the usual rule for differentiating a product applies also to vector
products, *provided* that the order of the two factors is preserved.) Now,
the first term in (3.20) is zero, because it is the vector product of a vector
with itself. By (1.1), the second term is simply $r \wedge F = G$. Thus we obtain

the important result that the rate of change of the angular momentum is equal to the moment of the applied force:

$$\dot{J} = G. \tag{3.21}$$

This should be compared with the equation $\dot{p} = F$ for the rate of change of the linear momentum.

Since the definition of the vector product (see Appendix, §A.3) depends on the choice of a right-hand screw convention, the directions of the vectors $G$ and $J$ also depend on this convention. A vector of this type is known as an *axial* vector. It is to be contrasted with an ordinary, or *polar*, vector. Axial vectors are often associated with rotation about an axis. What is specified physically is not the direction along the axis, but the sense of rotation about it (see Fig. 3.1).

### 3.4  Central Forces; Conservation of Angular Momentum

An external force is said to be *central* if it is always directed towards or away from a fixed point, called the *centre of force*. If we choose the origin to be this centre, this means that $F$ is always parallel to the position vector $r$. Since the vector product of two parallel vectors is zero, the condition for a force $F$ to be central is that its moment about the centre should vanish:

$$G = r \wedge F = 0. \tag{3.22}$$

From (3.21) it follows that if the force is central, the angular momentum is a constant:

$$J = \text{constant.} \tag{3.23}$$

This is the law of *conservation of angular momentum* in its simplest form. It really contains two statements: that the direction of $J$ is constant, and that its magnitude is constant. Let us look at these in turn.

The direction of $J$ is that of the normal to the plane of $r$ and $v = \dot{r}$. Hence, the statement that this direction is fixed implies that $r$ and $v$ must always lie in a fixed plane. In other words, the motion of the particle is confined to the plane containing the initial position vector and velocity vector (see Fig. 3.2). This is obvious physically; for, since the force is central, it has no component perpendicular to this plane, and, since the normal component of velocity is initially zero, it must always remain zero.

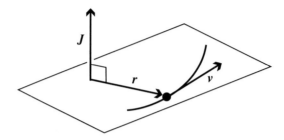

Fig. 3.2

To understand the meaning of the second part of the law, the constancy of the magnitude $J$ of $\boldsymbol{J}$, it is convenient to introduce polar co-ordinates $r$ and $\theta$ in the plane of the motion. In a short time interval, in which the co-ordinates change by amounts $dr$ and $d\theta$, the distances travelled in the radial and transverse directions are $dr$ and $rd\theta$, respectively (see Fig. 3.3). Thus the radial and transverse components of the velocity are

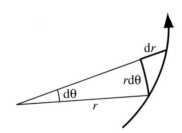

Fig. 3.3

$$v_r = \dot{r}, \qquad v_\theta = r\dot{\theta}. \qquad (3.24)$$

The magnitude of the angular momentum is then

$$J = mrv_\theta = mr^2\dot{\theta}. \qquad (3.25)$$

It is now easy to find a geometrical interpretation of the statement that $J$ is a constant. We note that when the angle $\theta$ changes by an amount $d\theta$, the radius vector sweeps out an area

$$dA = \tfrac{1}{2}r^2 d\theta.$$

Thus the rate of sweeping out area is

$$\frac{\mathrm{d}A}{\mathrm{d}t} = \tfrac{1}{2}r^2\dot\theta = \frac{J}{2m} = \text{constant.} \tag{3.26}$$

One of the most important applications is to planetary motion, and in that context, the constancy of the rate of sweeping out area is generally referred to as *Kepler's second law*, although as we have seen it applies more generally to motion under any central force. It applies even if the force is non-conservative — for example, to a particle attached to the end of a string which is gradually being wound in towards the centre. An equivalent statement of the law, which follows at once from (3.25), is that the transverse component of the velocity, $v_\theta$, varies inversely with the radial distance $r$.

When the force is both central and conservative, the two conservation laws, (3.5) and (3.23), together give us a great deal of information about the motion of the particle. This important special case is the subject of the next chapter.

## 3.5  Polar Co-ordinates

In problems with particular symmetries, it is often convenient to use non-Cartesian co-ordinates. In particular, in the case of axial or spherical symmetry, we may use cylindrical polar co-ordinates $\rho, \varphi, z$, or spherical polar co-ordinates $r, \theta, \varphi$. These are related to Cartesian co-ordinates by

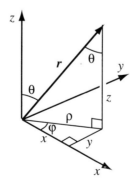

Fig. 3.4

$$x = \rho \cos \varphi = r \sin \theta \cos \varphi,$$
$$y = \rho \sin \varphi = r \sin \theta \sin \varphi, \qquad (3.27)$$
$$z = z \qquad = r \cos \theta.$$

(See Fig. 3.4.) The relation between the two is clearly

$$z = r \cos \theta$$
$$\rho = r \sin \theta, \qquad (3.28)$$
$$\varphi = \varphi.$$

It is easy to find the components of velocity in the various co-ordinate directions, analogous to (3.24). In the cylindrical polar case, the co-ordinates $\rho$ and $\varphi$ are just plane polar co-ordinates in the $xy$-plane, and we obviously have

$$v_\rho = \dot\rho, \qquad v_\varphi = \rho \dot\varphi \qquad v_z = \dot z. \qquad (3.29)$$

In the spherical polar case, a curve along which $\theta$ varies, while $r$ and $\varphi$ are fixed, is a circle of radius $r$ (corresponding on the Earth to a meridian of longitude). A curve along which only $\varphi$ varies is a circle of radius $\rho = r \sin \theta$ (corresponding to a parallel of latitude). Thus the elements of length in the three co-ordinate directions are $dr$, $r \, d\theta$ and $r \sin \theta \, d\varphi$, and we get

$$v_r = \dot r, \qquad v_\theta = r\dot\theta, \qquad v_\varphi = r \sin \theta \, \dot\varphi. \qquad (3.30)$$

For use later in this chapter, it will be very useful to have expressions for the kinetic energy in terms of polar co-ordinates. These co-ordinates, though curvilinear, are still *orthogonal* in the sense that the three co-ordinate directions at each point are mutually perpendicular. Thus we can write, for example,

$$\boldsymbol{v}^2 = v_\rho^2 + v_\varphi^2 + v_z^2.$$

(This would not be true if the co-ordinates were oblique.) Hence from (3.29) and (3.30) we find

$$T = \tfrac{1}{2}m(\dot\rho^2 + \rho^2\dot\varphi^2 + \dot z^2), \qquad (3.31)$$

and

$$T = \tfrac{1}{2}m(\dot r^2 + r^2\dot\theta^2 + r^2 \sin^2 \theta \, \dot\varphi^2). \qquad (3.32)$$

Note the characteristic difference between these expressions and the corresponding one for the case of Cartesian co-ordinates: here $T$ depends not only on the time derivatives of the co-ordinates, but also on the co-ordinates themselves.

We could now go on and find expressions for the acceleration in terms of these co-ordinates, and hence write down equations of motion. While it is not particularly hard to do this directly in these simple cases, it is very convenient to have a general prescription for writing down the equations of motion in arbitrary co-ordinates. A method of doing this that we shall use extensively later is due to the eighteenth-century mathematician Joseph–Louis Lagrange. It can be done as soon as we have found the expression for the kinetic energy, like (3.31) or (3.32).

The simplest — though not the only — way of deriving Lagrange's equations is to use what is known as a *variational principle*. This is a principle which states that some quantity has a minimum value, or, more generally, a stationary value. Principles of this kind are used in many branches of physics, particularly in quantum mechanics and in optics. It may therefore be useful, before discussing the particular principle we shall require, to consider the general kind of problem in which such methods are used. This we shall do in the next section.

## 3.6   The Calculus of Variations

Let us begin with what may seem a ridiculously simple example: what is the shortest path between two points in a plane? Of course we know the answer already, but the method we shall use to derive it can also be applied to less trivial examples — for instance, to find the shortest path between two points on a curved surface.

Suppose the two points are $(x_0, y_0)$ and $(x_1, y_1)$. Any curve joining them is represented by an equation $y = y(x)$ such that the function satisfies the boundary conditions

$$y(x_0) = y_0, \qquad y(x_1) = y_1. \tag{3.33}$$

Consider two neighbouring points on this curve. The distance $\mathrm{d}l$ between them is given by

$$\mathrm{d}l = \sqrt{\mathrm{d}x^2 + \mathrm{d}y^2} = \sqrt{1 + y'^2}\,\mathrm{d}x,$$

where $y' = dy/dx$. Thus the total length of the curve is

$$l = \int_{x_0}^{x_1} \sqrt{1 + y'^2}\,dx. \tag{3.34}$$

The problem, therefore, is to find that function $y(x)$, subject to the boundary conditions (3.33), that will make this integral a minimum.

This problem differs from the more familiar kind of minimum-value problem in that what we have to vary is not just a single variable or set of variables, but a *function* $y(x)$. However, we can still apply the same criterion: when the integral has a minimum value, it must be unchanged to first order by making a small variation in the function $y(x)$. (We shall not be concerned with the problem of distinguishing maxima from minima. All we shall do is to find the stationary values.)

More generally, we may be interested in finding the stationary values of an integral of the form

$$I = \int_{x_0}^{x_1} f(y, y')\,dx, \tag{3.35}$$

where $f(y, y')$ is a specified function of $y$ and its first derivative. We shall solve this general problem, and then apply the result to the integral (3.34).

Consider a small variation $\delta y(x)$ in the function $y(x)$, subject to the condition that the values of $y$ at the end-points are unchanged (see Fig. 3.5):

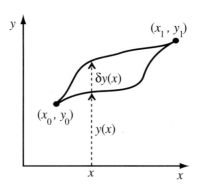

Fig. 3.5

$$\delta y(x_0) = 0, \qquad \delta y(x_1) = 0. \tag{3.36}$$

To first order, the variation in $f(y, y')$ is

$$\delta f = \frac{\partial f}{\partial y}\delta y + \frac{\partial f}{\partial y'}\delta y', \qquad \text{where} \qquad \delta y' = \frac{\mathrm{d}(\delta y)}{\mathrm{d}x}.$$

Thus the variation of the integral $I$ is

$$\delta I = \int_{x_0}^{x_1} \left( \frac{\partial f}{\partial y}\delta y + \frac{\partial f}{\partial y'}\frac{\mathrm{d}(\delta y)}{\mathrm{d}x} \right) \mathrm{d}x.$$

In the second term, we may integrate by parts. The integrated term, namely

$$\left[ \frac{\partial f}{\partial y'}\delta y \right]_{x_0}^{x_1},$$

vanishes because of the conditions (3.36). Hence we obtain

$$\delta I = \int_{x_0}^{x_1} \left[ \frac{\partial f}{\partial y} - \frac{\mathrm{d}}{\mathrm{d}x}\left( \frac{\partial f}{\partial y'} \right) \right] \delta y(x)\, \mathrm{d}x. \tag{3.37}$$

Now, in order that $I$ be stationary, this variation $\delta I$ must vanish for an *arbitrary* small variation $\delta y(x)$ (subject only to the boundary conditions (3.36)). This is only possible if the integrand vanishes identically. Thus we require

$$\frac{\partial f}{\partial y} - \frac{\mathrm{d}}{\mathrm{d}x}\left( \frac{\partial f}{\partial y'} \right) = 0. \tag{3.38}$$

This is known as the *Euler–Lagrange equation*. It is in general a second-order differential equation for the unknown function $y(x)$, whose solution contains two arbitrary constants that may be determined from the boundary conditions.

We can now return to solving the problem we started with, of finding the curves of shortest length (called *geodesics*) in a plane, pretending for the moment that we don't know the answer!

### *Example:* Geodesics on a plane

What is the shortest path between two given points in a plane?

In this case, comparing (3.34) with (3.35), we have to choose $f = \sqrt{1 + y'^2}$ and, therefore,

$$\frac{\partial f}{\partial y} = 0, \qquad \frac{\partial f}{\partial y'} = \frac{y'}{\sqrt{1 + y'^2}}.$$

Thus the Euler–Lagrange equation (3.38) reads

$$\frac{\mathrm{d}}{\mathrm{d}x}\left(\frac{y'}{\sqrt{1+y'^2}}\right) = 0.$$

This equation states that the expression inside the parentheses is a constant, and hence that $y'$ is a constant. Its solutions are therefore the straight lines

$$y = ax + b \qquad (a, b = \text{constants}).$$

Thus we have proved that the shortest path between two points in a plane is a straight line! The constants $a$ and $b$ are of course fixed by the conditions (3.33).

So far, we have used $x$ as the independent variable, but in the applications we consider later we shall be concerned instead with functions of the time $t$. It is easy to generalize the discussion to the case of a function $f$ of $n$ variables $q_1, q_2, \ldots, q_n$ and their time derivatives $\dot{q}_1, \dot{q}_2, \ldots, \dot{q}_n$. In order that the integral

$$I = \int_{t_0}^{t_1} f(q_1, \ldots, q_n, \dot{q}_1, \ldots, \dot{q}_n)\mathrm{d}t$$

be stationary, it must be unchanged to first order by a variation in any one of the functions $q_i(t)$ $(i = 1, 2, \ldots, n)$, subject to the conditions $\delta q_i(t_0) = \delta q_i(t_1) = 0$. Thus we require the $n$ Euler–Lagrange equations

$$\frac{\partial f}{\partial q_i} - \frac{\mathrm{d}}{\mathrm{d}t}\left(\frac{\partial f}{\partial \dot{q}_i}\right) = 0, \qquad i = 1, 2, \ldots, n. \tag{3.39}$$

These $n$ second-order partial differential equations determine the $n$ functions $q_i(t)$ to within $2n$ arbitrary constants of integration.

## 3.7  Hamilton's Principle; Lagrange's Equations

We shall now show that the equations of motion for a particle moving under a conservative force can be written as the Euler–Lagrange equations corresponding to a suitable integral. We define the *Lagrangian function*

$$L = T - V = \tfrac{1}{2}m(\dot{x}^2 + \dot{y}^2 + \dot{z}^2) - V(x, y, z). \tag{3.40}$$

(Note the minus sign — this is *not* the total energy.) Its derivatives are

$$\frac{\partial L}{\partial \dot{x}} = m\dot{x} = p_x, \qquad \frac{\partial L}{\partial x} = -\frac{\partial V}{\partial x} = F_x, \tag{3.41}$$

with similar equations for the $y$ and $z$ components. Thus the equation of motion $\dot{p}_x = F_x$ can be written in the form

$$\frac{\mathrm{d}}{\mathrm{d}t}\left(\frac{\partial L}{\partial \dot{x}}\right) = \frac{\partial L}{\partial x}. \tag{3.42}$$

But this has precisely the form of an Euler–Lagrange equation for the integral

$$I = \int_{t_0}^{t_1} L \, \mathrm{d}t. \tag{3.43}$$

This is generally known as the *action integral*. Strictly speaking, this terminology does not accord with the original, in which the term 'action' was used for a related but different quantity (the integral of twice the kinetic energy), which is relatively little used. However, it is more convenient, and has now become standard practice, to use this simple name for the commonly occurring function (3.43).

Thus we have arrived at *Hamilton's principle* (named after the nineteenth-century Irish mathematician and physicist Sir William Rowan Hamilton): the action integral $I$ is stationary under arbitrary variations $\delta x, \delta y, \delta z$ which vanish at the limits of integration $t_0$ and $t_1$.

The importance of this principle lies in the fact that it can immediately be applied to any set of co-ordinates. If, instead of using Cartesian co-ordinates $x, y, z$, we employ a set of curvilinear co-ordinates $q_1, q_2, q_3$, then we can express the Lagrangian function $L = T - V$ in terms of $q_1, q_2, q_3$ and their time derivatives, $\dot{q}_1, \dot{q}_2, \dot{q}_3$. The action integral must then be stationary with respect to arbitrary variations $\delta q_1, \delta q_2, \delta q_3$, subject to the conditions that the variations are zero at the limits $t_0$ and $t_1$. Thus we obtain *Lagrange's equations*:

$$\frac{\mathrm{d}}{\mathrm{d}t}\left(\frac{\partial L}{\partial \dot{q}_i}\right) = \frac{\partial L}{\partial q_i}, \qquad i = 1, 2, 3. \tag{3.44}$$

These are the equations of motion in terms of the co-ordinates $q_1, q_2, q_3$.

By analogy with (3.41), we may define the *generalized momenta* and *generalized forces*,

$$p_i = \frac{\partial L}{\partial \dot{q}_i} \qquad \text{and} \qquad F_i = \frac{\partial L}{\partial q_i}, \tag{3.45}$$

so that the equations (3.44) read

$$\dot{p}_i = F_i. \tag{3.46}$$

It must be emphasized, however, that in general the $p_i$ and $F_i$ are *not* components of the momentum vector $\boldsymbol{p}$ and the force vector $\boldsymbol{F}$ — we shall see this explicitly in the examples below.

To see how this works in a simple case, let us return to the simple pendulum discussed in §2.1.

---

*Example:* **Simple pendulum revisited**

Find the equation of motion for a simple pendulum.

Using the potential energy function (2.9), we find

$$L = T - V = \tfrac{1}{2}ml^2\dot{\theta}^2 - mgl(1 - \cos\theta).$$

Thus

$$\frac{\partial L}{\partial \dot{\theta}} = ml^2\dot{\theta}, \qquad \frac{\partial L}{\partial \theta} = -mgl\sin\theta,$$

and Lagrange's equation yields

$$ml^2\ddot{\theta} = -mgl\sin\theta.$$

Note that to find this equation of motion, we have not needed expressions for the polar components of acceleration, as we would to derive it directly from Newton's laws; nor have we needed to introduce the tension in the rod.

---

We can also apply these equations to the polar coordinates discussed in §3.5. In the cylindrical polar case, the derivatives of the kinetic energy function (3.31) are

$$\frac{\partial T}{\partial \dot{\rho}} = m\dot{\rho}, \qquad \frac{\partial T}{\partial \dot{\varphi}} = m\rho^2 \dot{\varphi}, \qquad \frac{\partial T}{\partial \dot{z}} = m\dot{z},$$

$$\frac{\partial T}{\partial \rho} = m\rho\dot{\varphi}^2, \qquad \frac{\partial T}{\partial \varphi} = 0, \qquad \frac{\partial T}{\partial z} = 0.$$

Hence Lagrange's equations (3.44) take the form

$$\frac{\mathrm{d}}{\mathrm{d}t}(m\dot{\rho}) = m\rho\dot{\varphi}^2 - \frac{\partial V}{\partial \rho},$$

$$\frac{\mathrm{d}}{\mathrm{d}t}(m\rho^2\dot{\varphi}) = -\frac{\partial V}{\partial \varphi}, \tag{3.47}$$

$$\frac{\mathrm{d}}{\mathrm{d}t}(m\dot{z}) = -\frac{\partial V}{\partial z}.$$

The generalized momenta $m\dot{\rho}$ and $m\dot{z}$ corresponding to the co-ordinates $\rho$ and $z$ are the components of the momentum vector $\boldsymbol{p}$ in the $\rho$ and $z$ directions. However, the generalized momentum $m\rho^2\dot{\varphi}$ is an *angular* momentum; in fact, the angular momentum $J_z$ about the $z$-axis. It is easy to see that, whenever $q_i$ is an angle, the corresponding generalized momentum $p_i$ has the dimensions of angular, rather than linear, momentum. Note that when the potential energy is independent of $\varphi$, then the second of Eqs. (3.47) expresses the fact that the component $J_z$ of angular momentum is conserved. One must be careful, when using the notation $p_\varphi$ to specify whether this is the $\varphi$ component of the momentum vector $\boldsymbol{p}$ or the $\varphi$ generalized momentum, $J_z$.

Proceeding in a similar way, we can write down the equations of motion in spherical polars, using the derivatives of the kinetic energy function (3.32). We find

$$\frac{\mathrm{d}}{\mathrm{d}t}(m\dot{r}) = mr(\dot{\theta}^2 + \sin^2\theta\,\dot{\varphi}^2) - \frac{\partial V}{\partial r},$$

$$\frac{\mathrm{d}}{\mathrm{d}t}(mr^2\dot{\theta}) = mr^2\sin\theta\cos\theta\,\dot{\varphi}^2 - \frac{\partial V}{\partial \theta}, \tag{3.48}$$

$$\frac{\mathrm{d}}{\mathrm{d}t}(mr^2\sin^2\theta\,\dot{\varphi}) = -\frac{\partial V}{\partial \varphi}.$$

Once again, the generalized momentum $mr^2\sin^2\theta\,\dot{\varphi}$ is the component of angular momentum $J_z$ about the $z$-axis, and is a constant if $V$ is independent of $\varphi$. The generalized momentum $mr^2\dot{\theta}$ also has the dimensions of angular momentum. Since it is equal to the radial distance $r$ multiplied by the $\theta$ component of momentum, $mr\dot{\theta}$, it is in fact the $\varphi$ component of angular momentum, $J_\varphi$. Note, however, that even when $V$ is independent of both angular variables $\theta$ and $\varphi$, as is the case for a central force, $J_\varphi$ is not in general a constant. This is because the $\varphi$ direction is not a fixed direction in space, like the $z$-axis, but changes according to the position of the particle. Thus, even when $\boldsymbol{J}$ is a fixed vector, the component $J_\varphi$ may vary.

Before concluding this section, it will be useful to discuss briefly the special case of a *central* conservative force, which is treated in detail in the following chapter. In spherical polars, the components of the force $\boldsymbol{F}$ (*not* the generalized forces) are given in terms of the potential by

$$F_r = -\frac{\partial V}{\partial r}, \qquad F_\theta = -\frac{1}{r}\frac{\partial V}{\partial \theta}, \qquad F_\varphi = -\frac{1}{r\sin\theta}\frac{\partial V}{\partial \varphi}.$$

(See (A.50).) Thus the force will be purely radial if, and only if, $V$ is independent of both angular variables $\theta$ and $\varphi$. The potential energy function (3.9) of the three-dimensional harmonic oscillator is an example. Then we have

$$\boldsymbol{F} = -\hat{\boldsymbol{r}}\frac{\mathrm{d}V(r)}{\mathrm{d}r}. \tag{3.49}$$

This shows incidentally that the magnitude of a central, conservative force depends only on $r$, not on $\theta$ or $\varphi$.

We know from the discussion of §3.4 that the angular momentum vector $\boldsymbol{J}$ in this case is a constant, and that the motion is confined to a plane. The equations of motion are considerably simplified by an appropriate choice of the polar axis $\theta = 0$. If it is chosen so that $\dot\varphi = 0$ initially, then by the third of Eqs. (3.48), $\dot\varphi$ will always be zero. The motion is then in a plane $\varphi = $ constant, through the polar axis, in which $r$ and $\theta$ play the role of plane polar co-ordinates. The angular momentum $J = J_\varphi = mr^2\dot\theta$ *is* then a constant, by the second of Eqs. (3.48).

Similarly, if the particle is originally moving in the 'equatorial' plane, $\theta = \frac{\pi}{2}$ with $\dot\theta = 0$, then from the second equation it follows that $\theta$ is always $\frac{\pi}{2}$. In this case, the plane polar co-ordinates are $r$ and $\varphi$, and the angular momentum is $J = J_z = mr^2\dot\varphi$.

## 3.8   Summary

The rate of change of kinetic energy of a particle is the rate at which the force does work, $\dot{T} = \dot{\boldsymbol{r}} \cdot \boldsymbol{F}$. The force is conservative if it has the form

$\boldsymbol{F} = -\boldsymbol{\nabla}V(\boldsymbol{r})$, and in that case the total energy $T + V$ is a constant. The condition for the existence of such a function is that $\boldsymbol{F}$ should be a function of position such that $\boldsymbol{\nabla} \wedge \boldsymbol{F} = \boldsymbol{0}$.

The rate of change of angular momentum of a particle is equal to the moment of the force, $\dot{\boldsymbol{J}} = \boldsymbol{r} \wedge \boldsymbol{F}$. When the force is central, the angular momentum is conserved. Then the motion is confined to a plane, and the rate of sweeping out area in this plane is a constant.

The use of these conservation laws greatly simplifies the treatment of any problem involving central or conservative forces. When the force is both central and conservative, they provide all the information we need to determine the motion of the particle, as we shall see in the following chapter.

Lagrange's equations are of great importance in advanced treatments of classical mechanics (and also in quantum mechanics). We have seen that they can be used to write down equations of motion in any system of coordinates, as soon as we have found the expressions for the kinetic energy and potential energy. In later chapters, we shall see that the method can readily and usefully be extended to more complicated systems than a single particle.

## Problems

1. Find which of the following forces are conservative, and for those that are find the corresponding potential energy function ($a$ and $b$ are constants, and $\boldsymbol{a}$ is a constant vector):
   (a)  $F_x = ax + by^2, \quad F_y = az + 2bxy, \quad F_z = ay + bz^2$;
   (b)  $F_x = ay, \quad F_y = az, \quad F_z = ax$;
   (c)  $F_r = 2ar \sin\theta \sin\varphi, \quad F_\theta = ar \cos\theta \sin\varphi, \quad F_\varphi = ar \cos\varphi$;
   (d)  $\boldsymbol{F} = \boldsymbol{a} \wedge \boldsymbol{r}$;
   (e)  $\boldsymbol{F} = r\boldsymbol{a}$;
   (f)  $\boldsymbol{F} = \boldsymbol{a}(\boldsymbol{a} \cdot \boldsymbol{r})$.

2. Given that the force is as in Problem 1(a), evaluate the work done in taking a particle from the origin to the point $(1, 1, 0)$: (i) by moving first along the $x$-axis and then parallel to the $y$-axis, and (ii) by going in a straight line. Verify that the result in each case is equal to minus the change in the potential energy function.

3. Repeat the calculations of Problem 2 for the force in 1(b).

4. Compute the work done in taking a particle around the circle $x^2 + y^2 =$

$a^2, z = 0$ if the force is (a) $\boldsymbol{F} = y\boldsymbol{i}$, and (b) $\boldsymbol{F} = x\boldsymbol{i}$. What do you conclude about these forces? (Use the parametrization $x = a\cos\varphi, y = a\sin\varphi, z = 0$.)

5. Evaluate the force corresponding to the potential energy function $V(\boldsymbol{r}) = cz/r^3$, where $c$ is a constant. Write your answer in vector notation, and also in spherical polars, and verify that it satisfies $\boldsymbol{\nabla}\wedge\boldsymbol{F} = \boldsymbol{0}$.

6. A projectile is launched with velocity $100\,\mathrm{m\,s^{-1}}$ at $60°$ to the horizontal. Atmospheric drag is negligible. Find the maximum height attained and the range. What other angle of launch would give the same range? Find the time of flight in each of the two cases.

7. *Find the equation for the trajectory of a projectile launched with velocity $v$ at an angle $\alpha$ to the horizontal, assuming negligible atmospheric resistance. Given that the ground slopes at an angle $\beta$, show that the range of the projectile (measured horizontally) is

$$x = \frac{2v^2}{g}\frac{\sin(\alpha - \beta)\cos\alpha}{\cos\beta}.$$

At what angle should the projectile be launched to achieve the maximum range?

8. *By expanding the logarithm in (3.17), find the approximate equation for the trajectory of a projectile subject to small atmospheric drag to first order in $\gamma$. (Note that this requires terms up to order $\gamma^3$ in the logarithm.) Show that to this order the range (on level ground) is

$$x = \frac{2uw}{g} - \frac{8\gamma uw^2}{3g^2},$$

and hence that to maximize the range for given launch speed $v$ the angle of launch should be chosen to satisfy $\cos 2\alpha = \sqrt{2}\gamma v/3g$. (*Hint*: In the term containing $\gamma$, you may use the zeroth-order approximation for the angle.) For a projectile whose terminal speed if dropped from rest (see Chapter 2, Problem 13) would be $500\,\mathrm{m\,s^{-1}}$, estimate the optimal angle and the range if the launch speed is $100\,\mathrm{m\,s^{-1}}$.

9. *Show that in the limit of strong damping (large $\gamma$) the time of flight of a projectile (on level ground) is approximately $t \approx (w/g + 1/\gamma)(1 - e^{-1-\gamma w/g})$. Show that to the same order of accuracy the range is $x \approx (u/\gamma)(1 - e^{-1-\gamma w/g})$. For a projectile launched at $800\,\mathrm{m\,s^{-1}}$ with $\gamma = 0.1\,\mathrm{s^{-1}}$, estimate the range for launch angles of $30°, 20°$ and $10°$.

10. A particle of mass $m$ is attached to the end of a light string of length $l$. The other end of the string is passed through a small hole and is

slowly pulled through it. Gravity is negligible. The particle is originally spinning round the hole with angular velocity $\omega$. Find the angular velocity when the string length has been reduced to $\frac{1}{2}l$. Find also the tension in the string when its length is $r$, and verify that the increase in kinetic energy is equal to the work done by the force pulling the string through the hole.

11. A particle of mass $m$ is attached to the end of a light spring of equilibrium length $a$, whose other end is fixed, so that the spring is free to rotate in a horizontal plane. The tension in the spring is $k$ times its extension. Initially the system is at rest and the particle is given an impulse that starts it moving at right angles to the spring with velocity $v$. Write down the equations of motion in polar co-ordinates. Given that the maximum radial distance attained is $2a$, use the energy and angular momentum conservation laws to determine the velocity at that point, and to find $v$ in terms of the various parameters of the system. Find also the values of $\ddot{r}$ when $r = a$ and when $r = 2a$.

12. *A light rigid cylinder of radius $2a$ is able to rotate freely about its axis, which is horizontal. A particle of mass $m$ is fixed to the cylinder at a distance $a$ from the axis and is initially at rest at its lowest point. A light string is wound on the cylinder, and a steady tension $F$ applied to it. Find the angular acceleration and angular velocity of the cylinder after it has turned through an angle $\theta$. Show that there is a limiting tension $F_0$ such that if $F < F_0$ the motion is oscillatory, but if $F > F_0$ it continues to accelerate. Estimate the value of $F_0$ by numerical approximation.

13. In the system of Problem 12, instead of a fixed tension applied to the string, a weight of mass $m/2$ is hung on it. Use the energy conservation equation to find the angular velocity as a function of $\theta$. Find also the angular acceleration and the tension in the string. (Compare your results with those in Problem 12.) Show that there is a point at which the tension falls to zero, and find the angle at which this occurs. What happens immediately beyond this point?

14. A wedge-shaped block of mass $M$ rests on a smooth horizontal table. A small block of mass $m$ is placed on its upper face, which is also smooth and inclined at an angle $\alpha$ to the horizontal. The system is released from rest. Write down the horizontal component of momentum, and the kinetic energy of the system, in terms of the velocity $v$ of the wedge and the velocity $u$ of the small block relative to it. Using conservation of momentum and the equation for the rate of change of kinetic energy,

find the accelerations of the blocks. Given that $M = 1\,\text{kg}$ and $m = 250\,\text{g}$, find the angle $\alpha$ that will maximize the acceleration of the wedge.

15. *A particle starts from rest and slides down a smooth curve under gravity. Find the shape of the curve that will minimize the time taken between two given points. [Take the origin as the starting point and the $z$ axis downwards. Show that the time taken is

$$\int_0^{z_1} \left[ \frac{1 + (\mathrm{d}x/\mathrm{d}z)^2}{2gz} \right]^{1/2} \mathrm{d}z,$$

and hence that for a minimum

$$\left( \frac{\mathrm{d}x}{\mathrm{d}z} \right)^2 = \frac{c^2 z}{1 - c^2 z},$$

where $c$ is an integration constant. To complete the integration, use the substitution $z = c^{-2} \sin^2 \theta$. This famous curve is known as the *brachistochrone*. It is in fact an example of a *cycloid*, the locus of a point on the rim of a circle of radius $1/2c^2$ being rolled beneath the $x$-axis.

This problem was first posed by Johann Bernoulli on New Year's Day 1697 as an open challenge. Newton's brilliant solution method initiated the calculus of variations. Bernoulli had an equally brilliant idea, using an optical analogy with refraction of a light ray through a sequence of plates and Fermat's principle of least time.]

16. *The position on the surface of a cone of semi-vertical angle $\alpha$ is specified by the distance $r$ from the vertex and the azimuth angle $\varphi$ about the axis. Show that the shortest path (or *geodesic*) along the surface between two given points is specified by a function $r(\varphi)$ obeying the equation

$$r \frac{\mathrm{d}^2 r}{\mathrm{d}\varphi^2} - 2 \left( \frac{\mathrm{d}r}{\mathrm{d}\varphi} \right)^2 - r^2 \sin^2 \alpha = 0.$$

Show that the solution is $r = r_0 \sec[(\varphi - \varphi_0) \sin \alpha]$, where $r_0$ and $\varphi_0$ are constants. [The equation may be solved by the standard technique of introducing a new dependent variable $u = \mathrm{d}r/\mathrm{d}\varphi$, and writing

$$\frac{\mathrm{d}^2 r}{\mathrm{d}\varphi^2} = \frac{\mathrm{d}u}{\mathrm{d}\varphi} = \frac{\mathrm{d}r}{\mathrm{d}\varphi} \frac{\mathrm{d}u}{\mathrm{d}r} = u \frac{\mathrm{d}u}{\mathrm{d}r}.$$

The substitution $u^2 = v$ then reduces the equation to a linear form that may be solved by using an integrating factor (see Chapter 2, Problem

13). Finally, to calculate $\varphi = \int(1/u)\mathrm{d}r$, one may use the substitution $r = 1/x$.]

17. *Find the geodesics on a sphere of unit radius. [*Hint*: Use $\theta$ as independent variable, and look for the path $\varphi = \varphi(\theta)$. To perform the integration, use the substitution $x = \cot\theta$.]

18. *Parabolic co-ordinates* $(\xi, \eta)$ in a plane are defined by $\xi = r + x, \eta = r - x$. Find $x$ and $y$ in terms of $\xi$ and $\eta$. Show that the kinetic energy of a particle of mass $m$ is

$$T = \frac{m}{8}(\xi + \eta)\left(\frac{\dot{\xi}^2}{\xi} + \frac{\dot{\eta}^2}{\eta}\right).$$

Hence find the equations of motion.

19. Write down the equations of motion in polar co-ordinates for a particle of unit mass moving in a plane under a force with potential energy function $V = -k\ln r + cr + gr\cos\theta$, where $k, c$ and $g$ are positive constants. Find the positions of equilibrium (a) if $c > g$, and (b) if $c < g$. By considering the equations of motion near these points, determine whether the equilibrium is stable (*i.e.*, will the particle, if given a small displacement, tend to return repeatedly?).

20. If $q_1, q_2, q_3$ are orthogonal curvilinear co-ordinates, and the element of length in the $q_i$ direction is $h_i \mathrm{d}q_i$, write down (a) the kinetic energy $T$ in terms of the generalized velocities $\dot{q}_i$, (b) the generalized momentum $p_i$ and (c) the component $\boldsymbol{e}_i \cdot \boldsymbol{p}$ of the momentum vector $\boldsymbol{p}$ in the $q_i$ direction. (Here $\boldsymbol{e}_i$ is a unit vector in the direction of increasing $q_i$.)

21. *By comparing the Euler–Lagrange equations with the corresponding components of the equation of motion $m\ddot{\boldsymbol{r}} = -\boldsymbol{\nabla}V$, show that the component of the acceleration vector in the $q_i$ direction is

$$\boldsymbol{e}_i \cdot \ddot{\boldsymbol{r}} = \frac{1}{mh_i}\left[\frac{\mathrm{d}}{\mathrm{d}t}\left(\frac{\partial T}{\partial \dot{q}_i}\right) - \frac{\partial T}{\partial q_i}\right].$$

Use this result to identify the components of the acceleration in cylindrical and spherical polars.

22. For the case of plane polar co-ordinates $r, \theta$, write the unit vectors $\boldsymbol{e}_r$ $(= \hat{\boldsymbol{r}})$ and $\boldsymbol{e}_\theta$ in terms of $\boldsymbol{i}$ and $\boldsymbol{j}$. Hence show that $\partial \boldsymbol{e}_r/\partial\theta = \boldsymbol{e}_\theta$ and $\partial \boldsymbol{e}_\theta/\partial\theta = -\boldsymbol{e}_r$. By starting with $\boldsymbol{r} = r\boldsymbol{e}_r$ and differentiating, rederive the expressions for the components of the velocity and acceleration vectors.

23. *Find the corresponding formulae for $\partial \boldsymbol{e}_i/\partial q_j$ for spherical polar co-ordinates, and hence verify the results obtained in Problem 21.

24. *The motion of a particle in a plane may be described in terms of *elliptic co-ordinates* $\lambda, \theta$ defined by

$$x = c \cosh \lambda \cos \theta, \qquad y = c \sinh \lambda \sin \theta, \qquad (\lambda \geq 0,\ 0 \leq \theta \leq 2\pi),$$

where $c$ is a positive constant. Show that the kinetic energy function may be written

$$T = \tfrac{1}{2} m c^2 (\cosh^2 \lambda - \cos^2 \theta)(\dot{\lambda}^2 + \dot{\theta}^2).$$

Hence write down the equations of motion.

25. *The method of *Lagrange multipliers* (see Appendix A, Problem 11) can be extended to the calculus of variations. To find the maxima and minima of an integral $I$, subject to the condition that another integral $J = 0$, we have to find the stationary points of the integral $I - \lambda J$ under variations of the function $y(x)$ *and* of the parameter $\lambda$. Apply this method to find the *catenary*, the shape in which a uniform heavy chain hangs between two fixed supports. [The required shape is the one that minimizes the total potential energy, subject to the condition that the total length is fixed. Show that this leads to a variational problem with

$$f(y, y') = (y - \lambda)\sqrt{1 + y'^2},$$

and hence to the equation

$$(y - \lambda)y'' = 1 + y'^2.$$

Solve this equation by introducing the new variable $u = y'$ and solving for $u(y)$.]

26. *A curve of given total length is drawn in a plane between the points $(\pm a, 0)$. Using the method of Problem 25, find the shape that will enclose the largest possible area between the curve and the $x$-axis.

# Chapter 4

# Central Conservative Forces

The most important forces of this type are the gravitational and electro-static forces, which obey the 'inverse square law'. Much of this chapter will therefore be devoted to this special case. We begin, however, by discussing a problem for which it is particularly easy to solve the equations of motion — the three-dimensional analogue of the harmonic oscillator discussed in §2.2.

## 4.1 The Isotropic Harmonic Oscillator

We consider a particle moving under the action of a central restoring force proportional to its distance from the origin,

$$\boldsymbol{F} = -k\boldsymbol{r}, \tag{4.1}$$

with $k$ constant. The corresponding equation of motion is

$$m\ddot{\boldsymbol{r}} + k\boldsymbol{r} = \boldsymbol{0},$$

or, in terms of components,

$$m\ddot{x} + kx = 0, \qquad m\ddot{y} + ky = 0, \qquad m\ddot{z} + kz = 0.$$

The equation of motion for each co-ordinate is thus identical with the equation (2.13) for the simple harmonic oscillator. This oscillator is called *isotropic* because all directions (not only those along the axes!) are equiv-alent. The *anisotropic* oscillator (which we shall not discuss) is described by similar equations, but with *different* constants in the three equations.

The general solution is again given by (2.19):

$$x = c_x \cos \omega t + d_x \sin \omega t,$$
$$y = c_y \cos \omega t + d_y \sin \omega t,$$
$$z = c_z \cos \omega t + d_z \sin \omega t,$$

where $\omega^2 = k/m$. In vector notation, the solution is

$$\boldsymbol{r} = \boldsymbol{c} \cos \omega t + \boldsymbol{d} \sin \omega t. \tag{4.2}$$

Clearly, the motion is periodic, with period $\tau = 2\pi/\omega$. (Other forms of the solution can be given, analogous to (2.20) or (2.23).)

As in the one-dimensional case, the arbitrary constant vectors $\boldsymbol{c}$ and $\boldsymbol{d}$ can be fixed by the initial conditions. If at $t = 0$ the particle is at $\boldsymbol{r}_0$ and moving with velocity $\boldsymbol{v}_0$, then

$$\boldsymbol{c} = \boldsymbol{r}_0, \qquad \boldsymbol{d} = \boldsymbol{v}_0/\omega. \tag{4.3}$$

We can easily verify the conservation laws for angular momentum and energy. From (4.2) it is clear that $\boldsymbol{r}$ always lies in the plane of $\boldsymbol{c}$ and $\boldsymbol{d}$, so that the direction of $\boldsymbol{J}$ is fixed. Using

$$\dot{\boldsymbol{r}} = -\omega \boldsymbol{c} \sin \omega t + \omega \boldsymbol{d} \cos \omega t,$$

we find explicitly

$$\boldsymbol{J} = m\boldsymbol{r} \wedge \dot{\boldsymbol{r}} = m\omega \boldsymbol{c} \wedge \boldsymbol{d} = m\boldsymbol{r}_0 \wedge \boldsymbol{v}_0, \tag{4.4}$$

which is obviously a constant.

The potential energy function corresponding to the force (4.1) — or (3.10) — is just (3.9), namely

$$V = \tfrac{1}{2}kr^2 = \tfrac{1}{2}k(x^2 + y^2 + z^2). \tag{4.5}$$

Thus, evaluating the energy, we find

$$E = \tfrac{1}{2}m\dot{r}^2 + \tfrac{1}{2}kr^2 = \tfrac{1}{2}k(c^2 + d^2) = \tfrac{1}{2}mv_0^2 + \tfrac{1}{2}kr_0^2, \tag{4.6}$$

which is again a constant.

To find the shape of the particle orbit, it is convenient to rewrite (4.2) in a slightly different, but equivalent, form. If $\theta$ is any fixed angle, we can write

$$\boldsymbol{r} = \boldsymbol{a} \cos(\omega t - \theta) + \boldsymbol{b} \sin(\omega t - \theta), \tag{4.7}$$

where

$$c = a \cos\theta - b \sin\theta,$$
$$d = a \sin\theta + b \cos\theta. \tag{4.8}$$

or, equivalently,

$$a = c \cos\theta + d \sin\theta,$$
$$b = -c \sin\theta + d \cos\theta.$$

We can now choose $\theta$ so that $a$ and $b$ are perpendicular. This requires

$$0 = a \cdot b = -(c^2 - d^2) \sin\theta \cos\theta + c \cdot d(\cos^2\theta - \sin^2\theta),$$

or

$$\tan 2\theta = \frac{2c \cdot d}{c^2 - d^2}.$$

We are now free to choose our axes so that the $x$-axis is in the direction of $a$ and the $y$-axis in the direction of $b$. Equation (4.7) then becomes

$$x = a \cos(\omega t - \theta), \qquad y = b \sin(\omega t - \theta), \qquad z = 0.$$

The equation of the orbit is obtained by eliminating the time from these equations. It is

$$\frac{x^2}{a^2} + \frac{y^2}{b^2} = 1, \qquad z = 0. \tag{4.9}$$

This is the well-known equation of an ellipse with centre at the origin, and semi-axes $a$ and $b$. (See Appendix B, Eq. (B.2), and Fig. 4.1.) The vectors $c$ and $d$ are what is known in geometry as a pair of conjugate semi-diameters of the ellipse.

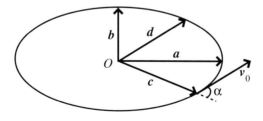

Fig. 4.1

The simplest way to determine the magnitudes of the semi-axes $a$ and $b$ from the initial conditions (that is, from $c$ and $d$) is to use the constancy of $E$ and $J$. Equating the expressions for these quantities, (4.6) and (4.4), to the corresponding expressions in terms of $a$ and $b$, we obtain

$$a^2 + b^2 = c^2 + d^2,$$
$$a \wedge b = c \wedge d.$$

(These equations may also be obtained directly from (4.8).) If $\alpha$ is the angle between $c$ and $d$, the second equation yields $ab = cd \sin \alpha$. Hence eliminating $b$ (or $a$) between these two equations, we obtain for $a^2$ (or $b^2$) the quadratic equation

$$a^4 - (c^2 + d^2)a^2 + c^2 d^2 \sin^2 \alpha = 0. \tag{4.10}$$

Conventionally, $a$ is taken to be the larger root, and called the *semi-major axis*.

## 4.2   The Conservation Laws

We now consider the general case of a central, conservative force. It corresponds, as we saw in §3.7, to a potential energy function $V(r)$ depending only on the radial co-ordinate $r$. There are two conservation laws, one for energy,

$$\tfrac{1}{2}m\dot{r}^2 + V(r) = E = \text{constant},$$

and one for angular momentum

$$m r \wedge \dot{r} = J = \text{constant}.$$

According to the discussion of §3.4, the second of these laws implies that the motion is confined to a plane, so that the problem is effectively two-dimensional. Introducing polar co-ordinates $r, \theta$ in this plane, we may write the two conservation laws in the form

$$\tfrac{1}{2}m(\dot{r}^2 + r^2\dot{\theta}^2) + V(r) = E,$$
$$m r^2 \dot{\theta} = J. \tag{4.11}$$

A great deal of information about the motion can be obtained directly from these equations, without actually having to solve them to find $r$ and

$\theta$ as functions of the time. We note that $\dot{\theta}$ may be eliminated to yield an equation involving only $r$ and $\dot{r}$:

$$\tfrac{1}{2}m\dot{r}^2 + \frac{J^2}{2mr^2} + V(r) = E. \tag{4.12}$$

We shall call this the *radial energy equation*. For a given value of $J$, it has precisely the same form as the one-dimensional energy equation with an *effective potential energy* function

$$U(r) = \frac{J^2}{2mr^2} + V(r). \tag{4.13}$$

It is easy to understand the physical significance of the extra term $J^2/2mr^2$ in this 'potential energy' function. It corresponds to a 'force' $J^2/mr^3$. This is precisely the 'centrifugal force' $mr\dot{\theta}^2$ (the first term on the right hand side of the radial equation in (3.48)), expressed in terms of the constant $J$ rather than the variable $\dot{\theta}$.

We can use the radial energy equation just as we did in §2.1. Since $\dot{r}^2$ is positive, the motion is limited to the range of values of $r$ for which

$$U(r) = \frac{J^2}{2mr^2} + V(r) \le E. \tag{4.14}$$

The maximum and minimum radial distances are given by the values of $r$ for which the equality holds.

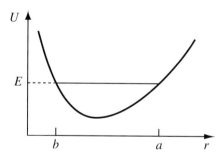

Fig. 4.2

As an example, let us consider again the case of the isotropic harmonic oscillator, for which $V(r) = \tfrac{1}{2}kr^2$. The corresponding effective potential energy function $U(r)$ is shown in Fig. 4.2. It has a minimum (corresponding to a position of stable equilibrium in the one-dimensional case) at

$$r = \left(\frac{J^2}{mk}\right)^{1/4}. \tag{4.15}$$

When the value of $E$ is equal to the minimum value of $U$, $\dot{r}$ must always be zero, and $r$ is fixed at the minimum position. In this case, the particle must move in a circle around the origin. It is interesting to note that we could also obtain (4.15) by equating the attractive force $kr$ to the centrifugal force in the circular orbit, $J^2/mr^3$.

For any larger value of $E$, the motion is confined to the region $b \leq r \leq a$ between the two limiting values of $r$, given by the solutions of the equality in (4.14). If the particle is initially at a distance $r_0$ from the origin, and moving with velocity $v_0$ in a direction making an angle $\alpha$ with the radius vector (as in Fig. 4.1), then the values of $E$ and $J$ are

$$E = \tfrac{1}{2}mv_0^2 + \tfrac{1}{2}kr_0^2, \qquad J = mr_0 v_0 \sin\alpha.$$

Thus the equation whose roots are $a$ and $b$ becomes (on multiplying by $2r^2/k$)

$$r^4 - \left(r_0^2 + \frac{m}{k}v_0^2\right)r^2 + \frac{m}{k}r_0^2 v_0^2 \sin^2\alpha = 0.$$

By (4.3), this is identical with the equation (4.10) found previously.

## 4.3   The Inverse Square Law

We now consider a force

$$\boldsymbol{F} = \frac{k}{r^2}\hat{\boldsymbol{r}},$$

where $k$ is a constant. The corresponding potential energy function is $V(r) = k/r$. The constant $k$ may be either positive or negative; in the first case, the force is repulsive, and in the second, attractive.

The radial energy equation for this case is

$$\tfrac{1}{2}m\dot{r}^2 + \frac{J^2}{2mr^2} + \frac{k}{r} = E. \tag{4.16}$$

It corresponds to the 'effective potential energy' function

$$U(r) = \frac{J^2}{2mr^2} + \frac{k}{r}.$$

## Repulsive case

We suppose first that $k > 0$. Then $U(r)$ decreases monotonically from $+\infty$ at $r = 0$ to $0$ at $r = \infty$ (see Fig. 4.3). Thus it has no minima, and circular

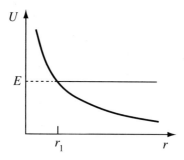

Fig. 4.3

motion is impossible, as is physically obvious. For any positive value of $E$, there is a minimum value of $r$, $r_1$ say, which is the unique positive root of $U(r) = E$, but no maximum value. If the radial velocity is initially inward, the particle must follow an orbit on which $r$ decreases to $r_1$ (at which point the velocity is purely transverse), and then increases again without limit. As is well known, and as we shall show in the next section, the orbit is actually a hyperbola.

Let us consider an example.

### *Example:* Distance of closest approach

A charged particle of charge $q$ moves in the field of a fixed point charge $q'$, with $qq' > 0$. Initially it is approaching the centre of force with velocity $v$ (at a large distance) along a path which, if continued in a straight line, would pass the centre at a distance $b$. Find the distance of closest approach.

This distance $b$ is known as the *impact parameter*, and will appear frequently in our future work. (See Fig. 4.4.) Since the particle starts from a great distance, its initial potential energy is negligible. (Note that we have chosen the arbitrary additive constant in $V$ so that $V(\infty) = 0$.) Thus the total energy is simply

$$E = \tfrac{1}{2}mv^2. \tag{4.17}$$

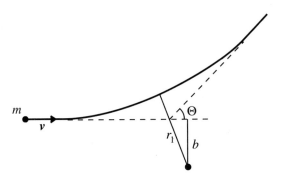

Fig. 4.4

Moreover, since the component of $\mathbf{r}$ perpendicular to $\mathbf{v}$ is $b$, the angular momentum is

$$J = mvb. \tag{4.18}$$

The distance of closest approach, $r_1$, is obtained by substituting these values, and $k = qq'/4\pi\epsilon_0$, into the radial energy equation, and then setting $\dot{r} = 0$. This yields

$$r_1^2 - 2ar_1 - b^2 = 0, \qquad \text{where} \qquad a = \frac{qq'}{4\pi\epsilon_0 mv^2}. \tag{4.19}$$

The required solution is the positive root

$$r_1 = a + \sqrt{a^2 + b^2}. \tag{4.20}$$

### Attractive case

We now suppose that $k < 0$, as in the gravitational case. It will be useful to define a quantity $l$, with the dimensions of length, by

$$l = \frac{J^2}{m|k|}. \tag{4.21}$$

Then the effective potential energy function is

$$U(r) = |k| \left( \frac{l}{2r^2} - \frac{1}{r} \right).$$

It is plotted in Fig. 4.5. Evidently, $U(\frac{1}{2}l) = 0$, and $U(r)$ has a minimum at $r = l$, with minimum value $U(l) = -|k|/2l$.

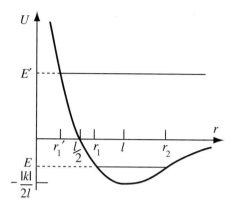

Fig. 4.5

Here, different types of motion are possible according to the value of $E$. We may distinguish four cases:

(a)   $E = -\dfrac{|k|}{2l}$.

This is the minimum value of $U$. Hence, $\dot{r}$ must always be zero, and the particle must move in a circle of radius $l$. It is easy to find the orbital velocity $v$ from the kinetic energy $T = E - V$. Since the potential energy is $V = -|k|/l$, we must have

$$T = \frac{|k|}{2l}, \qquad \text{whence} \qquad v = \sqrt{\frac{|k|}{ml}}. \tag{4.22}$$

This can also be found by equating the attractive force $|k|/l^2$ to the 'centrifugal force' $mv^2/l$.

Note the interesting result that for a circular orbit the potential energy is always twice as large in magnitude as the kinetic energy (see Problem 21).

(b)   $-\dfrac{|k|}{2l} < E < 0$.

This case is illustrated by the lower line $(E)$ in Fig. 4.5. The radial distance is limited between a minimum distance $r_1$ and a maximum distance $r_2$. As we shall see, the orbit is in fact an ellipse.

(c)   $E = 0$.

In this case, there is a minimum distance, $r_1 = \frac{1}{2}l$, but the maximum distance $r_2$ is infinite. Thus the particle has just enough energy to escape to infinity, with kinetic energy tending to zero at large distances. The orbit will be shown to be a parabola.

(d)   $E > 0$.

This corresponds to the upper line $(E')$ in Fig. 4.5. Again, there is a minimum distance but no maximum distance. Now, however, the particle can escape to infinity with non-zero limiting velocity. The orbit is in fact a hyperbola.

As a simple example of the use of these equations, we compute the escape velocity from the Earth.

### Example: Escape velocity

What is the minimum velocity with which a projectile launched from the surface of the Earth (taken to be a sphere of mass $M$ and radius $R$) can escape, and how does this depend on the angle of launch?

Suppose the projectile is launched with velocity $v$ at an angle $\alpha$ to the vertical. The energy and angular momentum are

$$E = \tfrac{1}{2}mv^2 - \frac{GMm}{R}, \qquad J = mRv\sin\alpha, \qquad (4.23)$$

where $G$ is Newton's gravitational constant. To express the energy in terms of more familiar quantities, we note that the gravitational force on a particle at the Earth's surface is

$$mg = \frac{GMm}{R^2},$$

from which we obtain the useful result

$$GM = R^2 g. \qquad (4.24)$$

Thus $E = \tfrac{1}{2}mv^2 - Rgm$. The projectile will escape to infinity provided that $E \geq 0$; or equivalently that its velocity exceeds the *escape velocity*,

$$v_e = \sqrt{2Rg}.$$

Note that this condition is *independent* of the angle of launch $\alpha$ (so long as we neglect atmospheric drag). Using the values $R = 6370\,\text{km}$, $g = 9.81\,\text{m s}^{-2}$, we find for the escape velocity from the Earth

$$v_e = 11.2\,\text{km s}^{-1}.$$

If the projectile is launched with a speed less than escape velocity, it will reach a maximum height and then fall back. (This is a generalization of the problem discussed in §3.2.) The problem of finding this distance provides another useful example.

### *Example:* Maximum height of projectile

If the launch speed $v$ is equal to the circular orbit speed $v_c$ in an orbit just above the Earth's surface, how high will the projectile reach for a given angle of launch $\alpha$?

The maximum distance $r_2$ from the centre is the larger root of the equation $U(r) = E$, or, with the values (4.23) for $E$ and $J$,

$$(2Rg - v^2)r^2 - 2R^2gr + R^2v^2 \sin^2 \alpha = 0.$$

The circular orbit velocity, by (4.22) and (4.24), is

$$v_c = \sqrt{Rg} = 7.9\,\text{km s}^{-1}.$$

Thus the equation reduces to $r^2 - 2Rr + R^2 \sin^2 \alpha = 0$, so that

$$r_2 = R(1 + \cos \alpha).$$

For a vertical launch, the maximum distance is $2R$. For any other angle, it is less, and in the limit of an almost horizontal launch, the orbit is nearly a circle of radius $R$.

This example illustrates the kind of problem which may readily be solved by using the radial energy equation. It is particularly useful when we are interested only in $r$, and not in the polar angle $\theta$ or the time $t$.

### *Energy levels of the hydrogen atom*

This is not a problem that can be solved by the methods of classical mechanics alone. However, the energy levels derived from quantum mechanics can also be obtained by imposing on the classical orbits an *ad hoc* 'quantization rule'. Indeed, historically, this is how they were first obtained. According to Bohr's 'old quantum theory', the electron in an atom cannot occupy any orbit, but only a certain discrete set of orbits. In the case of circular orbits, the quantization rule he imposed was that the angular momentum $J$ be an integer multiple of $\hbar$ (Planck's constant divided by $2\pi$). The constant $k$ in this case is $k = -e^2/4\pi\epsilon_0$, where $e$ is the electric charge on the proton, or minus the charge on the electron. Thus for $J = n\hbar$ the radius $a_n$ of the orbit is, by (4.21),

$$a_n = \frac{4\pi\epsilon_0 J^2}{me^2} = n^2 a_1,$$

where $a_1$, the radius of the first Bohr orbit (usually denoted by $a_0$), is

$$a_1 = \frac{4\pi\epsilon_0 \hbar^2}{me^2} = 5.29 \times 10^{-11} \, \text{m}.$$

The corresponding energy levels are

$$E_n = -\frac{e^2}{8\pi\epsilon_0 a_n} = -\frac{1}{n^2} \frac{e^2}{8\pi\epsilon_0 a_1}.$$

These values agree well with the energies of atomic transitions, as determined from the spectrum of hydrogen.

## 4.4   Orbits

We now turn to the problem of determining the *orbit* of a particle moving under a central conservative force. This can be done by eliminating the time from the two conservation equations, (4.11), to obtain an equation relating $r$ and $\theta$. The simplest way of doing this is to work not with $r$ itself, but with the variable $u = 1/r$, and look for an equation determining $u$ as a function of $\theta$. Now

$$\frac{du}{d\theta} = -\frac{1}{r^2} \frac{dr}{d\theta},$$

whence

$$\dot{r} = \frac{dr}{d\theta}\dot{\theta} = -r^2 \dot{\theta} \frac{du}{d\theta} = -\frac{J}{m} \frac{du}{d\theta}.$$

Thus substituting for $\dot{r}$ in the radial energy equation (4.12), we obtain

$$\frac{J^2}{2m}\left(\frac{\mathrm{d}u}{\mathrm{d}\theta}\right)^2 + \frac{J^2}{2m}u^2 + V = E, \tag{4.25}$$

in which of course $V$ is to be regarded as a function of $1/u$. This equation can be integrated (numerically, if not analytically) to give the equation of the orbit.

We shall consider explicitly only the case of the inverse square law, for which $V = ku$. We shall treat both cases $k > 0$ and $k < 0$ together, by writing $V = \pm|k|u$. It will be useful, as in the preceding section, to define the length parameter

$$l = \frac{J^2}{m|k|} \tag{4.26}$$

Then, multiplying (4.25) by $2/|k|$, we obtain

$$l\left(\frac{\mathrm{d}u}{\mathrm{d}\theta}\right)^2 + lu^2 \pm 2u = \frac{2E}{|k|},$$

where the upper sign refers to the repulsive case, $k > 0$, and the lower to the attractive case, $k < 0$.

To solve this equation, we multiply both sides by $l$ and add 1 to 'complete the square'. We now introduce the new variable

$$z = lu \pm 1, \qquad \text{so that} \qquad \frac{\mathrm{d}z}{\mathrm{d}\theta} = l\frac{\mathrm{d}u}{\mathrm{d}\theta}.$$

Then we may write the equation as

$$\left(\frac{\mathrm{d}z}{\mathrm{d}\theta}\right)^2 + z^2 = \frac{2El}{|k|} + 1 = e^2, \tag{4.27}$$

say, defining a new dimensionless constant $e$. Note that since the left hand side is a sum of squares, it can have a solution only when the the right hand side is positive, in agreement with our earlier result that the minimum value of $E$ is $-|k|/2l$.

The general solution of this equation is

$$z = lu \pm 1 = e\cos(\theta - \theta_0),$$

where $\theta_0$ is an arbitrary constant of integration. Thus finally, replacing $u$ by $1/r$ and multiplying by $r$, we find that the orbit equation is, in the repulsive case

$$r[e\cos(\theta - \theta_0) - 1] = l, \qquad (k > 0) \qquad (4.28)$$

and in the attractive case

$$r[e\cos(\theta - \theta_0) + 1] = l. \qquad (k < 0) \qquad (4.29)$$

These are the polar equations of *conic sections*, referred to a *focus* as origin (see Appendix B). The constant $e$, the *eccentricity*, determines the shape of the orbit; $l$, called the *semi-latus rectum*, determines its scale; and $\theta_0$ its orientation relative to the coordinate axes.

In the repulsive case, $e$ must be greater than unity, since otherwise the square bracket in (4.28) is always negative. Therefore, by (4.27), $E$ must be positive.

In the attractive case, $e = 0$ when $E$ has its minimum value, $-|k|/2l$; the orbit is then the circle $r = l$. So long as $e < 1$, or $E < 0$, the square bracket in (4.29) is always positive, and there is a value of $r$ for every $\theta$. Thus the orbit is closed. When $e \geq 1$, or $E \geq 0$, $r$ can become infinite when the square bracket vanishes. This is in agreement with our previous conclusion that the particle will escape to infinity if and only if its total energy is positive.

Note that $r$ takes its minimum value when $\theta = \theta_0$. Thus $\theta_0$ specifies the direction of the point of closest approach. In the attractive case, the constant $l$ also has a simple geometrical meaning. It is the radial distance at right angles to this direction; that is, $r = l$ when $\theta = \theta_0 \pm \frac{1}{2}\pi$.

### Elliptic orbits ($E < 0$, $0 \leq e < 1$)

For most applications, it is better to use the orbit equation directly in polar form. It is, however, straightforward to rewrite it in the possibly more familiar Cartesian form (see Appendix B). If we choose the axes so that $\theta_0 = 0$, we obtain after some algebra

$$\frac{(x + ae)^2}{a^2} + \frac{y^2}{b^2} = 1,$$

where

$$a = \frac{l}{1 - e^2} = \frac{|k|}{2|E|} \qquad \text{and} \qquad b^2 = al = \frac{J^2}{2m|E|} \qquad (4.30)$$

This is the equation of an *ellipse* with centre at $(-ae, 0)$, and semi-axes $a$ and $b$. (See Fig. 4.6) It is useful to note that the *semi-major axis $a$* is

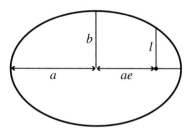

Fig. 4.6

determined by the value of the energy $E$, while the semi-latus rectum $l$ is fixed by the angular momentum $J$.

The time taken by the particle to traverse any part of its orbit may be found from the relation (3.26) between angular momentum and rate of sweeping out area,

$$\frac{\mathrm{d}A}{\mathrm{d}t} = \frac{J}{2m}.$$

All we have to do is to evaluate the area swept out by the radius vector, and multiply by $2m/J$. In particular, since the area of the ellipse is $A = \pi ab$, the orbital period is $\tau = 2m\pi ab/J$. Thus by (4.26) and (4.30),

$$\left(\frac{\tau}{2\pi}\right)^2 = \frac{m^2 a^2 b^2}{J^2} = \frac{m^2 a^3 l}{m|k|l},$$

and thus

$$\left(\frac{\tau}{2\pi}\right)^2 = \frac{m}{|k|} a^3. \tag{4.31}$$

(It is easy to verify directly from (4.22) that this gives the correct period for a circular orbit of radius $a$.) For a planet or satellite orbiting a central body of mass $M$,

$$\left(\frac{\tau}{2\pi}\right)^2 = \frac{a^3}{GM}. \tag{4.32}$$

This yields *Kepler's third law* of planetary motion: the square of the orbital period is proportional to the cube of the semi-major axis.

*Example:* **Time spent in the two halves of Earth's orbit**

If the Earth's orbit is divided in two by its minor axis, how much longer does it spend in one half than the other?

The eccentricity of Earth's orbit is $e = 0.0167$. It is clear from Fig. 4.6 that the areas swept out by the position vector of the Earth in the two halves are equal to half the area of the ellipse $\pm$ the area of a triangle with base $2b$ and height $ae$, namely $\frac{1}{2}\pi ab \pm aeb$. Thus the times taken are $(\frac{1}{2} \pm e/\pi)$ years. The difference between the two is $2e/\pi$ years, or 3.88 days.

### *Hyperbolic orbits* ($E > 0$, $e > 1$)

For both the repulsive and attractive cases, the Cartesian equation of the orbit is

$$\frac{(x - ae)^2}{a^2} - \frac{y^2}{b^2} = 1,$$

where

$$a = \frac{l}{e^2 - 1} = \frac{|k|}{2E} \qquad \text{and} \qquad b^2 = al = \frac{J^2}{2mE}. \tag{4.33}$$

This is the equation of a *hyperbola* with centre at $(+ae, 0)$ and semi-axes $a$ and $b$. (See Fig. 4.7.) One branch of the hyperbola (on the left in the figure) corresponds to the orbit in the attractive case and the other to the orbit in the repulsive case. As before, $a$ is determined by the energy $E$, and $l$ by

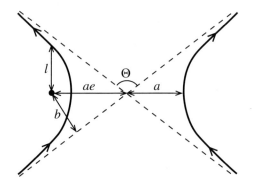

Fig. 4.7

the angular momentum $J$. Note that by (4.17) and (4.18), the *semi-minor axis b* is identical with the impact parameter introduced earlier.

The directions in which $r$ becomes infinite are, in the repulsive case, $\theta = \pm\arccos(1/e)$, and, in the attractive case, $\theta = \pm[\pi - \arccos(1/e)]$. In both cases, therefore, the angle through which the particle is deflected from its original direction of motion is

$$\Theta = \pi - 2\arccos(1/e). \tag{4.34}$$

This angle $\Theta$ is called the *scattering angle*. For later use, it will be helpful to find the relation between the impact parameter $b$, the scattering angle $\Theta$, and the limiting velocity $v$ at infinity. From (4.34), we have $e = \sec\frac{1}{2}(\pi - \Theta)$. Substituting in (4.33), this gives

$$b^2 = a^2(e^2 - 1) = a^2[\sec^2\tfrac{1}{2}(\pi - \Theta) - 1] = a^2\cot^2\tfrac{1}{2}\Theta.$$

But also from (4.33), using $E = \frac{1}{2}mv^2$, we have $a = |k|/mv^2$. Thus we obtain

$$b = \frac{|k|}{mv^2}\cot\tfrac{1}{2}\Theta. \tag{4.35}$$

Let us complete this section by considering a rather special orbit.

### Example: Orbit of a comet

The minimum distance of a comet from the Sun is half the radius of the Earth's orbit (assumed circular), and its velocity at that point is twice the orbital velocity of the Earth. Find its velocity when it crosses the Earth's orbit, and the angle at which the orbits cross. Will the comet subsequently escape from the solar system? What kind of orbit does it follow?

At perihelion, $r = a_E/2$ and $v = 2v_{c,E}$. Hence the energy is

$$E = 2mv_{c,E}^2 - \frac{2GM_Sm}{a_E} = 0,$$

using (4.22). It follows that the comet's velocity $v$ when it crosses the Earth's orbit is given by

$$\tfrac{1}{2}mv^2 = \frac{GM_Sm}{a_E} = mv_{c,E}^2,$$

whence $v = \sqrt{2}v_{c,E}$. The angular momentum is $J = ma_E v_{c,E}$, whence by (4.26), $l = a_E$. Thus the angle $\alpha$ at which the orbits cross is given by

$$ma_E v \cos \alpha = J = ma_E v_{c,E},$$

which yields $\alpha = 45°$.

Since $E = 0$, the comet has just enough energy to escape from the solar system. Its orbit is a parabola.

## 4.5 Scattering Cross-sections

One of the most important ways of obtaining information about the structure of small bodies (for example atomic nuclei) is to bombard them with particles and measure the number of particles scattered in various directions. The angular distribution of scattered particles will depend on the shape of the target and on the nature of the forces between the particles and the target. To be able to interpret the results of such an experiment, we must know how to calculate the expected angular distribution when the forces are known.

We shall consider first a particularly simple case. We suppose that the target is a fixed, hard (that is, perfectly elastic) sphere of radius $R$, and that a uniform, parallel beam of particles impinges on it. Let the particle flux in the beam, that is, the number of particles crossing unit area normal to the beam direction per unit time, be $f$. Then the number of particles which strike the target in unit time is

$$w = f\sigma, \tag{4.36}$$

where $\sigma$ is the cross-sectional area presented by the target, namely

$$\sigma = \pi R^2. \tag{4.37}$$

Now let us consider one of these particles. We suppose that it impinges on the target with velocity $v$ and impact parameter $b$. Then, as is clear from Fig. 4.8, it will hit the target at an angle $\alpha$ to the normal given by

$$b = R \sin \alpha.$$

The force on the particle is an impulsive central conservative force, corresponding to a potential $V(r)$ which is zero for $r > R$, and rises very sharply in the neighbourhood of $r = R$. Thus the kinetic energy and angular momentum must be the same before and after the collision. Let us take the

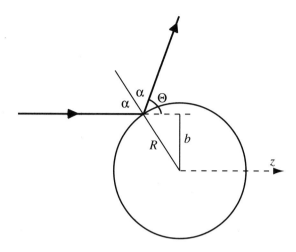

Fig. 4.8

positive $z$ direction ($\theta = 0$) to be the direction of motion of the incoming particles. Then, by the axial symmetry of the problem, the particle must move in a plane $\varphi = $ constant. From energy conservation, its velocity must be the same in magnitude before and after the collision. Then, from angular momentum conservation, it follows that the particle will bounce off the sphere at an angle to the normal equal to the incident angle $\alpha$. Thus the particle is deflected through an angle $\theta = \pi - 2\alpha$, related to the impact parameter by

$$b = R \cos \tfrac{1}{2}\theta. \tag{4.38}$$

We can now calculate the number of particles scattered in a direction specified by the polar angles $\theta, \varphi$, within angular ranges $\mathrm{d}\theta, \mathrm{d}\varphi$. The particles scattered through angles between $\theta$ and $\theta + \mathrm{d}\theta$ are those that came in with impact parameters between $b$ and $b + \mathrm{d}b$, where

$$\mathrm{d}b = -\tfrac{1}{2} R \sin \tfrac{1}{2}\theta \, \mathrm{d}\theta.$$

(Note that in our case $\mathrm{d}b$ is actually negative.)

Consider now a cross-section of the incoming beam. The particles we are interested in are those which cross a small element of area

$$\mathrm{d}\sigma = b \, |\mathrm{d}b| \, \mathrm{d}\varphi. \tag{4.39}$$

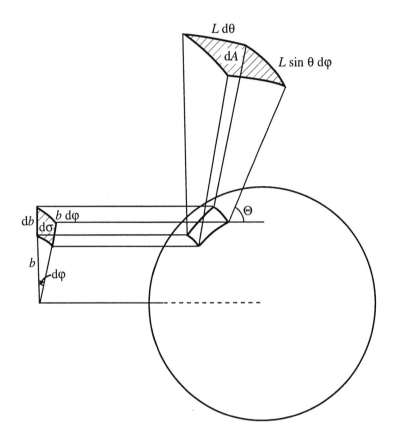

Fig. 4.9

(See Fig. 4.9.) Inserting the values of $b$ and $db$, we find

$$d\sigma = \tfrac{1}{4}R^2 \sin\theta \, d\theta \, d\varphi. \tag{4.40}$$

The rate at which particles cross this area, and therefore the rate at which they emerge in the given angular range, is

$$dw = f d\sigma. \tag{4.41}$$

In order to measure this rate, we may place a small detector at a large distance from the target in the specified direction. We therefore want to express our result in terms of the cross-sectional area $dA$ of the detector, and its distance $L$ from the target (which we assume to be much greater than $R$). Now from the discussion of §3.5 (see the paragraph preceding (3.30)), we see that the element of area on a sphere of radius $L$ is

$$dA = L\,d\theta \times L\sin\theta\,d\varphi.$$

We define the *solid angle* subtended at the origin by the area $dA$ to be

$$d\Omega = \sin\theta\,d\theta\,d\varphi, \tag{4.42}$$

so that

$$dA = L^2 d\Omega. \tag{4.43}$$

The solid angle is measured in *steradians* (sr). It plays the same role for a sphere as does the angle in radians for a circle; Eq. (4.43) is the analogue of the equation $ds = L\,d\theta$ for a circle of radius $L$. Just as the total angle subtended by an entire circle is $2\pi$, so the total solid angle subtended by an entire sphere is

$$\iint d\Omega = \frac{1}{L^2}\iint dA = \int_0^\pi \sin\theta\,d\theta \int_0^{2\pi} d\varphi = 4\pi.$$

The important quantity is not the cross-sectional area $d\sigma$ itself, but rather the ratio $d\sigma/d\Omega$, which is called the *differential cross-section*. By (4.41) and (4.43), the rate $dw$ at which particles enter the detector is

$$dw = f\frac{d\sigma}{d\Omega}\frac{dA}{L^2}. \tag{4.44}$$

It is useful to note an alternative definition of the differential cross-section, which is applicable even if we cannot follow the trajectory of each individual particle, and therefore cannot say just which of the incoming particles are those that emerge in a particular direction. We may *define* $d\sigma/d\Omega$ to be the ratio of scattered particles per unit solid angle to the number of incoming particles per unit area. Then the rate at which particles are detected is obtained by multiplying the differential cross-section by the flux of incoming particles, and by the solid angle subtended at the target by the detector, as in (4.44). Note that $d\sigma/d\Omega$ has the dimensions of area (solid angle, like angle, is dimensionless); it is measured in square metres per steradian $(\mathrm{m^2\,sr^{-1}})$.

In the particular case of scattering from a hard sphere, the differential cross-section is, by (4.40) and (4.42),

$$\frac{d\sigma}{d\Omega} = \tfrac{1}{4}R^2. \tag{4.45}$$

It has the special feature of being *isotropic*, or independent of the scattering angle. Thus the rate at which particles enter the detector is, in this case, independent of the direction in which it is placed. We note that the total cross-section (4.37) is correctly given by integrating (4.45) over all solid angles; in this case, we have merely to multiply by the total solid angle, $4\pi$.

## 4.6    Mean Free Path

The total cross-section $\sigma$ is useful in discussing the attenuation of a beam of particles passing through matter. Let us consider first a particle moving through a material containing $n$ atoms per unit volume, and suppose that the total cross-section for scattering by a single atom is $\sigma$. Consider a cylinder whose axis is the line of motion of the particle, and whose cross-sectional area is $\sigma$. Then the particle will collide with any atoms whose centres lie within the cylinder. Now, the number of such atoms in a length $x$ of the cylinder is $n\sigma x$. This is therefore the average number of collisions made by the particle when it travels a distance $x$. Thus the mean distance travelled between collisions — the *mean free path* $\lambda$ — is $x/n\sigma x$, i.e.

$$\lambda = \frac{1}{n\sigma}. \tag{4.46}$$

Now consider a beam of particles, with flux $f$, impinging normally on a wall. How far will they penetrate? We need to calculate the flux of particles that penetrate to a depth $x$ without suffering a collision. Let this flux be $f(x)$, and consider a thin slice of wall of thickness $\mathrm{d}x$ and area $A$ at the depth $x$ (see Fig. 4.10). The rate at which unscattered particles enter the slice is $Af(x)$, and the rate at which they emerge on the other side is $Af(x + \mathrm{d}x)$. The difference between the two must be the rate at which collisions occur within the slice. Now the total number of atoms in the slice is $nA\mathrm{d}x$, and the total cross-sectional area presented by all of them is $\sigma nA\,\mathrm{d}x$ (assuming $\mathrm{d}x$ is small enough that none of them overlap). Thus the rate at which collisions occur in the slice is $f(x)\sigma nA\,\mathrm{d}x$. Equating the two quantities, we have

$$Af(x) - Af(x + \mathrm{d}x) = f(x)\sigma nA\,\mathrm{d}x = f(x)\frac{A}{\lambda}\mathrm{d}x,$$

by (4.46). Equivalently,

$$\frac{\mathrm{d}f(x)}{\mathrm{d}x} = -\frac{1}{\lambda}f(x).$$

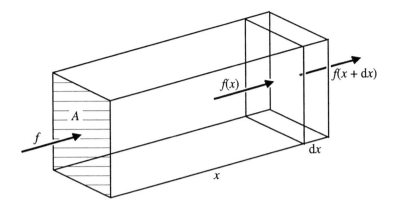

Fig. 4.10

This equation may be integrated to give

$$f(x) = Ce^{-x/\lambda},$$

where $C$ is an arbitrary constant. But $f(0) = f$, so finally

$$f(x) = fe^{-x/\lambda}. \tag{4.47}$$

Thus the number of particles in the beam decreases by a factor of $1/e$ in one mean free path.

Very often, when we wish to study the structure of a small object, like an atom, it would be quite impractical to use a target consisting of only a single atom. Instead, we have to use a target containing a large number $N$ of atoms. If the thickness of the target is $x$, the rate at which collisions occur within the target will be

$$Af(0) - Af(x) = A(1 - e^{-x/\lambda})f.$$

If $x$ is small compared to the mean free path, then we may retain only the linear term in the expansion of the exponential. In this case, the rate is approximately $Axf/\lambda = Axn\sigma f$. Since $N = nAx$, this shows that for a thin target the number of particles scattered will be just $N$ times the number scattered by a single atom. For a thick target, it would be less, because the atoms would effectively screen each other.

When the target is thin, the probability of multiple collisions, in which a particle strikes several target atoms in succession, will be small. If we assume that it is negligible, then we can conclude that the angular distribution of scattered particles will also be the same as that from a single target atom. This will be the case if all the dimensions of the target are small in comparison with $\lambda$. If we use a detector whose distance from the target, $L$, is large compared to the target size (so that the scattering angle does not depend appreciably on which of the atoms in the target is struck), then the rate at which particles enter the detector will be

$$dw = Nf\frac{d\sigma}{d\Omega}\frac{dA}{L^2},\tag{4.48}$$

that is, just $N$ times the rate for a single target atom.

## 4.7   Rutherford Scattering

We discuss in this section a problem which was of crucial importance in obtaining an understanding of the structure of the atom. In a classic experiment, performed in 1911, Rutherford bombarded atoms with $\alpha$-particles (helium nuclei). Because these particles are much heavier than electrons, they are deflected only very slightly by the electrons in the atom (see Chapter 7), and can therefore be used to study the heavy atomic nucleus. From observations of the angular distribution of the scattered $\alpha$-particles, Rutherford was able to show that the law of force between $\alpha$-particle and nucleus is the inverse square law down to very small distances. Thus he concluded that the positive charge is concentrated in a very small nuclear volume rather than being spread out over the whole volume of the atom.

Let us now calculate the differential cross-section for the scattering of a particle of charge $q$ and mass $m$ by a fixed point charge $q'$. The impact parameter $b$ is related to the scattering angle $\theta$ by (4.35), *i.e.*

$$b = a\cot\tfrac{1}{2}\theta, \qquad a = \frac{qq'}{4\pi\epsilon_0 mv^2}.\tag{4.49}$$

Thus

$$db = -\frac{a\,d\theta}{2\sin^2\tfrac{1}{2}\theta},$$

so that, substituting in (4.39), we obtain

$$d\sigma = \frac{a^2\cos\tfrac{1}{2}\theta\,d\theta\,d\varphi}{2\sin^3\tfrac{1}{2}\theta}.$$

Dividing by the solid angle (4.42), we find for the differential cross-section

$$\frac{d\sigma}{d\Omega} = \frac{a^2}{4\sin^4 \frac{1}{2}\theta}. \tag{4.50}$$

This is the *Rutherford scattering cross-section.*

We note that, in contrast to the differential cross-section for hard-sphere scattering, this cross-section is strongly dependent both on the velocity of the incoming particle and on the scattering angle. It also increases rapidly with increasing charge. For scattering of an $\alpha$-particle on a nucleus of atomic number $Z$, $qq' = 2Ze^2$ (where $-e$ is the electronic charge). Thus we expect the number of particles scattered to increase like $Z^2$ with increasing atomic number.

We saw in §4.3 that the minimum distance of approach is given by (4.20). Thus to investigate the structure of the atom at small distances, we must use high-velocity particles, for which $a$ is small, and examine large-angle scattering, corresponding to particles with small impact parameter $b$. The cross-section (4.50) is large for small values of the scattering angle, but physically it is the large-angle scattering which is of most interest. For, the fact that particles *can* be scattered through large angles is an indication that there are very strong forces acting at short distances. If the positive nuclear charge were spread out over a large volume, the force would be inverse-square-law only down to the radius of the charge distribution. Beyond that point, it would decrease as we go to even smaller distances. Consequently, the particles that penetrate to within this distance would experience a weaker force than the inverse square law predicts, and would be scattered through smaller angles.

A peculiar feature of the differential cross-section (4.50) is that the corresponding *total* cross-section is infinite. This is a consequence of the infinite range of the Coulomb force. However far away from the nucleus a particle may be, it still experiences some force, and is scattered through a non-zero (though small) angle. Thus the total number of particles scattered through any angle, however small, is indeed infinite. We can easily calculate the number of particles scattered through any angle greater than some small lower limit $\theta_0$. These are the particles which had impact parameters $b$ less than $b_0 = a \cot \frac{1}{2}\theta_0$. The corresponding cross-section is, therefore,

$$\sigma(\theta > \theta_0) = \pi b_0^2 = \pi a^2 \cot^2 \frac{1}{2}\theta_0. \tag{4.51}$$

(This may also be verified by direct integration of (4.50).)

## 4.8  Summary

For a particle moving under any central, conservative force, information about the radial motion may be obtained from the radial energy equation, which results from eliminating $\dot{\theta}$ between the conservation equations for energy and angular momentum. The values of $E$ and $J$ can be determined from the initial conditions, and this equation then tells us the radial velocity at any value of $r$.

When information about the angle $\theta$ is needed, we must find the equation of the orbit. For the inverse square law, the orbit is an ellipse or a hyperbola, according as $E < 0$ or $E > 0$. The semi-major axis is fixed by $E$, and the semi-*latus rectum* by $J$.

If we are concerned with finding the time taken to traverse part of the orbit, we can use the relation between the angular momentum and the rate of sweeping out area.

When a beam of particles strikes a target, the angular distribution of scattered particles may be found from the differential cross-section $\mathrm{d}\sigma/\mathrm{d}\Omega$. This may be calculated from a knowledge of the relation between the scattering angle and the impact parameter. The attenuation of the beam is related to the total cross-section $\sigma$, obtained by integrating $\mathrm{d}\sigma/\mathrm{d}\Omega$ over all solid angles.

---

## Problems

1. The orbits of synchronous communications satellites have been chosen so that viewed from the Earth they appear to be stationary. Find the radius of the orbits.

2. Find the radii of synchronous orbits about Jupiter and about the Sun. [Their mean rotation periods are 10 hours and 27 days, respectively. The mass of Jupiter is 318 times that of the Earth. The semi-major axis of the Earth's orbit, or *astronomical unit* (AU) is $1.50 \times 10^8$ km.]

3. The semi-major axis of Jupiter's orbit is 5.20 AU. Find its orbital period in years, and its mean (time-averaged) orbital speed. (Mean orbital speed of Earth $= 29.8\,\mathrm{km\,s^{-1}}$.)

4. The orbit of an asteroid extends from the Earth's to Jupiter's, just touching both. Find its orbital period. (Treat the planetary orbits as circular and coplanar.)

5. Find the maximum and minimum orbital speeds of the asteroid in Problem 4.

6. The Moon's mass and radius are $0.0123\,M_{\mathrm{E}}$ and $0.273\,R_{\mathrm{E}}$ (E = Earth). For Jupiter the corresponding figures are $318\,M_{\mathrm{E}}$ and $11.0\,R_{\mathrm{E}}$. Find in each case the gravitational acceleration at the surface, and the escape velocity.

7. Calculate the period of a satellite in an orbit just above the Earth's atmosphere (whose thickness may be neglected). Find also the periods for close orbits around the Moon and Jupiter.

8. The Sun has an orbital speed of about $220\,\mathrm{km\,s^{-1}}$ around the centre of the Galaxy, whose distance is $28\,000$ light years. Estimate the total mass of the Galaxy in solar masses.

9. A particle of mass $m$ moves under the action of a harmonic oscillator force with potential energy $\frac{1}{2}kr^2$. Initially, it is moving in a circle of radius $a$. Find the orbital speed $v$. It is then given a blow of impulse $mv$ in a direction making an angle $\alpha$ with its original velocity. Use the conservation laws to determine the minimum and maximum distances from the origin during the subsequent motion. Explain your results physically for the two limiting cases $\alpha = 0$ and $\alpha = \pi$.

10. Write down the effective potential energy function $U(r)$ for the system described in Chapter 3, Problem 11. Initially, the particle is moving in a circular orbit of radius $2a$. Find the orbital angular velocity $\omega$ in terms of the natural angular frequency $\omega_0$ of the oscillator when not rotating. If the motion is lightly disturbed, the particle will execute small oscillations about the circular orbit. By considering the effective potential energy function $U(r)$ near its minimum, find the angular frequency $\omega'$ of small oscillations. Hence describe the disturbed orbit qualitatively.

11. Show that the comet discussed at the end of §4.4 crosses the Earth's orbit at opposite ends of a diameter. Find the time it spends inside the Earth's orbit. (To evaluate the area required, write the equation of the orbit in Cartesian co-ordinates. See Appendix B.)

12. A star of mass $M$ and radius $R$ is moving with velocity $v$ through a cloud of particles of density $\rho$. If all the particles that collide with the star are trapped by it, show that the mass of the star will increase at a rate

$$\frac{\mathrm{d}M}{\mathrm{d}t} = \pi\rho v\left(R^2 + \frac{2GMR}{v^2}\right).$$

Given that $M = 10^{31}$ kg and $R = 10^8$ km, find how the effective cross-sectional area compares with the geometric cross-section $\pi R^2$ for velocities of $1000 \, \mathrm{km \, s^{-1}}$, $100 \, \mathrm{km \, s^{-1}}$ and $10 \, \mathrm{km \, s^{-1}}$.

13. *Find the polar equation of the orbit of an isotropic harmonic oscillator by solving the differential equation (4.25), and verify that it is an ellipse with centre at the origin. (*Hint*: Change to the variable $v = u^2$.) Check also that the period is given correctly by $\tau = 2mA/J$.

14. Discuss qualitatively the orbits of a particle under a repulsive force with potential energy function $V = \frac{1}{2}kr^2$ where $k$ is *negative*, using the effective potential energy function $U$. How would the orbit equation, as found in Problem 13, differ in this case? What shape is the orbit?

15. *If the Earth's orbit is divided in two by the *latus rectum*, show that the difference in time spent in the two halves, in years, is

$$\frac{2}{\pi} \left( e\sqrt{1 - e^2} + \arcsin e \right),$$

and hence for small $e$ about twice as large as the difference computed in the example in §4.4. (*Hint*: Use Cartesian co-ordinates to evaluate the required area. The identity $\pi/2 - \arcsin \sqrt{1 - e^2} = \arcsin e$ may be useful.)

16. A spacecraft is to travel from Earth to Jupiter along an elliptical orbit that just touches each of the planetary orbits (*i.e.*, the orbit of the asteroid in Problem 4). Use the results of Problems 3, 4 and 5 to find the relative velocity of the spacecraft with respect to the Earth just after launching and that with respect to Jupiter when it nears that planet, neglecting in each case the gravitational attraction of the planet. Where in its orbit must Jupiter be at the time of launch, relative to the Earth? Where will the Earth be when it arrives?

[This semi-elliptical trajectory is known as a *Hohmann transfer* and it is energy-efficient for interplanetary travel using high-thrust rockets in that a discrete boost is required at the beginning and at the end of the journey in order respectively to leave Earth and to arrive at the target planet. A similarly careful choice of timing for initiating a return to Earth is necessary, so that there is an inevitable minimum time that must be spent in the region of (here) Jupiter before this can be done. It is an interesting and important exercise to compare timings for round trips of planetary exploration using this form of transfer. For Mars, an obvious target in the short term, the time required is about 32 Earth months with about 15 months spent at Mars. There are of course other ways of effecting transfer, but at rather greater cost, either in fuel or in time spent at the destination!]

17. *Suppose that the asteroid of Problems 4 and 5 approaches the Earth with an impact parameter of $5R_E$, where $R_E$ = Earth's radius, moving in the same plane and overtaking it. (This is an improbably close encounter for a large asteroid; however the spectacular impact of comet Shoemaker–Levy 9 with Jupiter in July 1994 should prevent us from being too complacent about this threat!) Find the distance of closest approach and the angle through which the asteroid is scattered, in the frame of reference in which the Earth is at rest. (Assume that the asteroid is small enough to have negligible effect on the Earth's orbit.) What is its new velocity $v$ relative to the Sun? Show that the semi-major axis of its new orbit is $a_E v_E^2/(2v_E^2 - v^2)$, where $a_E$ and $v_E$ are the Earth's orbital radius and orbital velocity. Find the asteroid's new orbital period.

18. *Show that the position of a planet in its elliptical orbit can be expressed, using a frame with $x$-axis in the direction of *perihelion* (point of closest approach to the Sun), in terms of an angular parameter $\psi$ by $x = a(\cos\psi - e)$, $y = b\sin\psi$. (See Problem B.1. In the literature, $\psi$ is sometimes called the *eccentric anomaly*, while the polar angle $\theta$ is the *true anomaly*.) Show that $r = a(1 - e\cos\psi)$, and that the time from perihelion is given by $t = (\tau/2\pi)(\psi - e\sin\psi)$ (Kepler's equation).

19. Use the parametrization of Problem 18 to calculate the time-averaged values of the kinetic and potential energies $T$ and $V$ over a complete period. Hence verify the *virial theorem*, $\langle V \rangle_{\text{av}} = -2\langle T \rangle_{\text{av}}$.

20. *Find a parametrization similar to that of Problem 18 for a hyperbolic orbit, using hyperbolic functions.

21. On reaching the vicinity of Jupiter, the spacecraft in Problem 16 is swung around the planet by its gravitational attraction — a 'slingshot' manoeuvre. Consider this encounter in the frame of reference in which Jupiter is at rest. What is the magnitude and direction of the spacecraft's velocity before scattering? What is its magnitude after scattering? If the scattering angle in this frame is $90°$, what must be the impact parameter? What is the distance of closest approach to the planet, in terms of Jupiter radii? ($M_J = 318M_E$, $R_J = 11.0R_E$.)

22. *If the manoeuvre in Problem 21 is in the orbital plane, so that the final velocity of the spacecraft relative to Jupiter is radially away from the Sun, what is its velocity in magnitude and direction relative to the Sun? Use the radial energy equation to determine the spacecraft's

farthest distance from the Sun (its *aphelion* distance) in astronomical units. Find also its new orbital period. When it returns, what will be its perihelion distance?

23. An alternative to the manoeuvre described in Problem 22 is for the spacecraft to be scattered out of the orbital plane. Assume that relative to Jupiter its velocity after scattering is directed normal to the orbital plane. What is its velocity relative to the Sun? What will be its aphelion distance and orbital period? How far from the orbital plane will it reach? (*Hint*: Immediately after scattering, the radial component of its velocity is zero. This is therefore the perihelion point of the new orbit. The farthest point from the original orbital plane will occur when it is at one end of the semi-minor axis of the orbit.)

24. *A ballistic rocket (one that moves freely under gravity after its initial launch) is fired from the surface of the Earth with velocity $v < \sqrt{Rg}$ at an angle $\alpha$ to the vertical. (Ignore the Earth's rotation.) Find the equation of its orbit. Express the range $2R\theta$ (measured along the Earth's surface) in terms of the parameters $l$ and $a$, and hence show that to maximize the range, we should choose $\alpha$ so that $l = 2a - R$. (*Hint*: A sketch may help.) Deduce that the maximum range is $2R\theta$ where $\sin\theta = v^2/(2Rg - v^2)$. Given that the maximum range is 3600 nautical miles, find the launch velocity and the angle at which the rocket should be launched. (*Note*: 1 nautical mile = 1 minute of arc over the Earth's surface.)

25. Discuss the possible types of orbit for a particle moving under a central inverse-cube-law force, described by the potential energy function $V = k/2r^2$. For the repulsive case ($k > 0$), show that the orbit equation is $r\cos n(\theta - \theta_0) = b$, where $n, b$ and $\theta_0$ are constants. Show that for the attractive case ($k < 0$), the nature of the orbit depends on the signs of $J^2 + mk$ and of $E$. Find the equation of the orbit for each possible type. (Include the cases where one of these parameters vanishes.)

26. Show that the scattering angle for particles of mass $m$ and initial velocity $v$ scattered by a repulsive inverse-cube-law force is $\pi - \pi/n$ (see Problem 25). Hence find the differential cross-section.

27. *The potential energy of a particle of mass $m$ is $V(r) = k/r + c/3r^3$, where $k < 0$ and $c$ is a small constant. (The gravitational potential energy in the equatorial plane of the Earth has approximately this form, because of its flattened shape — see Chapter 6.) Find the angular velocity $\omega$ in a circular orbit of radius $a$, and the angular frequency $\omega'$ of small radial oscillations about this circular orbit. Hence show that a

nearly circular orbit is approximately an ellipse whose axes precess at an angular velocity $\Omega \approx (c/|k|a^2)\omega$.

28. A beam of particles strikes a wall containing $2 \times 10^{29}$ atoms per $m^3$. Each atom behaves like a sphere of radius $3 \times 10^{-15}$ m. Find the thickness of wall that exactly half the particles will penetrate without scattering. What thickness would be needed to stop all but one particle in $10^6$?

29. An $\alpha$-particle of energy $4\,keV$ ($1\,eV = 1.6 \times 10^{-19}$ J) is scattered by an aluminium atom through an angle of $90°$. Calculate the distance of closest approach to the nucleus. (Atomic number of $\alpha$-particle $= 2$, atomic number of Al $= 13$, $e = 1.6 \times 10^{-19}$ C.) A beam of such particles with a flux of $3 \times 10^8\,m^{-2}\,s^{-1}$ strikes a target containing $50\,mg$ of aluminium. A detector of cross-sectional area $400\,mm^2$ is placed $0.6\,m$ from the target in a direction at right angles to the beam direction. Find the rate of detection of $\alpha$-particles. (Atomic mass of Al $= 27\,u$; $1\,u = 1.66 \times 10^{-27}$ kg.)

30. *It was shown in §3.4 that Kepler's second law of planetary motion implies that the force is central. Show that his first law — that the orbit is an ellipse with the Sun at a focus — implies the inverse square law. (*Hint*: By differentiating the orbit equation $l/r = 1 + e\cos\theta$, and using (3.26), find $\dot{r}$ and $\ddot{r}$ in terms of $r$ and $\theta$. Hence calculate the radial acceleration.)

31. Show that Kepler's third law, $\tau \propto a^{3/2}$, implies that the force on a planet is proportional to its mass.
    [This law was originally expressed by Kepler as $\tau \propto \bar{r}^{3/2}$, where $\bar{r}$ is a 'mean value' of $r$. For an ellipse, the mean over *angle* $\theta$ is in fact $b$; the mean over *time* is actually $a(1 + \frac{1}{2}e^2)$; it is the mean over *arc length* — or the median — which is given by $a$! Of course, for most planets in our Solar System these values are almost equal.]

# Chapter 5

# Rotating Frames

Hitherto, we have always used inertial frames, in which the laws of motion take on the simple form expressed in Newton's laws. There are, however, a number of problems that can most easily be solved by using a non-inertial frame. For example, when discussing the motion of a particle near the Earth's surface, it is often convenient to use a frame which is rigidly fixed to the Earth, and rotates with it. In this chapter, we shall find the equations of motion with respect to such a frame, and discuss some applications of them.

## 5.1 Angular Velocity; Rate of Change of a Vector

Let us consider a solid body which is rotating with constant angular velocity $\omega$ about a fixed axis. Let $\boldsymbol{n}$ be a unit vector along the axis, whose direction is defined by the right-hand rule: it is the direction in which a right-hand-thread screw would move when turned in the direction of the rotation. Then we define the *vector angular velocity* $\boldsymbol{\omega}$ to be a vector of magnitude $\omega$ in the direction of $\boldsymbol{n}$: $\boldsymbol{\omega} = \omega\boldsymbol{n}$. Clearly, angular velocity, like angular momentum, is an *axial* vector (see §3.3).

For example, for the Earth, $\boldsymbol{\omega}$ is a vector pointing along the polar axis, towards the north pole. Its magnitude is equal to $2\pi$ divided by the length of the *sidereal* day (the rotation period with respect to the fixed stars, which is less than that with respect to the Sun by one part in 365), that is

$$\omega = \frac{2\pi}{86164}\,\mathrm{s}^{-1} = 7.292 \times 10^{-5}\,\mathrm{s}^{-1}. \tag{5.1}$$

If we take the origin to lie on the axis of rotation, then the velocity of a point of the body at position $r$ is given by the simple formula

$$v = \omega \wedge r. \qquad (5.2)$$

To prove this, we note that the point moves with angular velocity $\omega$ around a circle of radius $\rho = r \sin \theta$ (see Fig. 5.1). Thus its speed is

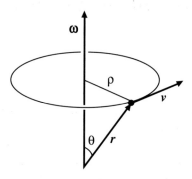

Fig. 5.1

$$v = \omega \rho = \omega r \sin \theta = |\omega \wedge r|.$$

Moreover, the direction of $v$ is that of $\omega \wedge r$; for clearly, $v$ is perpendicular to the plane containing $\omega$ and $r$, and it is easy to see from the figure that its sense is given correctly by the right-hand rule. Thus (5.2) is correct both as regards magnitude and direction.

It is not necessary that $r$ should be the position vector of a point of the rotating body. If $a$ is *any* vector fixed in the rotating body, then by the same argument,

$$\frac{da}{dt} = \omega \wedge a. \qquad (5.3)$$

In particular, if $i, j, k$ are orthogonal unit vectors fixed in the body, then

$$\frac{di}{dt} = \omega \wedge i, \qquad \frac{dj}{dt} = \omega \wedge j, \qquad \frac{dk}{dt} = \omega \wedge k. \qquad (5.4)$$

For example, if $\omega$ is in the $k$ direction, then

$$\frac{di}{dt} = \omega j, \qquad \frac{dj}{dt} = -\omega i, \qquad \frac{dk}{dt} = 0.$$

(Compare Fig. 5.2.)

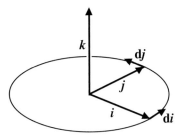

Fig. 5.2

Now consider a vector $\boldsymbol{a}$ specified with respect to the rotating axes $\boldsymbol{i}, \boldsymbol{j}, \boldsymbol{k}$ by the components $a_x, a_y, a_z$, so that

$$\boldsymbol{a} = a_x\boldsymbol{i} + a_y\boldsymbol{j} + a_z\boldsymbol{k}.$$

(Our discussion applies quite generally, but it may be helpful to think of $\boldsymbol{a}$ as the position vector of a particle with respect to the Earth.) We must distinguish two kinds of 'rates of change', and it will be convenient to suspend for the moment the convention whereby $\mathrm{d}\boldsymbol{a}/\mathrm{d}t$ and $\dot{\boldsymbol{a}}$ mean the same thing. We shall denote by $\mathrm{d}\boldsymbol{a}/\mathrm{d}t$ the rate of change of $\boldsymbol{a}$ as measured by an inertial observer at rest relative to the origin, and by $\dot{\boldsymbol{a}}$ the rate of change as measured by an observer rotating with the solid body. We wish to find the relation between these two rates of change.

Now, although our two observers differ about the rate of change of a vector, they will always agree about the rate of change of any scalar quantity, and in particular about the rates of change of the three components $a_x, a_y, a_z$. (Note that for the inertial observer, these are *not* the components of $\boldsymbol{a}$ with respect to his own set of axes. An observer outside the Earth will always agree with one on the Earth about the latitude and longitude of a small body moving on the surface, though these are not the co-ordinates he would naturally use himself to describe its position.) Hence we can write

$$\frac{\mathrm{d}a_x}{\mathrm{d}t} = \dot{a}_x, \qquad \frac{\mathrm{d}a_y}{\mathrm{d}t} = \dot{a}_y, \qquad \frac{\mathrm{d}a_z}{\mathrm{d}t} = \dot{a}_z. \tag{5.5}$$

According to the observer on the rotating body, the rate of change of $\boldsymbol{a}$ is fully described by the rates of change of its three components, so that

$$\dot{\boldsymbol{a}} = \dot{a}_x\boldsymbol{i} + \dot{a}_y\boldsymbol{j} + \dot{a}_z\boldsymbol{k}. \tag{5.6}$$

However, to the inertial observer, the axes $i, j, k$ are themselves changing with time according to (5.4), so that

$$\frac{\mathrm{d}\boldsymbol{a}}{\mathrm{d}t} = \left(\frac{\mathrm{d}a_x}{\mathrm{d}t}\boldsymbol{i} + \frac{\mathrm{d}a_y}{\mathrm{d}t}\boldsymbol{j} + \frac{\mathrm{d}a_z}{\mathrm{d}t}\boldsymbol{k}\right) + \left(a_x\frac{\mathrm{d}\boldsymbol{i}}{\mathrm{d}t} + a_y\frac{\mathrm{d}\boldsymbol{j}}{\mathrm{d}t} + a_z\frac{\mathrm{d}\boldsymbol{k}}{\mathrm{d}t}\right)$$

$$= (\dot{a}_x\boldsymbol{i} + \dot{a}_y\boldsymbol{j} + \dot{a}_z\boldsymbol{k}) + \boldsymbol{\omega} \wedge (a_x\boldsymbol{i} + a_y\boldsymbol{j} + a_z\boldsymbol{k}),$$

by (5.5) and (5.4). Thus, finally, we obtain

$$\frac{\mathrm{d}\boldsymbol{a}}{\mathrm{d}t} = \dot{\boldsymbol{a}} + \boldsymbol{\omega} \wedge \boldsymbol{a}. \tag{5.7}$$

Applied to the position vector $\boldsymbol{r}$, this result is almost obvious, for it states that the velocity with respect to the inertial observer is the sum of the velocity $\dot{\boldsymbol{r}}$ with respect to the rotating frame and the velocity $\boldsymbol{\omega} \wedge \boldsymbol{r}$ of a particle at $\boldsymbol{r}$ rotating with the body. We have given a detailed proof because of the central importance of this result, and because its application to other vectors is much less obvious.

We shall frequently encounter in our later work equations similar to (5.3): $\mathrm{d}\boldsymbol{a}/\mathrm{d}t = \boldsymbol{\omega} \wedge \boldsymbol{a}$. It is important to realize that one can reverse the argument that led up to it. If $\boldsymbol{a}$ satisfies this equation, then it must be a vector of constant length rotating with angular velocity $\omega$ about the direction of $\boldsymbol{\omega}$. For, if we introduce a frame of reference rotating with angular velocity $\boldsymbol{\omega}$, then according to (5.7), $\dot{\boldsymbol{a}} = \boldsymbol{0}$, so that $\boldsymbol{a}$ is fixed in the rotating frame.

## 5.2   Particle in a Uniform Magnetic Field

A particle of charge $q$ moving with velocity $\boldsymbol{v}$ in a magnetic field $\boldsymbol{B}(\boldsymbol{r})$ experiences a force proportional to its velocity, and perpendicular to it,

$$\boldsymbol{F} = q\boldsymbol{v} \wedge \boldsymbol{B}. \tag{5.8}$$

The equation of motion is then

$$m\frac{\mathrm{d}\boldsymbol{v}}{\mathrm{d}t} = q\boldsymbol{v} \wedge \boldsymbol{B}. \tag{5.9}$$

Now let us suppose that the magnetic field is uniform (independent of position) and constant in time. Then (5.9) has precisely the form of (5.3),

$$\frac{\mathrm{d}\boldsymbol{v}}{\mathrm{d}t} = \boldsymbol{\omega} \wedge \boldsymbol{v}, \qquad \text{where} \qquad \boldsymbol{\omega} = -\frac{q}{m}\boldsymbol{B}. \tag{5.10}$$

Hence, by the argument of the preceding section, the velocity vector $v$ rotates about the direction of $B$ with a constant angular velocity. It follows that the path of the particle is a helix whose axis points along the field lines (see Fig. 5.3). If we take the direction of $B$ to be the $z$-direction, then $v$ has

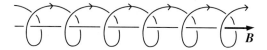

Fig. 5.3

a constant $z$-component, and a component in the $xy$-plane which rotates around a circle with angular velocity $\omega$.

In particular, if $v$ is initially parallel to $B$, the particle will continue to move with uniform velocity along a field line. The other limiting case, in which $v$ is perpendicular to $B$, is more interesting. In that case, the particle describes a circle of radius

$$r = \frac{v}{\omega} = \frac{mv}{qB}. \qquad (5.11)$$

This effect is used in accelerating particles in the *cyclotron*. In order to prevent the particles from escaping, a strong magnetic field is employed to constrain them to move in circles. To accelerate them, an alternating voltage is applied between two D-shaped pole pieces enclosing two halves of the circles (see Fig. 5.4). The angular frequency of this alternating voltage is chosen to coincide with the *cyclotron frequency*

$$\omega_{\mathrm{c}} = \frac{qB}{m}. \qquad (5.12)$$

Thus, by the time the particles have made a half-revolution, the direction of the electric field has reversed, and they experience an accelerating field each time they cross the gap. Each semicircle is therefore slightly larger than the preceding one, and the particles spiral outwards with increasing energy. Finally, one obtains a beam of high-velocity particles at the outer edge of the cyclotron.

This method can be used only so long as the speed remains small compared to the speed of light; for, according to relativity theory, the angular frequency $\omega_{\mathrm{c}}$ is not precisely constant when the speed is large, but decreases with increasing speed. Quite similar methods can be used, however, even at the highest energies.

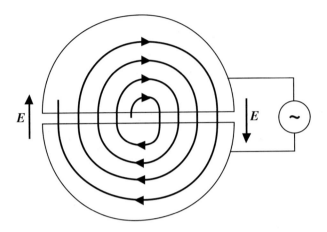

Fig. 5.4

If there are both electric and magnetic fields present, a charged particle experiences the Lorentz force, $\boldsymbol{F} = q(\boldsymbol{E} + \boldsymbol{v} \wedge \boldsymbol{B})$. An interesting case is that of *crossed* fields.

### Example: Crossed electric and magnetic fields

A particle of mass $m$ and charge $q$ moves in uniform perpendicular electric and magnetic fields, $\boldsymbol{E} = E\boldsymbol{j}$ and $\boldsymbol{B} = B\boldsymbol{k}$. Show that if it is initially moving in the $xy$-plane, it will continue to do so, and that it has a circular motion with a superimposed uniform drift velocity in a direction perpendicular to both $\boldsymbol{E}$ and $\boldsymbol{B}$. Find the solution for the case when the particle starts from rest at the origin.

The equation of motion $m\ddot{\boldsymbol{r}} = q\boldsymbol{E} + q\dot{\boldsymbol{r}} \wedge \boldsymbol{B}$ is, in terms of components,

$$m\ddot{x} = qB\dot{y}, \qquad m\ddot{y} = qE - qB\dot{x}, \qquad m\ddot{z} = 0.$$

Clearly, if $z = \dot{z} = 0$ initially, this will be true always. We can eliminate the term in $E$ by going over to a frame moving with relative velocity $(E/B)\boldsymbol{i}$ (see Chapter 1, Problem 5), in which the co-ordinates are given by

$$x' = x - (E/B)t, \qquad y' = y, \qquad z' = z.$$

Then $\dot{x}' = \dot{x} - E/B, \dot{y}' = \dot{y}, \dot{z}' = \dot{z}$. Thus the equations of motion become

$$m\ddot{x}' = qB\dot{y}', \qquad m\ddot{y}' = -qB\dot{x}', \qquad m\ddot{z}' = 0.$$

So in the moving frame, the particle moves in a circle, as in the discussion above. In the original frame, this means circular motion with a uniform velocity superimposed. The initial conditions yield $\dot{x}' = -E/B$, so the radius of the circular motion is $E/B\omega_c$. If the particle starts from the origin, the complete solution is

$$x = \frac{E}{B}\left(t - \frac{\sin\omega_c t}{\omega_c}\right), \qquad y = \frac{E}{B\omega_c}(1 - \cos\omega_c t), \qquad z = 0.$$

## 5.3 Acceleration; Apparent Gravity

We can use the formula (5.7) twice to obtain the relation between the absolute acceleration $\mathrm{d}^2\boldsymbol{r}/\mathrm{d}t^2$ of a particle and its acceleration $\ddot{\boldsymbol{r}}$ relative to a rotating frame.

The velocity of the particle relative to an inertial frame is

$$\boldsymbol{v} = \frac{\mathrm{d}\boldsymbol{r}}{\mathrm{d}t} = \dot{\boldsymbol{r}} + \boldsymbol{\omega} \wedge \boldsymbol{r}. \tag{5.13}$$

Applying the same formula to the rate of change of $\boldsymbol{v}$, we have

$$\frac{\mathrm{d}^2\boldsymbol{r}}{\mathrm{d}t^2} = \frac{\mathrm{d}\boldsymbol{v}}{\mathrm{d}t} = \dot{\boldsymbol{v}} + \boldsymbol{\omega} \wedge \boldsymbol{v}. \tag{5.14}$$

But, from (5.13), assuming that $\boldsymbol{\omega}$ is constant, we find

$$\dot{\boldsymbol{v}} = \ddot{\boldsymbol{r}} + \boldsymbol{\omega} \wedge \dot{\boldsymbol{r}},$$

and

$$\boldsymbol{\omega} \wedge \boldsymbol{v} = \boldsymbol{\omega} \wedge \dot{\boldsymbol{r}} + \boldsymbol{\omega} \wedge (\boldsymbol{\omega} \wedge \boldsymbol{r}).$$

Hence, substituting in (5.14), we obtain

$$\frac{\mathrm{d}^2\boldsymbol{r}}{\mathrm{d}t^2} = \ddot{\boldsymbol{r}} + 2\boldsymbol{\omega} \wedge \dot{\boldsymbol{r}} + \boldsymbol{\omega} \wedge (\boldsymbol{\omega} \wedge \boldsymbol{r}). \tag{5.15}$$

The second term on the right is called the *Coriolis* acceleration, and the third term is the *centripetal* acceleration. The latter is directed inwards towards the axis of rotation, and perpendicular to it, as may be seen by writing it in the form (see (A.16))

$$\omega \wedge (\omega \wedge r) = (\omega \cdot r)\omega - \omega^2 r.$$

The most important application of (5.15) is to a particle moving near the surface of the Earth. For a particle moving under gravity, and under an additional, mechanical force $F$, the equation of motion is

$$m\frac{d^2 r}{dt^2} = mg + F,$$

where $g$ is a vector of magnitude $g$ pointing downward (towards the centre of the Earth). Using (5.15), and moving all the terms except the relative acceleration to the right side of the equation, we can write it as

$$m\ddot{r} = mg + F - 2m\omega \wedge \dot{r} - m\omega \wedge (\omega \wedge r). \qquad (5.16)$$

The last two terms are *apparent* (or *fictitious*) forces, which arise because of the non-inertial nature of the frame of reference. We shall postpone discussion of the effects of the third term — the *Coriolis force* — to the next section, and consider here the last term, which is the familiar *centrifugal force*.

This force is a slowly varying function of position, proportional, like gravity, to the mass of the particle. When we make a laboratory measurement of the acceleration due to gravity, what we actually measure is not $g$ but

$$g^* = g - \omega \wedge (\omega \wedge r). \qquad (5.17)$$

In particular, a plumb line does not point directly towards the centre of the Earth, but is swung outwards through a small angle by the centrifugal force. (See Fig. 5.5, in which this effect is greatly exaggerated.) Let us consider a point in colatitude $(\pi/2 - \text{latitude})$ $\theta$. Then

$$|\omega \wedge (\omega \wedge r)| = \omega|\omega \wedge r| = \omega^2 r \sin \theta.$$

Thus the horizontal and vertical components of the effective gravitational acceleration $g^*$ (using 'vertical' to mean towards the Earth's centre) are

$$g_h^* = \omega^2 r \sin \theta \cos \theta, \qquad g_v^* = g - \omega^2 r \sin^2 \theta.$$

(There are further corrections, to be discussed later, due to the non-spherical shape of the Earth.)

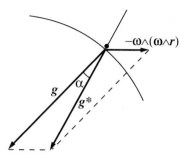

Fig. 5.5

The magnitude of the centrifugal force may be found by inserting the value (5.1) for $\omega$, and for $r$ the mean radius of the Earth, 6371 km. We find

$$\omega^2 r = 34 \, \text{mm s}^{-2}. \tag{5.18}$$

Since $\omega^2 r \ll g$, the angle $\alpha$ between the apparent and true verticals is approximately

$$\alpha \approx \frac{g_h^*}{g_v^*} \approx \frac{\omega^2 r}{g} \sin\theta \cos\theta.$$

The maximum value occurs at $\theta = 45°$, and is about $0°6'$.

At the pole, there is no centrifugal force, and $g^* = g$. On the equator, $g^* = g - \omega^2 r$. Thus we might expect the measured value of the acceleration due to gravity to be larger at the pole by $34 \, \text{mm s}^{-2}$. The actual measured difference is somewhat larger than this,

$$\Delta g^* = g_{\text{pole}}^* - g_{\text{equator}}^* = 52 \, \text{mm s}^{-2}. \tag{5.19}$$

This discrepancy arises from the fact that the Earth is not a perfect sphere, but more nearly spheroidal in shape, flattened at the poles. Thus the gravitational acceleration $g$, even excluding the centrifugal term, is itself larger at the pole than on the equator. These two effects are not really independent, for the flattening of the Earth is a consequence of its rotation. We shall discuss this point in the next chapter.

As another example of the effect of the centrifugal force, we shall consider the problem of a liquid in a rotating vessel.

*Example:* **Surface of a rotating liquid**

Liquid in a vessel rotating with angular velocity $\omega$ is rotating with the same angular velocity. Find the equation for its surface.

We use a frame rotating with the vessel, so that in the rotating frame the liquid is at rest, but subject to the centrifugal force. Now this force is conservative, and corresponds to a potential energy

$$V_{\text{cent}} = -\tfrac{1}{2}m\omega^2\rho^2 = -\tfrac{1}{2}m\omega^2(x^2 + y^2). \qquad (5.20)$$

Consider a particle on the surface of the liquid. Unless the surface is one on which potential energy is constant, the particle will tend to move towards regions of lower potential energy. Hence if the liquid is in equilibrium under the gravitational and centrifugal forces, its surface must be an equipotential surface,

$$gz - \tfrac{1}{2}\omega^2(x^2 + y^2) = \text{constant}.$$

This is a paraboloid of revolution about the $z$-axis.

This effect is used in constructing some high-powered optical telescopes. Because a paraboloid is the ideal shape to focus light from directly above it, the mirror is made by spinning a thin film of liquid, for example mercury. This is very much cheaper than grinding glass mirrors to a precise shape — though it has the disadvantage that the telescope can only look at objects near the zenith.

## 5.4 Coriolis Force

The *Coriolis force*, $-2m\boldsymbol{\omega} \wedge \dot{\mathbf{r}}$, is an apparent, velocity-dependent force arising from the Earth's rotation. To understand its physical origin, it may be helpful to consider a flat, rotating disc. Suppose that a particle moves across the disc under no forces, so that an inertial observer sees it moving diametrically across in a straight line. (See Fig. 5.6(a).) Then, because the disc is rotating, an observer on the disc will see the particle crossing successive radii, following a curved track as in Fig. 5.6(b). If he is unaware that the disc is rotating, he will ascribe this curvature to a force acting on the particle at right angles to its velocity. This is the Coriolis force.

We may regard Fig. 5.6 as representing the Earth viewed from the north pole. Thus we see that the effect of the Coriolis force is to make a particle

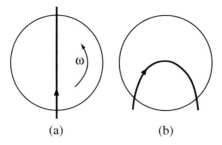

Fig. 5.6

deviate to the right in the northern hemisphere, and to the left in the southern. (It also has a vertical component, in general.) It is responsible for some well-known meteorological phenomena, which we shall discuss below.

First, however, we examine some effects of the Coriolis force which are observable in the laboratory. Let us consider a particle moving near the surface of the Earth in colatitude $\theta$. We shall suppose that the distance through which it moves is sufficiently small for both the gravitational and centrifugal forces to be effectively constant. Then we may combine them in a constant effective gravitational acceleration $\boldsymbol{g}^*$. For convenience, we shall drop the star, and write this simply as $\boldsymbol{g}$. The equation of motion is then

$$m\ddot{\boldsymbol{r}} = m\boldsymbol{g} + \boldsymbol{F} - 2m\boldsymbol{\omega} \wedge \dot{\boldsymbol{r}}. \tag{5.21}$$

As an example, let us take a small body falling freely under gravity towards the Earth's surface.

### *Example:* **Freely falling body**

A body is dropped from rest at a height $h$ above the ground. How far from the point directly below its starting point does it land?

To analyze this problem, it is convenient to choose our axes so that $\boldsymbol{i}$ is east, $\boldsymbol{j}$ is north, and $\boldsymbol{k}$ up. (See Fig. 5.7.) Here 'up' means opposite to $\boldsymbol{g}^*$, and the colatitude $\theta$ is the angle between that direction and the Earth's axis. The angular velocity vector then has the components

$$\boldsymbol{\omega} = (0, \omega \sin \theta, \omega \cos \theta).$$

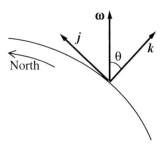

Fig. 5.7

Hence the Coriolis force is

$$-2m\boldsymbol{\omega} \wedge \dot{\boldsymbol{r}} = 2m\omega(\dot{y}\cos\theta - \dot{z}\sin\theta, -\dot{x}\cos\theta, \dot{x}\sin\theta). \qquad (5.22)$$

Neglecting the Coriolis force, the motion is described by

$$x = 0, \qquad y = 0, \qquad z = h - \tfrac{1}{2}gt^2.$$

We shall calculate the effect of the Coriolis force only to first order; that is, we neglect terms of order $\omega^2$. Since the Coriolis force (5.22) contains an explicit factor of $\omega$, this means that we may substitute for $\dot{\boldsymbol{r}}$ the zeroth-order value $\dot{\boldsymbol{r}} = (0, 0, -gt)$. The equations of motion (5.21) then read

$$m\ddot{x} = 2m\omega gt\sin\theta, \qquad m\ddot{y} = 0, \qquad m\ddot{z} = -mg.$$

The solution with appropriate initial conditions is, therefore,

$$x = \tfrac{1}{3}\omega gt^3\sin\theta, \qquad y = 0, \qquad z = h - \tfrac{1}{2}gt^2.$$

Thus the body will hit the ground, $z = 0$, at a point east of that vertically below its point of release, at a distance

$$x = \frac{\omega}{3}\left(\frac{8h^3}{g}\right)^{1/2}\sin\theta. \qquad (5.23)$$

For example, if a particle is dropped from a height of $100\,\mathrm{m}$ in latitude $45°$, the deviation is about $16\,\mathrm{mm}$.

It is instructive to consider how an inertial observer would describe this experiment. Since the particle is dropped from rest relative to the Earth, it has a component of velocity towards the east relative to the inertial

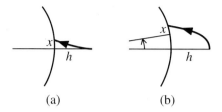

(a)        (b)

Fig. 5.8

observer. As it falls, the angular momentum about the Earth's axis remains constant, and therefore its angular velocity increases, so that it gets ahead of the ground beneath it. Figs. 5.8(a) and 5.8(b) show the experiment as it appears (but much exaggerated) to an observer on the Earth and to an inertial observer, respectively.

### Foucault's pendulum

Another way of observing the effect of the Coriolis force is to use *Foucault's pendulum*. This is simply an ordinary pendulum free to swing in any direction, and carefully arranged to be perfectly symmetric, so that its periods of oscillation in all directions are precisely equal. (It must also be long, and fairly heavy, so that it will go on swinging freely for several hours at least, despite the resistance of the air.) Examples of the Foucault pendulum are to be found in many science museums.

If the amplitude is small, the pendulum equation is simply the equation of two-dimensional simple harmonic motion. The vertical component of the Coriolis force is negligible, for it is merely a small correction to $g$, whose sign alternates on each half-period. (Even for a velocity $\dot{x}$ as large as $10\,\mathrm{m\,s^{-1}}$, we have $2\omega\dot{x} = 1.5\,\mathrm{mm\,s^{-2}} \ll g$.) The important components are the horizontal ones. For small amplitude, the velocity of the pendulum bob is almost horizontal, so that $\dot{z} \approx 0$. Thus the equations of motion for the $x$ and $y$ co-ordinates are

$$\ddot{x} = -\frac{g}{l}x + 2\omega\dot{y}\cos\theta, \qquad \ddot{y} = -\frac{g}{l}y - 2\omega\dot{x}\cos\theta, \qquad (5.24)$$

or, in vector notation,

$$\ddot{\boldsymbol{r}} = -\frac{g}{l}\boldsymbol{r} - 2\omega\cos\theta\,\boldsymbol{k}\wedge\dot{\boldsymbol{r}}. \qquad (5.25)$$

Let us first suppose that the pendulum is at the north pole ($\theta = 0$). Then it is clear from the equation of motion in a non-rotating frame that it must swing in a fixed direction in space, while the Earth rotates beneath it. Thus, relative to the Earth, its oscillation plane must rotate around the vertical with angular velocity $-\omega$. Now, at any other latitude, the only difference in Eq. (5.25) is that in place of the angular velocity $\boldsymbol{\omega}$ we have only its vertical component, $\omega \cos \theta \, \boldsymbol{k}$. Hence, we should expect that the oscillation plane of the pendulum rotates with angular velocity $-\Omega = -\omega \cos \theta$ around the vertical. In effect, we may regard the Earth's surface in colatitude $\theta$ as rotating about the vertical with angular velocity $\omega \cos \theta$ (and also with angular velocity $\omega \sin \theta$ about a horizontal north-south axis, but this component leads only to a vertical Coriolis force, which does not appreciably affect the motion.)

We can verify this conclusion by obtaining an explicit solution of Eqs. (5.24). A neat way of doing this is to combine the two equations by introducing a complex variable $z = x + iy$ (not of course to be confused with the vertical co-ordinate). Then $\dot{y} - i\dot{x} = -i\dot{z}$, so that, adding i times the second equation to the first, we get

$$\ddot{z} + 2i\Omega\dot{z} + \omega_0^2 z = 0,$$

where $\Omega = \omega \cos \theta$ and $\omega_0^2 = g/l$. We now look for solutions of the form $z = Ae^{pt}$ and, as in §2.5, find for $p$ the equation

$$p^2 + 2i\Omega p + \omega_0^2 = 0.$$

The roots of this equation are $p = -i\Omega \pm i\omega_1$, where $\omega_1^2 = \omega_0^2 + \Omega^2$. Hence, the general solution of the equation for $z$ is

$$z = Ae^{-i(\Omega - \omega_1)t} + Be^{-i(\Omega + \omega_1)t}.$$

In particular, if the pendulum is released from rest at position $(a, 0)$ at $t = 0$, we must choose $A = B = a/2$, so that the solution is

$$z = ae^{-i\Omega t} \cos \omega_1 t,$$

or, in terms of $x$ and $y$,

$$x = a \cos \Omega t \cos \omega_1 t, \qquad y = -a \sin \Omega t \cos \omega_1 t.$$

So at time $t$, since $\Omega \ll \omega_0$ and $\omega_0 \simeq \omega_1$, the pendulum bob oscillates between the extreme points $\pm a(\cos\Omega t, \sin\Omega t)$. The solution represents an oscillation with amplitude $a$ in a plane rotating with angular velocity $-\Omega$.

It should be noted that in deriving (5.24) we neglected terms of order $\Omega^2$ (in particular, the variation in the centrifugal force is of this order). Thus the difference between the angular frequency $\omega_1$ in our solution and $\omega_0$ is not significant. The period of the pendulum will not be substantially affected.

At the pole, the plane of oscillation makes a complete revolution in just 24 hours. At any other latitude, the period is greater than this, and is in fact $2\pi/\omega\cos\theta$. In latitude $45°$, it is about 34 hours, while on the equator (where $\boldsymbol{\omega}$ is purely horizontal), it is infinite.

### Cyclones and trade winds

There are also some important large-scale effects of the Coriolis force. Suppose that for some reason a region of low pressure develops in the northern hemisphere. The air around it will be pushed inwards by the force of the pressure gradient. As it starts to move, however, the Coriolis force cause it to curve to the right. Thus an anticlockwise rotation is set up around the low-pressure zone. The process will continue until an approximate equilibrium is established between the pressure force, acting inwards, and the Coriolis force (plus the centrifugal force of the rotation) acting outwards. This configuration is a *cyclone*, or *depression*, familiar to those who live in temperate latitudes. (As in the case of Foucault's pendulum, there is no effect of this kind on the equator.) More generally, in these latitudes, the wind velocity is not directly from regions of high to low pressure, but more nearly along the isobars, keeping the low pressure on the left in the northern hemisphere.

The same effect is responsible, on a yet larger scale, for the trade winds. The heating of the Earth's surface near the equator causes the air to rise, and be replaced by cooler air flowing in from higher latitudes. However, because of the Coriolis force, it does not flow directly north or south, but is made to deviate towards the west. Thus we have the north-east trade winds in the northern hemisphere, and the south-east trade winds in the southern.

For reasons that are too complex to discuss in detail here, this pattern does not extend to high latitudes. Further from the equator there are belts of high pressure, which, like the equatorial region itself, are characterized by light and variable winds. Beyond these, the direction of the pressure gradient is reversed. Thus around $40°$ to $50°$ N or S the prevailing wind direction is westerly rather than easterly. Near the poles, there is a further reversal, and the circumpolar winds blow from the east.

## 5.5   Larmor Effect

As a rather different example of the use of rotating frames, we shall consider the effect of a magnetic field on a particle of charge $q$ moving in an orbit around a fixed point charge $-q'$. The equation of motion, including the magnetic force (5.8), is

$$m\frac{d^2\boldsymbol{r}}{dt^2} = -\frac{k}{r^2}\hat{\boldsymbol{r}} + q\frac{d\boldsymbol{r}}{dt} \wedge \boldsymbol{B},$$

where $k = qq'/4\pi\epsilon_0$. Let us rewrite this equation in terms of a rotating frame of reference, using (5.7) and (5.15). We obtain

$$\ddot{\boldsymbol{r}} + 2\boldsymbol{\omega} \wedge \dot{\boldsymbol{r}} + \boldsymbol{\omega} \wedge (\boldsymbol{\omega} \wedge \boldsymbol{r}) = -\frac{k}{mr^2}\hat{\boldsymbol{r}} + \frac{q}{m}(\dot{\boldsymbol{r}} + \boldsymbol{\omega} \wedge \boldsymbol{r}) \wedge \boldsymbol{B}.$$

Now, if we choose $\boldsymbol{\omega} = -(q/2m)\boldsymbol{B}$, then the terms in $\dot{\boldsymbol{r}}$ drop out. The last term on the left only cancels half the last term on the right, however, so we obtain

$$\ddot{\boldsymbol{r}} = -\frac{k}{mr^2}\hat{\boldsymbol{r}} + \left(\frac{q}{2m}\right)^2 \boldsymbol{B} \wedge (\boldsymbol{B} \wedge \boldsymbol{r}).$$

We now assume that the magnetic field is sufficiently weak for the quadratic term in $\boldsymbol{B}$ to be negligible in comparison to the electrostatic force term. The necessary condition for this is that

$$\omega^2 = \left(\frac{qB}{2m}\right)^2 \ll \frac{k}{mr^3} = \frac{qq'}{4\pi\epsilon_0 mr^3} \approx \omega_0^2, \tag{5.26}$$

where we have used (4.31) to express the right side in terms of the mean angular velocity $\omega_0$ of the particle in its orbit.

When this condition is satisfied, we obtain the approximate equation

$$\ddot{\boldsymbol{r}} = -\frac{k}{mr^2}\hat{\boldsymbol{r}}.$$

But this is just the usual equation of motion for a particle in an inverse-square-law field. Consequently, a bounded orbit in the rotating frame is an ellipse. In the original, non-rotating frame, it is a slowly precessing ellipse, precessing with angular velocity $\boldsymbol{\omega}$. (See Fig. 5.9, which is drawn for the special case in which the orbit lies in the plane normal to $\boldsymbol{B}$. In general, it

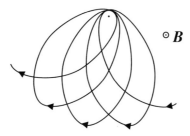

Fig. 5.9

will be inclined, and the plane of the orbit will precess about the direction of $\boldsymbol{B}$.) Note that, in view of the restriction (5.26), the axes of the ellipse will precess only through a small angle on each revolution.

This effect is known as the *Larmor effect*, and the angular velocity of precession,

$$\omega_{\mathrm{L}} = \frac{qB}{2m},$$

is called the *Larmor frequency*. Note that it is just half the cyclotron frequency (5.12). The difference arises from the factor of 2 in the Coriolis acceleration term of (5.15).

This effect leads to observable changes in the spectra emitted by atoms in the presence of a magnetic field, since the Bohr energy level corresponding to a given angular momentum is slightly shifted in the presence of a magnetic field. (The shift in spectral lines is known as the *Zeeman effect*.)

When the weak magnetic field approximation is not applicable then the resulting motion can be very complex — see §14.4.

## 5.6 Angular Momentum and the Larmor Effect

The Larmor effect is a particular example of a more general, and very important, phenomenon. As we shall see later in a number of different examples, the effect of a small force on a rotating system is often to make the axis of rotation precess — that is, revolve about some fixed direction.

It will therefore be useful to consider here an alternative treatment of the problem which is of more general applicability.

If the magnetic field is weak, then to a first approximation the particle must move in an inverse-square-law orbit whose characteristics change slowly with time. Since the magnetic force is perpendicular to the velocity, it does no work. Hence the energy is constant, and therefore so is the semi-major axis of the orbit. However the angular momentum $\boldsymbol{J}$ does change with time, according to the equation

$$\frac{\mathrm{d}\boldsymbol{J}}{\mathrm{d}t} = \boldsymbol{r} \wedge \boldsymbol{F} = q\boldsymbol{r} \wedge (\boldsymbol{v} \wedge \boldsymbol{B}) = q[(\boldsymbol{r} \cdot \boldsymbol{B})\boldsymbol{v} - (\boldsymbol{r} \cdot \boldsymbol{v})\boldsymbol{B}], \qquad (5.27)$$

using (A.16).

To be specific, let us suppose that the magnetic field is in the $z$-direction, $\boldsymbol{B} = B\boldsymbol{k}$, and that the particle is moving in a circular orbit of radius $r$ in a plane inclined to the $xy$-plane at an angle $\alpha$. (See Fig. 5.10.) It will be convenient to introduce three orthogonal unit vectors, $\boldsymbol{n}$ normal to the

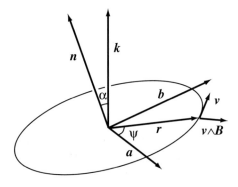

Fig. 5.10

plane of the orbit, $\boldsymbol{a}$ in the direction of $\boldsymbol{k} \wedge \boldsymbol{n}$ and $\boldsymbol{b} = \boldsymbol{n} \wedge \boldsymbol{a}$. Then the angular momentum is

$$\boldsymbol{J} = m\boldsymbol{r} \wedge \boldsymbol{v} = mrv\boldsymbol{n}. \qquad (5.28)$$

If we specify the position of the particle in its orbit by the angle $\psi$ between $\boldsymbol{a}$ and $\boldsymbol{r}$, then in terms of the axes $\boldsymbol{a}, \boldsymbol{b}, \boldsymbol{n}$, the components of $\boldsymbol{r}$, $\boldsymbol{v}$ and $\boldsymbol{B}$

are

$$\begin{aligned} \boldsymbol{r} &= (r\cos\psi, r\sin\psi, 0), \\ \boldsymbol{v} &= (-v\sin\psi, v\cos\psi, 0), \\ \boldsymbol{B} &= (0, B\sin\alpha, B\cos\alpha). \end{aligned}$$

(There is also a small contribution to $\boldsymbol{v}$ from the fact that the plane of the orbit is changing, but for small magnetic fields this will be negligible.)

Now, in a circular orbit, $\boldsymbol{r} \cdot \boldsymbol{v}$ is always zero. Hence the right hand side of (5.27) is

$$qBr\sin\alpha\sin\psi\,\boldsymbol{v} = qBrv\sin\alpha(-\sin^2\psi, \sin\psi\cos\psi, 0). \tag{5.29}$$

Since the magnetic field is weak, the change in $\boldsymbol{J}$ in a single orbit will be small. Thus the oscillatory term in (5.29) is unimportant: it leads only to a very small periodic oscillation in $\boldsymbol{J}$. The important term is the one which leads to a *secular* variation of $\boldsymbol{J}$ (a steady change in one direction). It follows that we can replace (5.29) by its average value over a complete oscillation period, *i.e.*, over all values of $\psi$ from 0 to $2\pi$. Thus we replace $\sin^2\psi = \frac{1}{2}(1 - \cos 2\psi)$ by $\frac{1}{2}$, and $\sin\psi\cos\psi$ by 0. This yields the equation

$$\frac{\mathrm{d}\boldsymbol{J}}{\mathrm{d}t} = \langle \boldsymbol{r} \wedge \boldsymbol{F} \rangle_{\mathrm{av}} = -\tfrac{1}{2}qBrv\sin\alpha\,\boldsymbol{a}. \tag{5.30}$$

Thus the vector $\boldsymbol{J}$ will move in the direction of $-\boldsymbol{a}$. Since $\mathrm{d}\boldsymbol{J}/\mathrm{d}t$ is perpendicular to $\boldsymbol{J}$, its magnitude remains unchanged, while its direction precesses around the direction of $\boldsymbol{B}$. In fact, (5.30) can be seen to be another example of an equation of the form (5.3). From (5.28), we have

$$\boldsymbol{k} \wedge \boldsymbol{J} = mrv\boldsymbol{k} \wedge \boldsymbol{n} = mrv\sin\alpha\,\boldsymbol{a}.$$

Thus (5.30) may be written

$$\frac{\mathrm{d}\boldsymbol{J}}{\mathrm{d}t} = -\frac{qB}{2m}\boldsymbol{k} \wedge \boldsymbol{J}. \tag{5.31}$$

This is the equation of a vector $\boldsymbol{J}$ rotating with angular velocity

$$\boldsymbol{\omega} = -\frac{q\boldsymbol{B}}{2m}.$$

Since the direction of $J$ is that of the normal to the orbital plane, this agrees with our previous conclusion that the orbital plane precesses around the direction of the magnetic field, with a precessional angular velocity equal to the angular Larmor frequency.

### Comparison with a current loop

It is instructive to contrast this behaviour of a charged particle with the superficially similar case of a current loop. A circular current loop in the same position as the particle orbit would experience a similar force, producing a moment about the $-a$-axis. The effect of the force, however, would be quite different. Since the loop itself (that is, its *mass*) is not rotating, the effect would be simply to rotate it about this axis, so that its normal tends to become aligned with the direction of $B$. (This is the familiar fact that a current loop behaves like a magnetic dipole.)

To understand why a particle moving in an orbit behaves so differently, it may be helpful to consider, in place of a continuously acting force, a small impulsive blow delivered once every orbit, say at the farthest point above the $xy$-plane. The effect of such a blow is to give the particle a small component of velocity perpendicular to its original orbit. The next orbit will be in a plane slightly tilted with respect to the original one, but reaching just as far from the $xy$-plane. The net effect is to make the orbit swing around the $z$-axis. The essential point here is that it is the *velocity* of the particle, rather than its position, that is changed instantaneously. Thus the effect of a small blow is not seen in a shift of the orbit at the point of the blow, but rather at a point 90° later.

We shall see in Chapter 9 that a rigid body displays a very similar type of behaviour. If one applies a small force downwards on the rim of a rapidly rotating wheel, it is not that point that moves down, but a point 90° later. (This is easy to verify with a freely spinning bicycle wheel.)

## 5.7   Summary

In problems involving a rotating body — particularly the Earth — it is often convenient to use a rotating frame of reference. The equations of motion in such a frame contain additional terms representing apparent forces which arise because the frame is non-inertial. These are the centrifugal force, directed outwards from the axis of rotation, and the velocity-dependent

Coriolis force, $-2m\boldsymbol{\omega} \wedge \dot{\boldsymbol{r}}$. Because the magnetic force on a charged particle tends to produce rotation about the direction of the magnetic field, rotating frames are also useful in many problems involving a magnetic field.

---

## Problems

1. Find the centrifugal acceleration at the equator of the planet Jupiter and of the Sun. In each case, express your answer also as a fraction of the surface gravity. (The rotation periods are 10 hours and 27 days, respectively, the radii $7.1 \times 10^4$ km and $7.0 \times 10^5$ km, and the masses $1.9 \times 10^{27}$ kg and $2.0 \times 10^{30}$ kg.)

2. Water in a rotating container of radius 50 mm is 30 mm lower in the centre than at the edge. Find the angular velocity of the container.

3. The water in a circular lake of radius 1 km in latitude $60°$ is at rest relative to the Earth. Find the depth by which the centre is depressed relative to the shore by the centrifugal force. For comparison, find the height by which the centre is *raised* by the curvature of the Earth's surface. (Earth radius = 6400 km.)

4. Find the velocity relative to an inertial frame (in which the centre of the Earth is at rest) of a point on the Earth's equator. An aircraft is flying above the equator at $1000 \, \mathrm{km \, h^{-1}}$. Assuming that it flies straight and level (*i.e.*, at a constant altitude above the surface) what is its velocity relative to the inertial frame (a) if it flies north, (b) if it flies west, and (c) if it flies east?

5. The apparent weight of the aircraft in Problem 4 when on the ground at the equator is 100 t weight. What is its apparent weight in each of the three cases (a)–(c)?

6. A bird of mass 2 kg is flying at $10 \, \mathrm{m \, s^{-1}}$ in latitude $60°$N, heading due east. Find the horizontal and vertical components of the Coriolis force acting on it.

7. The wind speed in colatitude $\theta$ is $v$. By considering the forces on a small volume of air, show that the pressure gradient required to balance the horizontal component of the Coriolis force, and thus to maintain a constant wind direction, is $dp/dx = 2\omega\rho v \cos\theta$, where $\rho$ is the density of the air. Evaluate this gradient in $\mathrm{mbar \, km^{-1}}$ for a wind speed of $50 \, \mathrm{km \, h^{-1}}$ in latitude $30°$N. (1 bar = $10^5$ Pa; density of air = $1.3 \, \mathrm{kg \, m^{-3}}$.)

8. An aircraft is flying at $800 \, \text{km} \, \text{h}^{-1}$ in latitude 55°N. Find the angle through which it must tilt its wings to compensate for the horizontal component of the Coriolis force.

9. An orbiting space station may be made to rotate to provide an artificial gravity. Given that the radius is 25 m, find the rotation period required to produce an apparent gravity equal to $0.7g$. A man whose normal weight is 75 kg weight runs around the station in one direction and then the other (*i.e.*, on a circle on the inside of the cylindrical wall) at $5 \, \text{m} \, \text{s}^{-1}$. Find his apparent weight in each case. What effects will he experience if he climbs a ladder to a higher level (*i.e.*, closer to the axis), climbing at $1 \, \text{m} \, \text{s}^{-1}$?

10. A beam of particles of charge $q$ and velocity $v$ is emitted from a point source, roughly parallel with a magnetic field $\boldsymbol{B}$, but with a small angular dispersion. Show that the effect of the field is to focus the beam to a point at a distance $z = 2\pi m v / |q| B$ from the source. Calculate the focal distance for electrons of kinetic energy $500 \, \text{eV}$ in a magnetic field of $0.01 \, \text{T}$. (Charge on electron $= -1.6 \times 10^{-19} \, \text{C}$, mass $= 9.1 \times 10^{-31} \, \text{kg}$, $1 \, \text{eV} = 1.6 \times 10^{-19} \, \text{J}$.)

11. *Write down the equation of motion for a charged particle in uniform, *parallel* electric and magnetic fields, both in the $z$-direction, and solve it, given that the particle starts from the origin with velocity $(v, 0, 0)$. A screen is placed at $x = a$, where $a \ll mv/qB$. Show that the locus of points of arrival of particles with given $m$ and $q$, but different speeds $v$, is approximately a parabola. How does this locus depend on $m$ and $q$?

12. A beam of particles with velocity $(v, 0, 0)$ enters a region containing crossed electric and magnetic fields, as in the example at the end of §5.2. Show that if the ratio $E/B$ is correctly chosen the particles are undeviated, while particles with other speeds follow curved trajectories. Suppose the particles have velocities equal to $v$ in magnitude, but with a small angular dispersion. Show that if the path length $l$ is correctly chosen, all such particles are focussed onto a line parallel to the $z$-axis. (Thus a slit at that point can be used to select particles with a given speed.) For electrons of velocity $10^8 \, \text{m} \, \text{s}^{-1}$ in a magnetic field of $0.02 \, \text{T}$, find the required electric field, and the correct (smallest possible) choice for $l$.

13. The angular velocity of the electron in the lowest Bohr orbit of the hydrogen atom is approximately $4 \times 10^{16} \, \text{s}^{-1}$. What is the largest

magnetic field which may be regarded as small in this case, in the sense of §5.5? Determine the Larmor frequency in a field of 2 T.

14. *The orbit of an electron (charge $-e$) around a nucleus (charge $Ze$) is a circle of radius $a$ in a plane perpendicular to a uniform magnetic field $\boldsymbol{B}$. By writing the equation of motion in a frame rotating with the electron, show that the angular velocity $\omega$ is given by one of the roots of the equation

$$m\omega^2 - eB\omega - Ze^2/4\pi\epsilon_0 a^3 = 0.$$

Verify that for small values of $B$, this agrees with §5.5. Evaluate the two roots if $B = 10^5\,\mathrm{T}$, $Z = 1$ and $a = 5.3 \times 10^{-11}\,\mathrm{m}$. (Note, however, that in reality $a$ would be changed by the field.)

15. *A projectile is launched due north from a point in colatitude $\theta$ at an angle $\pi/4$ to the horizontal, and aimed at a target whose distance is $y$ (small compared to Earth's radius $R$). Show that if no allowance is made for the effects of the Coriolis force, the projectile will miss its target by a distance

$$x = \omega \left(\frac{2y^3}{g}\right)^{1/2} (\cos\theta - \tfrac{1}{3}\sin\theta).$$

Evaluate this distance if $\theta = 45°$ and $y = 40\,\mathrm{km}$. Why is it that the deviation is to the east near the north pole, but to the west both on the equator and near the south pole? (Neglect atmospheric resistance.)

16. *Solve the problem of a particle falling from height $h$ above the equator by using an inertial frame, and verify that the answer agrees with that found using a rotating frame. (*Hint*: Use equations (3.48). Recall Fig. 5.8.)

17. Find the equations of motion for a particle in a frame rotating with *variable* angular velocity $\boldsymbol{\omega}$, and show that there is another apparent force of the form $-m\dot{\boldsymbol{\omega}} \wedge \boldsymbol{r}$. Discuss the physical origin of this force.

18. Find the equation of motion for a particle in a *uniformly accelerated* frame, with acceleration $\boldsymbol{a}$. Show that for a particle moving in a uniform gravitational field, and subject to other forces, the gravitational field may be eliminated by a suitable choice of $\boldsymbol{a}$.

19. *The co-ordinates $(x, y, z)$ of a particle with respect to a uniformly rotating frame may be related to those with respect to a fixed inertial

frame, $(x^*, y^*, z^*)$, by the transformation

$$\begin{bmatrix} x \\ y \\ z \end{bmatrix} = \begin{bmatrix} \cos\omega t & \sin\omega t & 0 \\ -\sin\omega t & \cos\omega t & 0 \\ 0 & 0 & 1 \end{bmatrix} \begin{bmatrix} x^* \\ y^* \\ z^* \end{bmatrix}.$$

(Here, we use matrix notation: this stands for three separate equations,

$$x = \cos\omega t \cdot x^* + \sin\omega t \cdot y^*,$$

*etc.*) Write down the inverse relation giving $(x^*, y^*, z^*)$ in terms of $(x, y, z)$. By differentiating with respect to $t$, rederive the relation (5.15) between $\mathrm{d}^2 \boldsymbol{r}/\mathrm{d}t^2$ and $\ddot{\boldsymbol{r}}$. [*Hint*: Note that $\ddot{\boldsymbol{r}} = (\ddot{x}, \ddot{y}, \ddot{z})$, while $\mathrm{d}^2 \boldsymbol{r}/\mathrm{d}t^2$ is the vector obtained by applying the above transformation to $(\ddot{x}^*, \ddot{y}^*, \ddot{z}^*)$.]

20. Another way of deriving the equation of motion (5.16) is to use Lagrange's equations. Express the kinetic energy $\frac{1}{2}m(\mathrm{d}\boldsymbol{r}/\mathrm{d}t)^2$ in terms of $(x, y, z)$, and show that Lagrange's equations (3.44) reproduce (5.16) for the case where the force is conservative.

# Chapter 6

# Potential Theory

This chapter is complementary to the preceding ones. In it we shall discuss not the problem of determining the motion of a particle under known forces, but the problem of finding the forces from a knowledge of the positions of other bodies. We deal specifically with gravitational and electrostatic forces obeying the inverse square law, which are determined by the positions of other masses and charges.

We return here to the convention whereby $\dot{\boldsymbol{r}} = \mathrm{d}\boldsymbol{r}/\mathrm{d}t$. We shall not use rotating frames again until Chapter 9.

## 6.1   Gravitational and Electrostatic Potentials

The gravitational potential energy of a particle of mass $m$ moving in the field of a fixed mass $m'$ at $\boldsymbol{r}'$ is $-Gmm'/|\boldsymbol{r} - \boldsymbol{r}'|$. If we have several masses $m_j$, located at the points $\boldsymbol{r}_j$, then the potential energy is the sum

$$V(\boldsymbol{r}) = -\sum_j \frac{Gmm_j}{|\boldsymbol{r} - \boldsymbol{r}_j|}.$$

(The fact that the potential energies add follows from the additive property of forces.)

Since the mass $m$ appears only as an overall factor, we may define the *gravitational potential* $\Phi(\boldsymbol{r})$ to be the potential energy per unit mass

$$V(\boldsymbol{r}) = m\Phi(\boldsymbol{r}), \tag{6.1}$$

so that

$$\Phi(\boldsymbol{r}) = -\sum_j \frac{Gm_j}{|\boldsymbol{r} - \boldsymbol{r}_j|}. \tag{6.2}$$

Note that $\Phi$ is always negative. (The potential has sometimes been defined to be *minus* the potential energy per unit mass, and therefore positive. We prefer, however, to retain the direct correspondence between potential and potential energy, so that particles tend to move towards regions of lower potential.)

The acceleration of a particle moving under gravitational forces is given by

$$m\ddot{\boldsymbol{r}} = -\boldsymbol{\nabla}V(\boldsymbol{r}) = -m\boldsymbol{\nabla}\Phi(\boldsymbol{r}).$$

Since this acceleration is independent of the mass $m$, we may define the *gravitational acceleration* or *gravitational field* $\boldsymbol{g}(\boldsymbol{r})$ by

$$\boldsymbol{g}(\boldsymbol{r}) = -\boldsymbol{\nabla}\Phi(\boldsymbol{r}), \tag{6.3}$$

Thus (6.2) and (6.3) are all that is needed to calculate the acceleration induced in a particle by a given distribution of masses.

The electrostatic case is very similar. We define the *electrostatic potential* $\phi(\boldsymbol{r})$ to be the electrostatic potential energy per unit charge,

$$V(\boldsymbol{r}) = q\phi(\boldsymbol{r}). \tag{6.4}$$

The potential due to charges $q_j$ at $\boldsymbol{r}_j$ is then

$$\phi(\boldsymbol{r}) = \sum_j \frac{q_j}{4\pi\epsilon_0|\boldsymbol{r}-\boldsymbol{r}_j|}. \tag{6.5}$$

The acceleration of a charged particle is given by

$$m\ddot{\boldsymbol{r}} = q\boldsymbol{E},$$

where the *electric field* $\boldsymbol{E}(\boldsymbol{r})$ is defined by

$$\boldsymbol{E}(\boldsymbol{r}) = -\boldsymbol{\nabla}\phi(\boldsymbol{r}), \tag{6.6}$$

Note that the acceleration depends on the charge-to-mass ratio $q/m$. (The corresponding ratio in the gravitational case is the ratio of gravitational to inertial mass, which is of course a constant, set equal to unity in conventional units. This proportionality is the content of the *equivalence principle*.)

Unlike the gravitational potential, the electrostatic potential $\phi$ may have either sign, because both positive and negative charges exist.

We can now forget about the particle at $\boldsymbol{r}$, and concentrate on the problem of calculating the potential from given information about the positions of the masses or charges.

## 6.2 The Dipole and Quadrupole

The *electric dipole* consists of two equal and opposite charges, $q$ and $-q$, placed close together, say at $\boldsymbol{a}$ and at the origin, respectively. (See Fig. 6.1.) The potential is

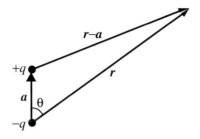

Fig. 6.1

$$\phi(\boldsymbol{r}) = \frac{q}{4\pi\epsilon_0|\boldsymbol{r} - \boldsymbol{a}|} - \frac{q}{4\pi\epsilon_0|\boldsymbol{r}|}. \tag{6.7}$$

We shall assume that $a \ll r$, so that we may expand by the binomial theorem. Since we shall need the result later, we evaluate the terms up to order $a^2$, but neglect $a^3$. If $\theta$ is the angle between $\boldsymbol{a}$ and $\boldsymbol{r}$, then

$$|\boldsymbol{r} - \boldsymbol{a}|^2 = r^2 - 2ar\cos\theta + a^2.$$

Hence

$$
\begin{aligned}
\frac{1}{|\boldsymbol{r} - \boldsymbol{a}|} &= \frac{1}{r}\left(1 - 2\frac{a}{r}\cos\theta + \frac{a^2}{r^2}\right)^{-1/2} \\
&= \frac{1}{r}\left[1 - \frac{1}{2}\left(-2\frac{a}{r}\cos\theta + \frac{a^2}{r^2}\right) + \frac{3}{8}\left(-2\frac{a}{r}\cos\theta + \frac{a^2}{r^2}\right)^2 - \cdots\right] \\
&= \frac{1}{r} + \frac{a}{r^2}\cos\theta + \frac{a^2}{r^3}\left(\tfrac{3}{2}\cos^2\theta - \tfrac{1}{2}\right) + \cdots .
\end{aligned}
\tag{6.8}
$$

(The general term in this series is of the form $(a^l/r^{l+1})P_l(\cos\theta)$, where $P_l$ is a polynomial, known as the *Legendre polynomial* of degree $l$. Here, however, we shall only need the terms up to $l = 2$.) In vector notation, we may write (6.8) as

$$\frac{1}{|r - a|} = \frac{1}{r} + \frac{a \cdot r}{r^3} + \frac{3(a \cdot r)^2 - a^2 r^2}{2r^5} + \cdots . \tag{6.9}$$

We now return to (6.7). Keeping only the linear term in $a$ and neglecting $a^2$, we can write it as

$$\phi(r) = \frac{d \cdot r}{4\pi\epsilon_0 r^3} = \frac{d \cos\theta}{4\pi\epsilon_0 r^2}, \tag{6.10}$$

where $d$ is the *electric dipole moment*, $d = qa$. The corresponding electric field, given by (6.6), has the spherical polar components (see (A.50))

$$\begin{aligned}
E_r &= -\frac{\partial \phi}{\partial r} & &= \frac{2d \cos\theta}{4\pi\epsilon_0 r^3}, \\
E_\theta &= -\frac{1}{r}\frac{\partial \phi}{\partial \theta} & &= \frac{d \sin\theta}{4\pi\epsilon_0 r^3}, \\
E_\varphi &= -\frac{1}{r\sin\theta}\frac{\partial \phi}{\partial \varphi} & &= 0.
\end{aligned} \tag{6.11}$$

(To avoid any possible risk of confusion, we denote the potential and the azimuth angle by the distinct symbols $\phi$ and $\varphi$.) This field is illustrated in Fig. 6.2, in which the solid lines are field lines, drawn in the direction of $E$, and the dashed lines are the *equipotential* surfaces, $\phi = $ constant. Note that the two always intersect at right angles.

We can repeat the process of putting two charges (or *monopoles*) together to form a dipole, by putting two dipoles together to form a *quadrupole*. If we place oppositely oriented dipoles, with dipole moments $d$ and $-d$, at $a$ and at the origin, then the potential is

$$\phi(r) = \frac{d \cdot (r - a)}{4\pi\epsilon_0 |r - a|^3} - \frac{d \cdot r}{4\pi\epsilon_0 r^3}.$$

Now, neglecting terms of order $a^2$, we have

$$\frac{1}{|r - a|^3} \approx \frac{1}{r^3}\left(1 - 2\frac{a \cdot r}{r^2}\right)^{-3/2} \approx \frac{1}{r^3} + 3\frac{a \cdot r}{r^5}.$$

Hence, to this approximation

$$\phi(r) = \frac{3(d \cdot r)(a \cdot r) - (d \cdot a)r^2}{4\pi\epsilon_0 r^5}. \tag{6.12}$$

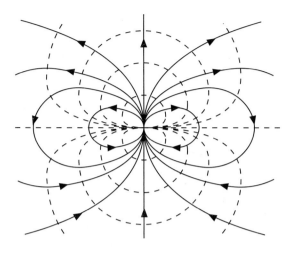

Fig. 6.2

In the special case, where the dipoles are placed end-on, so that $\boldsymbol{d}$ is parallel to $\boldsymbol{a}$, we may take this common direction to be the $z$-axis, and obtain

$$\phi(\boldsymbol{r}) = \frac{Q}{16\pi\epsilon_0 r^3}(3\cos^2\theta - 1), \tag{6.13}$$

where $Q$ is the *electric quadrupole moment*, $Q = 4da$. (The factor of 4 is purely conventional, and serves to simplify some of the later formulae.) The corresponding electric field is

$$
\begin{aligned}
E_r &= \frac{3Q}{16\pi\epsilon_0 r^4}(3\cos^2\theta - 1), \\
E_\theta &= \frac{3Q}{8\pi\epsilon_0 r^4}\cos\theta\sin\theta, \\
E_\varphi &= 0.
\end{aligned}
\tag{6.14}
$$

It is illustrated in Fig. 6.3.

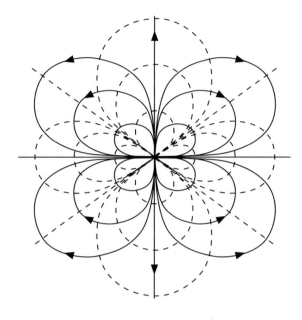

Fig. 6.3

## 6.3 Spherical Charge Distributions

When we have a continuous distribution of charge (or mass) we can replace
the sum in (6.5) by an integral:

$$\phi(\boldsymbol{r}) = \iiint \frac{\rho(\boldsymbol{r}')}{4\pi\epsilon_0|\boldsymbol{r} - \boldsymbol{r}'|}\, \mathrm{d}^3\boldsymbol{r}', \tag{6.15}$$

where $\rho(\boldsymbol{r}')$ is the *charge density*, and $\mathrm{d}^3\boldsymbol{r}'$ denotes the element of volume.
In spherical polars, the volume element is the product of the elements of
length in the three co-ordinate directions, namely (see §3.5)

$$\mathrm{d}^3\boldsymbol{r}' = \mathrm{d}x'\, \mathrm{d}y'\, \mathrm{d}z' = r'^2\mathrm{d}r'\sin\theta'\mathrm{d}\theta'\, \mathrm{d}\varphi'.$$

Note that $\theta'$ ranges from $0 \to \pi$, and $\varphi'$ from $0 \to 2\pi$.

We shall consider in this section spherically symmetric distributions of
charge, for which $\rho$ is a function only of the radial co-ordinate $r'$. Let us
first take a uniform thin spherical shell of charge density $\rho$, radius $a$ and
thickness $\mathrm{d}a$. Choosing the direction of $\boldsymbol{r}$ to be the $z$-axis, we can write

(6.15) as

$$\phi(\mathbf{r}) = \frac{\rho a^2 \mathrm{d}a}{4\pi\epsilon_0} \iint \frac{\sin\theta' \, \mathrm{d}\theta' \, \mathrm{d}\varphi'}{(r^2 - 2ar\cos\theta' + a^2)^{1/2}}.$$

The $\varphi'$ integration gives a factor of $2\pi$. The $\theta'$ integration can easily be performed by the substitution $u = \cos\theta'$, and yields

$$\phi(\mathbf{r}) = \rho a^2 \mathrm{d}a \frac{(r+a) - |r-a|}{2\epsilon_0 ar}.$$

We now have to consider separately the cases $r > a$ and $r < a$. In terms of the total charge $\mathrm{d}q = 4\pi\rho a^2 \mathrm{d}a$ of the spherical shell, we obtain

$$\phi(\mathbf{r}) = \begin{cases} \dfrac{\mathrm{d}q}{4\pi\epsilon_0 r}, & r > a, \\ \dfrac{\mathrm{d}q}{4\pi\epsilon_0 a}, & r < a. \end{cases} \tag{6.16}$$

Outside the shell, the potential is the same as that of a charge $\mathrm{d}q$ concentrated at the origin, and the electric field is $\mathbf{E} = \hat{\mathbf{r}} \, \mathrm{d}q/4\pi\epsilon_0 r^2$. Inside, the potential is a constant, and the electric field vanishes.

It is now easy to find the potential of any spherically symmetric distribution, by summing over all the spherical shells. It is clear that the electric field at a distance $r$ from the centre is equal to that of a point charge located at the origin, whose magnitude is the total charge enclosed within a sphere of radius $r$. (A similar result holds in the gravitational case.)

A simple example is a uniformly charged sphere.

*Example:* **Uniformly charged sphere**

What is the potential and electric field for a sphere of radius $a$, with constant charge density $\rho$ inside it?

For $r > a$ the potential is clearly just

$$\phi(\mathbf{r}) = \frac{q}{4\pi\epsilon_0 r}, \qquad \text{with} \qquad q = \frac{4}{3}\pi a^3 \rho.$$

Inside the sphere, we have to separate the contributions from the regions $r' < r$ and $r' > r$, and obtain

$$\phi(\mathbf{r}) = \int_0^r \frac{\rho r'^2}{\epsilon_0 r} \mathrm{d}r' + \int_r^a \frac{\rho r'}{\epsilon_0} \mathrm{d}r'.$$

Performing the integrations, we find

$$\phi(\boldsymbol{r}) = \begin{cases} \dfrac{q}{4\pi\epsilon_0 r}, & r > a, \\[2mm] \dfrac{q}{4\pi\epsilon_0}\left(\dfrac{3}{2a} - \dfrac{r^2}{2a^3}\right), & r < a. \end{cases} \tag{6.17}$$

This potential is illustrated in Fig. 6.4. The corresponding electric field is

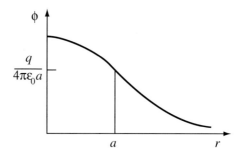

Fig. 6.4

$$\boldsymbol{E}(\boldsymbol{r}) = \begin{cases} \hat{\boldsymbol{r}}\,\dfrac{q}{4\pi\epsilon_0 r^2}, & r > a, \\[2mm] \hat{\boldsymbol{r}}\,\dfrac{qr}{4\pi\epsilon_0 a^3}, & r < a. \end{cases} \tag{6.18}$$

Thus, outside the sphere, the field obeys the inverse square law. Inside, it increases linearly from zero at the centre to the surface value $q/4\pi\epsilon_0 a^2$.

So far as its effect on outside bodies is concerned, any spherically symmetric distribution of charge or mass may be replaced by a point charge or mass located at the centre. This is the justification for treating the Sun, for example, as a point mass in discussing the motion of the planets. Of course, any deviation from spherical symmetry will lead to a modification in the inverse square law. We shall discuss the nature of this correction later.

## 6.4 Expansion of Potential at Large Distances

It is only for simple cases that we can calculate the potential exactly. In general, we must use approximation methods. In particular, we are often interested in the form of the potential at distances from the charge or mass distribution which are large compared to its dimensions. Then in (6.5) or (6.15), $r$ is much larger than $r_j$ or $r'$, and we may use the series expansion (6.8) to obtain an expansion of the potential in powers of $1/r$:

$$\phi(\boldsymbol{r}) = \phi_0(\boldsymbol{r}) + \phi_1(\boldsymbol{r}) + \phi_2(\boldsymbol{r}) + \cdots . \tag{6.19}$$

The leading ('monopole') term is

$$\phi_0(\boldsymbol{r}) = \frac{q}{4\pi\epsilon_0 r}, \tag{6.20}$$

where $q$ is the total charge

$$q = \sum_j q_j \quad \text{or} \quad q = \iiint \rho(\boldsymbol{r}')\,\mathrm{d}^3\boldsymbol{r}'. \tag{6.21}$$

Thus at very great distances, only the total charge is important, not the shape of the distribution. This is the potential we should get for an exactly spherically symmetric distribution. Thus the further terms in the series (6.19) may be regarded as measuring deviations from spherical symmetry.

The next term in (6.19) is

$$\phi_1(\boldsymbol{r}) = \frac{\boldsymbol{d} \cdot \boldsymbol{r}}{4\pi\epsilon_0 r^3}, \tag{6.22}$$

where $\boldsymbol{d}$ is the dipole moment, defined by

$$\boldsymbol{d} = \sum_j q_j \boldsymbol{r}_j \quad \text{or} \quad \boldsymbol{d} = \iiint \rho(\boldsymbol{r}')\boldsymbol{r}'\,\mathrm{d}^3\boldsymbol{r}'. \tag{6.23}$$

If we shift the origin through a distance $\boldsymbol{R}$, the total charge is clearly unaffected, but the dipole moment changes to

$$\boldsymbol{d} = \sum_j q_j(\boldsymbol{r}_j - \boldsymbol{R}) = \boldsymbol{d} - q\boldsymbol{R}. \tag{6.24}$$

Thus, if the total charge is non-zero, we can always make the dipole moment vanish by choosing the origin to lie at the *centre of charge*

$$\boldsymbol{R} = \frac{\sum q_j \boldsymbol{r}_j}{\sum q_j} = \frac{\boldsymbol{d}}{q}. \tag{6.25}$$

In this case, the dipole moment gives us information about the position of the centre of charge. However, if the total charge is zero, then by (6.24) the dipole moment is independent of the choice of origin — the electric dipole and quadrupole discussed in the preceding section are examples of this.

Note that in the gravitational case, the total mass can never vanish, so we can always ensure that the gravitational 'dipole moment' is zero by choosing our origin at the centre of mass.

The 'quadrupole' term in the potential (6.19), obtained by substituting the third term of (6.9) into (6.15), is

$$\phi_2(\boldsymbol{r}) = \iiint \rho(\boldsymbol{r}') \frac{3(\boldsymbol{r} \cdot \boldsymbol{r}')^2 - r^2 r'^2}{8\pi\epsilon_0 r^5} \, d^3\boldsymbol{r}'. \tag{6.26}$$

One can write this expression in a form similar to (6.12). However, we shall not consider the general case here (see Appendix A, Problem 17), but restrict the discussion to the special case of axial symmetry, where $\rho$ depends on $r'$ and $\theta'$ (or $\rho'$ and $z'$) but not on $\varphi'$. This case is of particular interest in connection with the gravitational potential of the Earth, which is to a good approximation axially symmetric, though flattened at the poles. We shall see that in the case of axial symmetry, (6.26) can be written in a form similar to (6.13) with a suitably defined quadrupole moment $Q$.

To do this, we examine the numerator of the integrand in (6.26). Written out in terms of components, it reads

$$x^2(2x'^2 - y'^2 - z'^2) + y^2(2y'^2 - x'^2 - z'^2) + z^2(2z'^2 - x'^2 - y'^2)$$
$$+ 6xyx'y' + 6xzx'z' + 6yzy'z'. \tag{6.27}$$

Now, because of axial symmetry, any integral involving an odd power of $x'$ or $y'$ will vanish, for example

$$\iiint \rho(\boldsymbol{r}')x'z' \, \mathrm{d}^3\boldsymbol{r}' = 0,$$

because the contributions from $(x', y', z')$ and $(-x', y', z')$ exactly cancel. Moreover, axial symmetry also implies that

$$\iiint \rho(\boldsymbol{r}')x'^2 \, \mathrm{d}^3\boldsymbol{r}' = \iiint \rho(\boldsymbol{r}')y'^2 \, \mathrm{d}^3\boldsymbol{r}',$$

since the $x$- and $y$-axes are in no way distinguished from one another. Hence we can replace $2x'^2 - y'^2 - z'^2$ in the integrand by $\frac{1}{2}(x'^2 + y'^2) - z'^2$ without affecting the value of the integral. When we do this, we see that the three remaining terms of (6.26) all involve the same integral, namely

$$Q = \iiint \rho(\mathbf{r}')(2z'^2 - x'^2 - y'^2)\, d^3\mathbf{r}'. \tag{6.28}$$

Thus (6.26) can be written

$$\phi_2(\mathbf{r}) = \frac{2z^2 - x^2 - y^2}{16\pi\epsilon_0 r^5} Q = \frac{Q}{16\pi\epsilon_0 r^3}(3\cos^2\theta - 1). \tag{6.29}$$

This is identical with the potential (6.12) of an axially symmetric quadrupole. The quantity $Q$ defined by (6.28) is the *quadrupole moment* of the distribution.

> *Example:* **Linear quadrupole**
>
> Two equal charges $q$ are positioned at the points $(0, 0, \pm a)$, and a charge $-2q$ at the origin. Find the quadrupole moment, and write down the potential and electric field at large distances.
>
> Here the central particle contributes nothing to $Q$, and the other two each give the same contribution, so $Q = 2q(2a^2) = 4qa^2$. The potential and electric field at large distances are given by (6.13) and (6.14) with this value of $Q$.

Thus we see that a better approximation than treating a charge distribution as a single point charge is to treat it as a point charge plus an electric quadrupole of moment given by (6.28). For a spherically symmetric charge distribution, $Q = 0$, because the integrals over $x'^2$, $y'^2$ and $z'^2$ all yield the same value. The value of $Q$ may be regarded as a measure of the flattening of the distribution. For a distribution with uniform positive charge density, $Q$ is positive if the shape is *prolate* (egg-shaped, with the $z$-axis longer than the others), and negative if it is *oblate*, like the Earth. In particular, for a *spheroid* (ellipsoid of revolution), with uniform charge density and semi-axes $a, a, c$, we can evaluate the integral in (6.28) explicitly. We shall have occasion to evaluate a very similar integral in connection with moments of inertia in Chapter 9, and therefore omit the details here. The result is

$$Q = \tfrac{2}{5}q(c^2 - a^2). \tag{6.30}$$

## 6.5   The Shape of the Earth

The Earth is approximately an oblate spheroid whose equatorial radius $a$ exceeds its polar radius $c$ by about $21.4\,\mathrm{km}$. Its *oblateness* is defined to be

$$\epsilon = \frac{a - c}{a} \approx \frac{1}{297}. \tag{6.31}$$

Its gravitational potential is therefore not precisely the inverse-square-law potential, $-GM/r$. The most important correction is the 'quadrupole' term,

$$\Phi_2(\boldsymbol{r}) = -\frac{GQ}{4r^3}(3\cos^2\theta - 1).$$

Since the Earth is oblate, $Q$ here is actually negative.

    If the density of the Earth were uniform, the quadrupole moment would be given simply by (6.30):

$$Q = -\tfrac{2}{5}M(a^2 - c^2) \approx -\tfrac{4}{5}Ma^2\epsilon.$$

Here we have used $c = a(1 - \epsilon)$, and dropped terms of order $\epsilon^2$. In fact, the density is considerably greater near the centre than at the surface, so that large values of $r'$ contribute proportionately less to the integral (6.28). Thus we must expect $Q$ to be rather smaller in magnitude than this value (see Problem 15). It will be convenient to write

$$Q = -2Ma^2 J_2,$$

where $J_2$ is a dimensionless parameter, somewhat less than $\tfrac{2}{5}\epsilon$, defining the magnitude of the quadrupole moment. (In general, $J_l$ parametrizes the term in $\Phi$ proportional to $P_l(\cos\theta)$ — see the note following Eq. (6.8).) Then the gravitational potential becomes

$$\Phi(\boldsymbol{r}) \approx -\frac{GM}{r} + \frac{GMa^2 J_2}{2r^3}(3\cos^2\theta - 1). \tag{6.32}$$

We shall see that both $\epsilon$ and $J_2$ can be determined from measurements of the acceleration due to gravity and the use of an equilibrium principle.

    The flattening of the Earth is in fact a consequence of its rotation. Over long periods of time, the Earth is capable of plastic deformation, and behaves more nearly like a liquid than a solid — and of course the larger part of the surface *is* covered by a liquid. Although its crust does have some rigidity, it could not preserve its shape for long unless it were at least approximately in equilibrium under the combined gravitational and centrifugal forces.

Now, as in the problem discussed at the end of §5.3, the surface of a liquid in equilibrium under conservative forces must be an equipotential surface, for otherwise there would be a tendency for the liquid to flow towards regions of lower potential. We can include the centrifugal force by adding to the potential a term

$$\Phi_{\text{cent}}(\boldsymbol{r}) = -\tfrac{1}{2}\omega^2(x^2 + y^2) = -\tfrac{1}{2}\omega^2 r^2 \sin^2\theta.$$

The equipotential equation is then

$$\Phi(\boldsymbol{r}) + \Phi_{\text{cent}}(\boldsymbol{r}) = \text{constant}, \qquad (6.33)$$

with $\Phi$ given by (6.32). Thus equating the values at the pole ($\theta = 0$) and on the equator ($\theta = \pi/2$), we obtain the equation

$$-\frac{GM}{c} + \frac{GMa^2 J_2}{c^3} = -\frac{GM}{a} - \frac{GM J_2}{2a} - \tfrac{1}{2}\omega^2 a^2.$$

Since $\epsilon$ and $J_2$ are both small quantities, we may neglect $\epsilon^2$ and $\epsilon J_2$. Thus, in the quadrupole term, we may ignore the difference between $c$ and $a$. In the inverse-square-law term, we can write $1/c \approx (1/a)(1+\epsilon)$. Hence to first order we find

$$-\frac{GM\epsilon}{a} + \frac{3GM J_2}{2a} = -\tfrac{1}{2}\omega^2 a^2.$$

Writing $g_0$ for the gravitational acceleration without the centrifugal term, $g_0 = GM/a^2$, and multiplying by $2/ag_0$, we obtain

$$\frac{\omega^2 a}{g_0} = 2\epsilon - 3J_2. \qquad (6.34)$$

If we made the approximation of treating the Earth as of uniform density, and therefore set $J_2 = \tfrac{2}{5}\epsilon$, then this equation would determine the oblateness in terms of the angular velocity. This yields $\epsilon \approx 1/230$, which is appreciably larger than the observed value.

On the other hand, if we regard $\epsilon$ and $J_2$ as independent, then we need another relation to fix them both. Such a relation can be found from the measured values of $g$. (See §5.3.) Now, the gravitational field $\boldsymbol{g} = -\boldsymbol{\nabla}\Phi$

corresponding to the potential (6.32) is

$$
\begin{aligned}
g_r &= -\frac{GM}{r^2} + \frac{3GMa^2 J_2}{2r^4}(3\cos^2\theta - 1), \\
g_\theta &= \frac{3GMa^2 J_2}{r^4}\cos\theta\sin\theta.
\end{aligned}
\tag{6.35}
$$

Note that $g_\theta$ is directed away from the poles, and towards the equator, as one might expect from thinking of it as due to the attraction of the equatorial bulge. The inward radial acceleration $-g_r$ is decreased at the poles, and increased at the equator, by the quadrupole term, so that the difference between the two values is not as large as the inverse square law would predict.

At the poles, and on the equator, $g_\theta = 0$, and the magnitude of $\boldsymbol{g}$ is equal to $-g_r$. Thus we find

$$
\begin{aligned}
g_{\text{pole}} &= \frac{GM}{c^2} - \frac{3GMa^2 J_2}{c^4} \approx g_0(1 + 2\epsilon - 3J_2) \\
g_{\text{eq}} &= \frac{GM}{a^2} + \frac{3GM J_2}{2a^2} \approx g_0(1 + \tfrac{3}{2}J_2),
\end{aligned}
$$

making the same approximations as before. Thus, for the difference, we find the expression

$$
\frac{\Delta g}{g_0} = \frac{g_{\text{pole}} - g_{\text{eq}}}{g_0} = 2\epsilon - \frac{9}{2}J_2.
\tag{6.36}
$$

For the value including the centrifugal term, $\Delta g^* = \Delta g + \omega^2 a$, we have to add (6.34) to (6.36). This yields

$$
\frac{\Delta g^*}{g_0} = 4\epsilon - \frac{15}{2}J_2.
\tag{6.37}
$$

Using the known values quoted in (5.18) and (5.19), we may solve the simultaneous equations (6.34) and (6.37) for $\epsilon$ and $J_2$. This yields

$$
\epsilon = 0.0034, \qquad J_2 = 0.0011.
$$

The value of $\epsilon$ is in good agreement with the the oblateness deduced from direct measurements of the polar and equatorial radii. As expected, the value of $J_2$ is somewhat less than $\tfrac{2}{5}\epsilon$. The relation between the two is determined by the density distribution within the Earth. (See Problem 15.)

One consequence of the fact that the Earth's surface is approximately an equipotential surface should be noted. Because of the general property that the force field is always perpendicular to the equipotentials, it means that the effective $g^*$, which is the field derived from $\Phi + \Phi_{\text{cent}}$, is always perpendicular to the Earth's surface. Thus, although a plumb line does not point towards the Earth's centre, it is perpendicular to the surface at that point — apart, of course, from minor deviations caused by mountains and the like. (Compare Chapter 1, Problem 10.)

### Satellite orbits

The 'quadrupole' term in the Earth's gravitational field has two important effects on the orbit of a close artificial satellite. Indeed, observations of such orbits provide the most reliable means of measuring its magnitude (and those of the even smaller higher-order correction terms). They have revealed quite substantial deviations from spheroidal shape and from hydrostatic equilibrium: the Earth's surface is only rather approximately an equipotential.

The first effect of the quadrupole term, arising mainly from the deviation of the radial component of $g$ from the inverse square law, is a precession of the major axis of the orbit within the orbital plane. (Compare Chapter 4, Problem 27.) The major axis precesses in the forward direction for orbits of small inclination to the equator, and in the retrograde direction for orbits with inclination greater than $\arcsin\sqrt{4/5} = 63.4°$. (The difference is a reflection of the differing sign of the quadrupole contribution to $g_r$ on the equator and near the poles.)

The second effect occurs because the force is no longer precisely central, so that the angular momentum changes with time according to

$$\frac{\mathrm{d}J}{\mathrm{d}t} = mr \wedge g. \tag{6.38}$$

Thus, as in the case of the Larmor effect discussed in §5.6, the orbital plane precesses around the direction of the Earth's axis. The rate of precession may be calculated by a method very similar to the one used there (see Problem 25). The precession is in fact greater for orbits of small inclination, and is zero for an orbit passing over the poles. For a close satellite with small inclination, it can be nearly 10° per day. It is always in a retrograde sense (opposite to the direction of revolution of the satellite in its orbit).

Both effects are strongly dependent on the radius of the satellite orbit. In fact, the rate of precession decreases like $r^{-7/2}$. At the radius of the Moon's orbit, the precessional angular velocity is only a few seconds of arc per year. (The Moon's orbit *does* precess, at about 19° per year, but this is a consequence of the non-uniformity of the Sun's gravitational field, not of the shape of the Earth. See Problem 26.)

### Planetary precession

It should be noted that a small oblateness of the Sun produces a slow precession of the Kepler orbits around it.

The major part of the precession of planetary orbits is due to the gravitational action of the planets on each other, leading for example in the case of Mercury to a predicted precession of the axes of the ellipse of about 531 seconds of arc every century. These effects can be modelled quite well for each planet by smearing the mass of each of the other planets into an appropriate coplanar uniform circular ring. The modifications to the inverse square law potential depend, of course, on whether the planetary distance being considered is inside or outside the ring.

For Mercury (alone) there is an observed residual precession of 43″ per century, not explainable by the Newtonian theory. In fact, general relativity does explain this precession — indeed that it does so is one of the celebrated tests of the theory.

## 6.6   The Tides

Tidal forces arise because the gravitational attraction of the Moon, and to a lesser extent of the Sun, is not uniform over the surface of the Earth. The attraction is stronger than average on the side of the Earth facing the Moon, and weaker than average on the far side, so there is a tendency for the Earth to be elongated along the line of centres.

Let $a$ be the position of the Moon relative to the Earth's centre, and consider a point $r$ on the Earth. The potential at this point is (see Fig. 6.5)

$$\Phi(r) = -\frac{Gm}{|r-a|},\qquad(6.39)$$

where $m$ is the mass of the Moon. Now, since $r \ll a$, we may expand in powers of $r/a$. (This is the reverse of the situation encountered in §6.2, where $r$ was much larger than $a$.) Taking the direction of the Moon to be

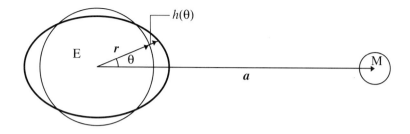

Fig. 6.5

the $z$-direction, and using (6.8), we obtain

$$\Phi(\boldsymbol{r}) = -Gm \left[ \frac{1}{a} + \frac{r}{a^2} \cos\theta + \frac{r^2}{a^3} (\tfrac{3}{2} \cos^2\theta - \tfrac{1}{2}) + \cdots \right].$$

The first term is a constant and does not yield any force. The second may be written

$$\Phi_1(\boldsymbol{r}) = -\frac{Gm}{a^2} z.$$

It yields a uniform gravitational acceleration $Gm/a^2$ directed towards the Moon. This term therefore describes the major effect of the Moon's gravitational force, which is to accelerate the Earth as a whole. It is irrelevant in discussing the phenomenon of the tides, since we are interested in motion relative to the centre of the Earth. In fact, we can eliminate it by going over to an *accelerated* frame, the frame in which the Earth's centre of mass is at rest. (See Chapter 5, Problem 18.)

The important term for our purposes is the quadratic term, which leads to a gravitational field

$$g_r = \frac{Gmr}{a^3} (3 \cos^2\theta - 1),$$
$$g_\theta = -\frac{3Gmr}{a^3} \cos\theta \sin\theta. \tag{6.40}$$

This field is directed outwards along the $z$-axis, towards and away from the Moon, and inwards in the $xy$-plane, as expected.

It should be noted that the field described by (6.40) is much weaker even than the corrections to the Earth's field discussed in the preceding section. (If it were not, it would show up in measurements of $g$.) It is easy to compute its magnitude. It is smaller than $g_0 = GM/r^2$ by the factor

$mr^3/Ma^3$. For the Moon, $m/M = 1/81.3$ and $r/a = 1/60.3$, so that this factor is

$$\frac{mr^3}{Ma^3} = 5.60 \times 10^{-8}. \tag{6.41}$$

If we denote the mass of the Sun by $m'$ and its mean distance, the semi-major axis of the Earth's orbit, by $a'$, then the corresponding values for the Sun are $m'/M = 3.33 \times 10^5$ and $r/a' = 4.26 \times 10^{-5}$. Thus the factor in that case is

$$\frac{m'r^3}{Ma'^3} = 2.57 \times 10^{-8}. \tag{6.42}$$

By a remarkable coincidence (unique in the solar system), these two fields are of roughly the same order of magnitude: the effect of the Sun is rather less than half that of the Moon.

The only reason why these very small fields can lead to significant effects is that they change with time. As the Earth rotates, the value of $\theta$ at any point on its surface varies. Thus the field described by (6.40) oscillates periodically with time. Because of the symmetry of (6.40) in the central plane, the main term has an oscillation period of 12 hours in the case of the Sun, and slightly more for the Moon, because of its changing position. This explains the most noticeable feature of the tides — their twice-daily periodicity. Unless the Moon is directly over the equator, however, a point on the Earth that passes directly beneath it will not also pass directly opposite, so that the two daily tides may be of unequal height. In other words, there is an additional term in the field with a period of 24 rather than 12 hours.

At new moon or full moon, the Sun and Moon are acting in the same direction, and the tides are unusually high; these are the *spring tides*. On the other hand, at the first and third quarters, when the Sun and Moon are at right angles to each other, their effects partially cancel, and we have the relatively low *neap tides*. (See Fig. 6.6.) Measurements of the relative heights of the tides at these times provided one of the earliest methods of estimating the mass of the Moon. (A more accurate method will be discussed in the next chapter.)

To get some idea of the order of magnitude of the effects produced by the tidal field, (6.40), let us make the simplest possible assumptions. We consider a perfectly rigid solid Earth, completely covered by ocean, and suppose that the natural periods of oscillation are short compared to the rotation period. In that case, the rotation is slow enough for the water

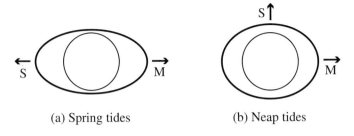

(a) Spring tides        (b) Neap tides

Fig. 6.6

to reach equilibrium under the combined forces of Earth and Moon. The problem is then one in hydrostatics rather than hydrodynamics.

In equilibrium, the surface of the water must be an equipotential surface. We can thus calculate the height $h(\theta)$ through which it is raised, as a function of the angle $\theta$ to the Moon's direction (see Fig. 6.5). Since $h(\theta)$ is certainly small, the change in the Earth's gravitational potential is approximately $g_0 h(\theta)$. This change must be balanced by the potential due to the Moon, so that

$$g_0 h(\theta) = \frac{Gmr^2}{a^3}(\tfrac{3}{2}\cos^2\theta - \tfrac{1}{2}).$$

Thus, using $g_0 = GM/r^2$, we find

$$h(\theta) = h_0(\tfrac{3}{2}\cos^2\theta - \tfrac{1}{2}), \qquad \text{where} \qquad h_0 = \frac{mr^4}{Ma^3}. \qquad (6.43)$$

This $h_0$ is just (6.41) times an extra factor of $r$. Thus, using $r = 6370\,\mathrm{km}$, we find, for the Moon, $h_0 = 0.36\,\mathrm{m}$, and for the Sun, $h_0' = 0.16\,\mathrm{m}$.

These figures must not be taken as more than rough order-of-magnitude estimates of the height of the tides. Even on a completely ocean-covered Earth, there would be important modifying effects. First, we have neglected the gravitational attraction of the ocean itself. The tidal bulges exert an attraction that tends to increase $h_0$ slightly, in fact by about 12 per cent (see Problem 19). More important, however, is the fact the Earth is not perfectly rigid, and is itself distorted by tidal forces. Because the observed tidal range refers not to absolute height but to the relative heights of sea surface and sea floor, this *reduces* the effective height by a substantial factor. Indeed, in the extreme case of a fluid Earth enclosed by a completely flexible crust, there would be essentially no observable tide at all.

The values obtained from (6.43) may seem rather small, especially when further reduced by the effect just mentioned. However, it is important to remember that they refer to the unrealistic case of an Earth without continents. The observed tidal range in mid-ocean (which can be measured by ranging from satellites) is in fact quite small, normally less than a metre. Large tides are a feature of continental shelf areas, and are strongly dependent on the local topography. Note too that the Moon's attraction at any point on the Earth's surface is a periodic force, whose effect may be greatly enhanced by the phenomenon of resonance, discussed in §2.6. The natural periods of oscillation of a body of water depend on a variety of factors, including its size, shape and depth, and there are critical values for which one of the periods is close to 12 hours. Then, because the damping produced by tidal friction is normally quite small except in very shallow water, very large tides can be set up.

Another effect of resonance is to delay the oscillations, so that they are not precisely in phase with the applied force. At resonance, the phase lag is a quarter period (about three hours in this case). Thus we should not normally expect the times of high tide to coincide with the times when the Moon is overhead, or directly opposite. In practice, the times are determined in a very complicated way by the actual shapes of the oceans.

## 6.7   The Field Equations

It is often convenient to obtain the electric or gravitational potential by solving a differential equation rather than performing an integration. Although we shall not in fact need to use this technique in this book, we include a short discussion of it because of its very important role in more advanced treatments of mechanics and electromagnetic theory.

To find the relevant equations, we consider first a single charge $q$ located at the origin. The electric field is then

$$\boldsymbol{E} = \frac{q}{4\pi\epsilon_0 r^2}\hat{\boldsymbol{r}}.$$

Now let us consider a closed surface $S$ surrounding the charge, and evaluate the surface integral of the component of $\boldsymbol{E}$ normal to the surface,

$$\iint_S \boldsymbol{E} \cdot \boldsymbol{n}\, \mathrm{d}S = \frac{q}{4\pi\epsilon_0} \iint_S \frac{\hat{\boldsymbol{r}} \cdot \boldsymbol{n}}{r^2}\, \mathrm{d}S,$$

where $\mathrm{d}S$ is the element of surface area, and $\boldsymbol{n}$ is a unit vector normal to the surface, directed outwards (see Fig. 6.7). If $\alpha$ is the angle between $\hat{\boldsymbol{r}}$

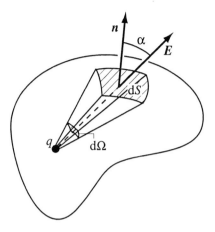

Fig. 6.7

and $\boldsymbol{n}$, then $\hat{\boldsymbol{r}} \cdot \boldsymbol{n} \, \mathrm{d}S = \mathrm{d}S \cos \alpha$ is the projection of the area $\mathrm{d}S$ on a plane normal to the radius vector $\boldsymbol{r}$. Hence it follows from the definition of solid angle (see (4.42)) that $\hat{\boldsymbol{r}} \cdot \boldsymbol{n} \, \mathrm{d}S / r^2$ is equal to the solid angle $\mathrm{d}\Omega$ subtended at the origin by the element of area $\mathrm{d}S$. Hence

$$\iint_S \boldsymbol{E} \cdot \boldsymbol{n} \, \mathrm{d}S = \frac{q}{4\pi\epsilon_0} \iint \mathrm{d}\Omega = \frac{q}{\epsilon_0}, \tag{6.44}$$

since the total solid angle subtended by a closed surface around the origin is $4\pi$. (It is to ensure that the factor of $4\pi$ here cancels that Coulomb's law in SI units is written with an explicit factor of $1/4\pi$.)

We have implicitly assumed that each radial line cuts the surface only once. However, we can easily remove this restriction. Evidently, a radial line must always cut it in an odd number of points. (See Fig. 6.8.) Moreover, the contributions to the surface integral from points where it goes into, rather than out of, the surface are all of opposite sign (because $\hat{\boldsymbol{r}} \cdot \boldsymbol{n} = \cos \alpha$ is negative at these points) and of equal magnitude. Thus all the terms but one will cancel, and we obtain the same answer (6.44).

Similarly, if we consider a surface $S$ which does *not* enclose the origin, we can see that the surface integral will vanish. For each radial line cuts $S$ in an *even* number of points, half in each direction, so that all contributions will cancel.

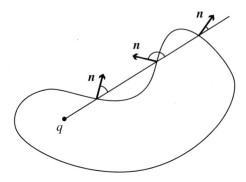

Fig. 6.8

It is now clear how we may generalize this result to an arbitrary distribution of charges. Any charge $q$ located anywhere within a closed surface $S$ will contribute to the surface integral an amount $q/\epsilon_0$, while charges outside contribute nothing. Therefore, the value of the integral is $1/\epsilon_0$ times the total charge enclosed within the surface. If we have a continuous distribution of charge, with charge density $\rho(\boldsymbol{r})$, then

$$\iint_S \boldsymbol{E} \cdot \boldsymbol{n}\,\mathrm{d}S = \frac{1}{\epsilon_0} \iiint_V \rho(\boldsymbol{r})\,\mathrm{d}^3\boldsymbol{r}, \qquad (6.45)$$

where $V$ is the volume enclosed by the surface $S$. This is often called *Gauss's law*.

Now, by the theorem of Gauss (see (A.36)), the surface integral is equal to the volume integral of the divergence of $\boldsymbol{E}$:

$$\iint_S \boldsymbol{E} \cdot \boldsymbol{n}\,\mathrm{d}S = \iiint_V \boldsymbol{\nabla} \cdot \boldsymbol{E}\,\mathrm{d}^3\boldsymbol{r}.$$

Hence (6.45) may be written

$$\iiint_V (\boldsymbol{\nabla} \cdot \boldsymbol{E} - \epsilon_0^{-1}\rho)\,\mathrm{d}^3\boldsymbol{r} = 0.$$

This equation must be true for an arbitrary volume $V$, and therefore the integrand itself must vanish:

$$\boldsymbol{\nabla} \cdot \boldsymbol{E} = \epsilon_0^{-1}\rho. \qquad (6.46)$$

This is the differential form of Gauss's law. The electric field is in fact completely determined (except for a possible additive uniform field extending over the whole of space) by this equation together with the condition that it be a conservative field,

$$\nabla \wedge \boldsymbol{E} = \boldsymbol{0}. \tag{6.47}$$

This latter equation guarantees the existence of the potential $\phi$ related to $\boldsymbol{E}$ by $\boldsymbol{E} = -\nabla\phi$. (See §6.6.) Substituting in (6.46), we obtain *Poisson's equation* for the potential,

$$\nabla^2\phi = -\epsilon_0^{-1}\rho. \tag{6.48}$$

The main importance of these equations lies in the fact that they provide *local* relations between the potential or field and the charge density. The expression (6.15) for $\phi$ in terms of $\rho$ is non-local, in the sense that it expresses the potential $\phi$ at one point as an integral involving the charge density at all points of space. To use this equation, we must know $\rho$ everywhere. Frequently, however, we are interested in the field in some restricted region of space, where we do know the value of $\rho$, while, instead of having information about $\rho$ outside this region, we have some conditions on the fields at the boundary. This type of problem may be solved by looking for solutions of (6.48) with appropriate boundary conditions. A particularly important special case is that in which $\rho$ vanishes inside the region of interest. Then (6.48) becomes *Laplace's equation*,

$$\nabla^2\phi = 0. \tag{6.49}$$

The gravitational case is of course entirely similar. The field equations are

$$\nabla \cdot \boldsymbol{g} = -4\pi G\rho, \tag{6.50}$$

where $\rho$ is now the *mass* density, and

$$\nabla \wedge \boldsymbol{g} = \boldsymbol{0}. \tag{6.51}$$

Note the explicit appearance of the factor of $4\pi$ on the right hand side of (6.50), which arises because the gravitational force is conventionally expressed in terms of *unrationalized* units, *i.e.*, there is no factor of $1/4\pi$ in (1.5). Note also the differing signs in (6.46) and (6.50), which again reflect the difference between (1.5) and (1.6).

Poisson's equation for the gravitational potential is

$$\nabla^2 \Phi = 4\pi G \rho. \tag{6.52}$$

## 6.8   Summary

The great advantage of calculating the potential, rather than the field itself directly, is that it is much easier to add scalar quantities than vector ones. Even in simple cases, like that of the uniformly charged sphere, a direct integration of the forces produced by each element of the sphere is difficult to perform. In fact, the introduction of the potential was one of the major advances in mechanics, which allowed many previously intractable problems to be handled relatively simply.

This chapter, in which we have discussed the method of determining the force on a particle from a knowledge of the positions of other masses or charges, is complementary to the previous chapters, where we discussed the motion of a particle under known forces. Together, they provide a method of solving most of the problems in which the object of immediate interest is (or may be taken to be) a single particle, moving under the action of conservative forces.

## Problems

1. Find the potential and electric field at points on the axis of symmetry of a uniformly charged flat circular disc of charge $q$ and radius $a$. What happens to the field if we keep the charge per unit area $\sigma = q/\pi a^2$ fixed, and let $a \to \infty$?

2. Calculate the quadrupole moment of the disc of Problem 1, and deduce the form of the field at large distance in any direction. Verify that on the axis it agrees with the exact value when $r/a$ is sufficiently large.

3. Write down the potential energy of a pair of charges, $q$ at $\boldsymbol{a}$ and $-q$ at the origin, in a field with potential $\phi(\boldsymbol{r})$. By considering the limit $a \to 0$, show that the potential energy of a dipole of moment $\boldsymbol{d}$ is $V = -\boldsymbol{d} \cdot \boldsymbol{E}$. If the electric field is uniform, when is this potential energy a minimum? Show that the dipole experiences a net moment, or *couple*, $\boldsymbol{G} = \boldsymbol{d} \wedge \boldsymbol{E}$, and that in a non-uniform field there is also a net force, $\boldsymbol{F} = (\boldsymbol{d} \cdot \nabla)\boldsymbol{E}$. (Take $\boldsymbol{d}$ in the $z$-direction, and show that $\boldsymbol{F} = d\,\partial \boldsymbol{E}/\partial z$.)

4. *Write the electric field of a dipole in vector notation. Using the result of Problem 3, find the potential energy of a dipole of moment $\boldsymbol{d}$ in the field of another dipole of moment $\boldsymbol{d'}$. (Take $\boldsymbol{d'}$ at the origin and $\boldsymbol{d}$ at position $\boldsymbol{r}$.) Find the forces and couples acting between the dipoles if they are placed on the $z$-axis and (a) both are pointing in the $z$-direction, (b) both are pointing in the $x$-direction, (c) $\boldsymbol{d}$ is in the $z$-direction, and $\boldsymbol{d'}$ in the $x$-direction, and (d) $\boldsymbol{d}$ is in the $x$-direction and $\boldsymbol{d'}$ in the $y$-direction.

5. Show that the work done in bringing two charges $q_1$ and $q_2$, initially far apart, to a separation $r_{12}$ is $q_1 q_2/4\pi\epsilon_0 r_{12}$. Write down the corresponding expression for a system of many charges. Show that the energy stored in the charge distribution is

$$V = \tfrac{1}{2} \sum_j q_j \phi_j(\boldsymbol{r}_j),$$

where $\phi_j(\boldsymbol{r}_j)$ is the potential at $\boldsymbol{r}_j$ due to all the other charges. Why does a factor of $\tfrac{1}{2}$ appear here, but not in the corresponding expression for the potential energy in an external potential $\phi(\boldsymbol{r})$?

6. Find the energy stored in a sphere of charge $q$ and radius $a$ with uniform charge density, and show that infinite energy is required to compress the sphere to a point. Find also the stored energy in the case where the charge is uniformly spread over the *surface* of the sphere.

7. Near the Earth's surface, there is normally a vertical electric field, typically of order $100\,\mathrm{V\,m^{-1}}$. Find the corresponding total charge on the Earth, and the stored energy per unit surface area.

8. Find the quadrupole moment of a distribution of charge on the surface of a sphere of radius $a$ with surface charge density $\sigma = \sigma_0(\tfrac{3}{2}\cos^2\theta - \tfrac{1}{2})$. Find the total energy stored in this distribution.

9. *Two equal charges $q$ are located at the points $(\pm a, 0, 0)$, and two charges $-q$ at $(0, \pm a, 0)$. Find the leading term in the potential at large distances, and the corresponding electric field.

10. Find the gravitational potential at large distances of a thin circular loop of radius $a$ and mass $m$, up to terms of order $r^{-3}$. Find also the potential, and the leading term in the gravitational field near the origin, at distances $r \ll a$.

11. *Six equal point masses $m$ are located at the points $\pm a\boldsymbol{i}$, $\pm a\boldsymbol{j}$ and $\pm a\boldsymbol{k}$. Show that the quadrupole term in the potential vanishes, and find the leading correction to the monopole term $-6Gm/r$. (*Note*: This requires expansion of the potential up to terms of order $a^4/r^5$.)

12. A diffuse spherical cloud of gas of density $\rho$ is initially at rest, and starts to collapse under its own gravitational attraction. Find the radial velocity of a particle which starts at a distance $a$ from the centre when it reaches the distance $r$. Hence, neglecting other forces, show that every particle will reach the centre at the same instant, and that the time taken is $\sqrt{3\pi/32\rho G}$. Evaluate this time in years if $\rho = 10^{-19}\,\mathrm{kg\,m^{-3}}$. (*Hint*: Assume that particles do not overtake those that start nearer the centre. Verify that your solution is consistent with this assumption. The substitution $r = a\sin^2\theta$ may be used to perform the integration.) Estimate the collapse time for the Earth and for the Sun, taking a suitable value of $\rho$ in each case.

13. Given that the mass of the gas cloud of Problem 12 is $10^{30}\,\mathrm{kg}$, and that its contraction is halted by the build-up of pressure when a star of radius $10^6\,\mathrm{km}$ has been formed, use the result of Problem 6 to find the total energy released, assuming that the density of the star is still uniform.

14. The rotation period of Jupiter is approximately 10 hours. Its mass and radius are $318M_E$ and $11.0R_E$, respectively ($E = $ Earth). Calculate approximately its oblateness, neglecting the variation in density. (The observed value is about $1/15$.)

15. *Assume that the Earth consists of a core of uniform density $\rho_c$, surrounded by a mantle of uniform density $\rho_m$, and that the boundary between the two is of similar shape to the outer surface, but with a radius only three-fifths as large. Find what ratio of densities $\rho_c/\rho_m$ is required to explain the observed quadrupole moment. (*Hint*: Treat the Earth as a superposition of two ellipsoids of densities $\rho_m$ and $\rho_c - \rho_m$. Note that in reality neither core nor mantle is of uniform density.)

16. The distance between the Earth and the Moon is gradually increasing (because of tidal friction — see §8.4). Estimate the height of the tides when the Moon was 10 Earth radii away. How far away will it be when the lunar and solar tides are equal in magnitude?

17. Suppose that at some time the Moon had been ocean-covered and rotating relative to the Earth. Find the ratio between the heights of the tides raised on the Moon by the Earth and on the Earth by the Moon. (The Moon's radius is $R_M = 0.27R_E$.) Estimate how high the tides on the Moon would have been when the Earth–Moon distance was $10R_E$.

18. Jupiter has a satellite Io only a little heavier than the Moon (mass $8.93 \times 10^{22}\,\mathrm{kg}$, as compared to $7.35 \times 10^{22}\,\mathrm{kg}$). The semi-major axis of its orbit is $4.22 \times 10^5\,\mathrm{km}$, compared with $3.84 \times 10^5\,\mathrm{km}$ for the Moon.

Estimate the height of the tides raised on Jupiter by Io. (See Problem 14 for Jovian data.)

19. *The tidal bulge in the ocean may be regarded as adding a surface mass distribution of the same form as the surface charge distribution of Problem 8, positive near $\theta = 0$ and $\pi$ and negative near $\theta = \pi/2$. Find the quadrupole moment of this distribution, and show that when it is included in (6.43) the effective height of the tides is *increased* by a factor $(1 - 3\rho_o/5\rho_E)^{-1}$, where $\rho_o$ is the mean density of the oceans and $\rho_E$ that of the Earth as a whole. Evaluate this factor, given that $\rho_E = 5.52 \times 10^3 \, \mathrm{kg\,m^{-3}}$ and $\rho_o = 1.03 \times 10^3 \, \mathrm{kg\,m^{-3}}$. (Note, however, that the effective height is decreased by a larger factor due to the non-rigidity of the Earth.)

20. *Two small identical uniform spheres of density $\rho$ and radius $r$ are orbiting the Earth in a circular orbit of radius $a$. Given that the spheres are just touching, with their centres in line with the Earth's centre, and that the only force between them is gravitational, show that they will be pulled apart by the Earth's tidal force if $a$ is less than $a_c = 2(\rho_E/\rho)^{1/3}R_E$, where $\rho_E$ is the mean density of the Earth and $R_E$ its radius. (This is an illustration of the existence of the *Roche limit*, within which small planetoids would be torn apart by tidal forces. The actual limit is larger than the one found here, because the spheres themselves would be distorted by the tidal force, thus enhancing the effect. It is $a_c = 2.45(\rho_E/\rho)^{1/3}R_E$. For the mean density of the Moon, for example, $\rho = 3.34 \times 10^3 \, \mathrm{kg\,m^{-3}}$, this gives $a_c = 2.89 R_E$.)

21. Verify that the potential of a uniformly charged sphere satisfies Poisson's equation, both inside and outside the sphere. (You will need (A.59) for this and subsequent problems.)

22. The potential $\phi(\boldsymbol{r}) = (q/4\pi\epsilon_0 r)\mathrm{e}^{-\mu r}$ may be regarded as representing the effect of screening of a charge $q$ at the origin by mobile charges in a plasma. Calculate the charge density $\rho$ (at points where $r \neq 0$) and find the total charge throughout space, excluding the origin.

23. By considering the equilibrium of a small volume element, show that in a fluid in equilibrium under pressure and gravitational forces, $\boldsymbol{\nabla} p = \rho \boldsymbol{g}$, where $\rho$ is the density and $p$ the pressure (the equation of *hydrostatic equilibrium*). Deduce that, for an incompressible fluid of uniform density $\rho$, $p + \rho\Phi$ is a constant. Use this result to obtain a rough estimate of the pressure at the centre of the Earth. (Mean density of Earth $= 5.5 \times 10^3 \, \mathrm{kg\,m^{-3}}$. Note that the pressure at the surface is essentially zero. The actual pressure at the centre is more than this estimate —

in fact about $3.6 \times 10^{11}$ Pa — because of the non-uniformity of $\rho$. It should be noted that — as indicated in Problem 12 — even very small departures from hydrostatic equilibrium would result in collapse on a short time scale, only checked by an increase in internal pressure via other processes.)

24. *Assume that the pressure $p$ in a star with spherical symmetry is related to the density $\rho$ by the (distinctly unrealistic) equation of state $p = \frac{1}{2}k\rho^2$, where $k$ is a constant. Use the fluid equilibrium equation obtained in Problem 23 to find a relation between $\rho$ and $\Phi$. Hence show that Poisson's equation yields

$$\frac{d^2[r\rho(r)]}{dr^2} = -\frac{4\pi G}{k} r\rho(r).$$

Solve this equation with the boundary conditions that $\rho$ is finite at $r = 0$ and vanishes at the surface of the star. Hence show that the radius $a$ of the star is determined solely by $k$ and is independent of its mass $M$. Show also that $M = (4/\pi)a^3\rho(0)$.

25. *Show that the moment of the Earth's gravitational force may be written in the form

$$m\boldsymbol{r} \wedge \boldsymbol{g} = \frac{3GMma^2 J_2}{r^5}(\boldsymbol{k} \cdot \boldsymbol{r})(\boldsymbol{k} \wedge \boldsymbol{r}).$$

Consider a satellite in a circular orbit of radius $r$ in a plane inclined to the equator at an angle $\alpha$. By introducing a pair of axes in the plane of the orbit, as in §5.6 (see Fig. 5.10), show that the average value of this moment is

$$\langle m\boldsymbol{r} \wedge \boldsymbol{g}\rangle_{\text{av}} = -\frac{3GMma^2 J_2}{2r^3}\cos\alpha\,(\boldsymbol{k} \wedge \boldsymbol{n}),$$

where $\boldsymbol{n}$ is the normal to the orbital plane. Hence show that the orbit precesses around the direction $\boldsymbol{k}$ of the Earth's axis at a rate $\Omega = -(3J_2a^2/2r^2)\omega\cos\alpha$, where $\omega$ is the orbital angular velocity. Evaluate this rate for an orbit 400 km above the Earth's surface, with an inclination of $30°$. Find also the precessional period.

26. *Show that the moment of the solar tidal force $m\boldsymbol{g}_{\text{S}}$ on an Earth satellite is

$$m\boldsymbol{r} \wedge \boldsymbol{g}_{\text{S}} = \frac{3GM_{\text{S}}m}{a_{\text{S}}^5}(\boldsymbol{r} \cdot \boldsymbol{a}_{\text{S}})(\boldsymbol{r} \wedge \boldsymbol{a}_{\text{S}}),$$

where $\boldsymbol{a}_S$ is the position vector of the Sun relative to the Earth. Consider the effect on the Moon, whose orbit is inclined at $\alpha = 5°$ to the ecliptic (the Earth's orbital plane). Introduce axes $\boldsymbol{i}$ and $\boldsymbol{j}$ in the plane of the ecliptic, with $\boldsymbol{i}$ in the direction where the Moon's orbit crosses it. Write down the positions of the Sun and Moon, relative to Earth, as functions of time. By averaging both over a month and over a year (*i.e.*, over the positions in their orbits of both bodies), show that the moment leads to a precession of the Moon's orbital plane, at a precessional angular velocity

$$\boldsymbol{\Omega} = -\frac{3\varpi^2}{4\omega} \cos \alpha \, \boldsymbol{k},$$

where $\varpi$ is the Earth's *orbital* angular velocity. Compute the precessional period. (Note that this calculation is only approximately correct.)

# Chapter 7

# The Two-Body Problem

We shall be mainly concerned in this chapter with an isolated system of two particles, subject only to the force between the two. However, as it is no harder to solve, and considerably extends the range of applicability of the results, we shall also allow the presence of a *uniform* gravitational field.

## 7.1 Centre-of-mass and Relative Co-ordinates

We denote the positions and masses of the two particles by $r_1, r_2$ and $m_1, m_2$. If the force on the first particle due to the second is $F$, then, by Newton's third law, that on the second due to the first is $-F$. Thus, in a uniform gravitational field $g$, the equations of motion are

$$m_1\ddot{r}_1 = m_1 g + F,$$
$$m_2\ddot{r}_2 = m_2 g - F. \tag{7.1}$$

It is convenient to introduce new variables in place of $r_1$ and $r_2$. We define the position of the *centre of mass*,

$$R = \frac{m_1 r_1 + m_2 r_2}{m_1 + m_2}, \tag{7.2}$$

and the *relative* position,

$$r = r_1 - r_2. \tag{7.3}$$

(See Fig. 7.1.) The centre of mass is often denoted by $\bar{r}$. This symbol is, however, typographically inconvenient when dots are used to denote time derivatives. From (7.1), we get the equation of motion for $R$ by adding:

$$M\ddot{R} = Mg, \quad \text{where} \quad M = m_1 + m_2. \tag{7.4}$$

159

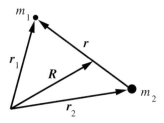

Fig. 7.1

One virtue of the notation $\boldsymbol{R}$ for the centre of mass is that it serves to emphasize the fact that $\boldsymbol{R}$ is associated with the total mass, $M$.

The equation for $\boldsymbol{r}$ is obtained by dividing by the masses and then subtracting:

$$\mu\ddot{\boldsymbol{r}} = \boldsymbol{F}, \qquad \text{where} \qquad \mu = \frac{m_1 m_2}{m_1 + m_2}. \tag{7.5}$$

The mass $\mu$ is called the *reduced* mass, because it is always less than either $m_1$ or $m_2$.

These two equations are now completely separate. Equation (7.4) shows that the centre of mass moves with uniform acceleration $\boldsymbol{g}$. In the case $\boldsymbol{g} = \boldsymbol{0}$ it is equivalent to the law of conservation of momentum,

$$M\dot{\boldsymbol{R}} = m_1\dot{\boldsymbol{r}}_1 + m_2\dot{\boldsymbol{r}}_2 = \boldsymbol{P} = \text{constant}. \tag{7.6}$$

The equation of motion (7.5) for the relative position is identical with that for a single particle of mass $\mu$ moving under the force $\boldsymbol{F}$. If we can solve this one-particle problem, then we can also solve the two-particle problem. When we have found $\boldsymbol{R}$ and $\boldsymbol{r}$ as functions of time, we can obtain the positions of the particles by solving the simultaneous equation (7.2) and (7.3):

$$\boldsymbol{r}_1 = \boldsymbol{R} + \frac{m_2}{M}\boldsymbol{r}, \qquad \boldsymbol{r}_2 = \boldsymbol{R} - \frac{m_1}{M}\boldsymbol{r}. \tag{7.7}$$

The separation between centre-of-mass and relative motion extends to the expressions for the total angular momentum and kinetic energy. We have

$$\boldsymbol{J} = m_1\boldsymbol{r}_1 \wedge \dot{\boldsymbol{r}}_1 + m_2\boldsymbol{r}_2 \wedge \dot{\boldsymbol{r}}_2$$

$$= m_1\left(\boldsymbol{R} + \frac{m_2}{M}\boldsymbol{r}\right) \wedge \left(\dot{\boldsymbol{R}} + \frac{m_2}{M}\dot{\boldsymbol{r}}\right) + m_2\left(\boldsymbol{R} - \frac{m_1}{M}\boldsymbol{r}\right) \wedge \left(\dot{\boldsymbol{R}} - \frac{m_1}{M}\dot{\boldsymbol{r}}\right),$$

by (7.7). It is easy to see that the cross terms between $\boldsymbol{R}$ and $\boldsymbol{r}$ cancel, and we are left with

$$\boldsymbol{J} = M\boldsymbol{R} \wedge \dot{\boldsymbol{R}} + \mu \boldsymbol{r} \wedge \dot{\boldsymbol{r}}. \tag{7.8}$$

Similarly, substituting (7.7) into

$$T = \tfrac{1}{2}m_1\dot{\boldsymbol{r}}_1^2 + \tfrac{1}{2}m_2\dot{\boldsymbol{r}}_2^2,$$

we find, after a little algebra,

$$T = \tfrac{1}{2}M\dot{\boldsymbol{R}}^2 + \tfrac{1}{2}\mu\dot{\boldsymbol{r}}^2. \tag{7.9}$$

It is instructive to examine this problem also from the Lagrangian point of view introduced in §3.7. Let us suppose that the internal, two-body force is conservative, and depends only on the relative co-ordinate $\boldsymbol{r}$. It then corresponds to a potential energy function $V_{\text{int}}(\boldsymbol{r})$. The total potential energy $V$ includes not only $V_{\text{int}}$ but also the potential energy of the external gravitational forces. Now the potential energy of a particle of mass $m$ in a uniform gravitational field $\boldsymbol{g}$ is $-m\boldsymbol{g}\cdot\boldsymbol{r}$. (If $\boldsymbol{g}$ is in the $-z$-direction, this is the familiar expression $mgz$.) Thus the Lagrangian function is

$$\begin{aligned} L &= T - V \\ &= \tfrac{1}{2}m_1\dot{\boldsymbol{r}}_1^2 + \tfrac{1}{2}m_2\dot{\boldsymbol{r}}_2^2 + m_1\boldsymbol{g}\cdot\boldsymbol{r}_1 + m_2\boldsymbol{g}\cdot\boldsymbol{r}_2 - V_{\text{int}}(\boldsymbol{r}_1 - \boldsymbol{r}_2). \end{aligned} \tag{7.10}$$

It is easy to verify that Lagrange's equations (3.44) yield the correct equations of motion (7.1).

Now, just as in §3.7, we may express $L$ in terms of any six independent co-ordinates in place of $x_1, y_1, z_1, x_2, y_2, z_2$. In particular, we may choose the coordinates of $\boldsymbol{R}$ and $\boldsymbol{r}$. Then, from (7.9) and (7.10), we find

$$L = \tfrac{1}{2}M\dot{\boldsymbol{R}}^2 + M\boldsymbol{g}\cdot\boldsymbol{R} + \tfrac{1}{2}\mu\dot{\boldsymbol{r}}^2 - V_{\text{int}}(\boldsymbol{r}). \tag{7.11}$$

The complete separation of the terms involving $\boldsymbol{R}$ from those involving $\boldsymbol{r}$ is now obvious. It is easy to check that Lagrange's equations for these co-ordinates are just (7.4) and (7.5).

In this form, it is clear that the crucial point is the separation of the *potential* energy. The kinetic energy may always be separated according to (7.9), but only for a *uniform* gravitational field does the potential energy separate in a similar way.

## 7.2    The Centre-of-mass Frame

It is often convenient to describe the motion of the system in terms of a frame of reference in which the centre of mass is at rest at the origin. (In a gravitational field, this is an accelerated, non-inertial frame, but it is still useful.) This is the *centre-of-mass* (*CM*) *frame*. We shall denote quantities referred to it by an asterisk.

The relative position $r$ is of course independent of the choice of origin, so, setting $R^* = 0$ in (7.7), we find

$$r_1^* = \frac{m_2}{M} r, \qquad r_2^* = -\frac{m_1}{M} r. \tag{7.12}$$

In this frame, the momenta of the two particles are equal and opposite:

$$m_1 \dot{r}_1^* = -m_2 \dot{r}_2^* = \mu \dot{r} = p^*, \tag{7.13}$$

say.

As we shall see explicitly later, it is often convenient to solve a problem first in the CM frame. To find the solution in some other frame, we then need the relations between the momenta in the two frames. Let us consider a frame in which the centre of mass is moving with velocity $\dot{R}$. Then the velocities of the two particles are

$$\dot{r}_1 = \dot{R} + \dot{r}_1^*, \qquad \dot{r}_2 = \dot{R} + \dot{r}_2^*.$$

Hence by (7.13), their momenta are

$$p_1 = m_1 \dot{r}_1 = m_1 \dot{R} + p^*, \qquad p_2 = m_2 \dot{r}_2 = m_2 \dot{R} - p^*. \tag{7.14}$$

From (7.8) and (7.9), it follows that the total momentum, angular momentum and kinetic energy in the CM frame are

$$P^* = 0,$$
$$J^* = \mu r \wedge \dot{r} = r \wedge p^*, \tag{7.15}$$
$$T^* = \tfrac{1}{2}\mu \dot{r}^2 \;\; = \frac{p^{*2}}{2\mu}.$$

Thus in any other frame we can write

$$P = M\dot{R},$$
$$J = M R \wedge \dot{R} + J^*, \tag{7.16}$$
$$T = \tfrac{1}{2} M \dot{R}^2 + T^*.$$

To obtain the values in any frame from those in the CM frame, we have only to add the contribution of a particle of mass $M$ located at the centre of mass $\boldsymbol{R}$. We shall see in the next chapter that this is a general conclusion, not restricted to two-particle systems.

As an example, let us consider two bodies moving under their mutual gravitational attraction, for instance the Moon and the Earth.

### *Example:* Motion of a satellite

How does the period of the Moon's orbit relate to the Earth–Moon distance? How far does the apparent position of the Sun oscillate because of the influence of the Moon on the Earth's position?

Equation (7.5) becomes

$$\mu \ddot{\boldsymbol{r}} = -\hat{\boldsymbol{r}} \frac{G m_1 m_2}{r^2} = -\hat{\boldsymbol{r}} \frac{G M \mu}{r^2}.$$

This is identical with the equation for a particle moving around a fixed point mass $M$. In particular, for an elliptic orbit, the period is given by (4.31),

$$\left( \frac{\tau}{2\pi} \right)^2 = \frac{a^3}{GM}.$$

Note that $a$ here is the semi-major axis of the *relative* orbit (the median distance between the bodies), and that $M$ is the *sum* of the masses, rather than the mass of the heavier body. Thus Kepler's third law is only approximately correct: the orbital period depends not only on the semi-major axis, but also on the mass. (See also Chapter 4, Problem 31.)

The only cases in the solar system for which the lighter mass is an appreciable fraction of the total are those of the Earth–Moon system, for which $m_1/m_2 = 1/81.3$, and the Pluto–Charon system, for which $m_1/m_2 = 1/8.5$. If we were to compute the period of the Moon's orbit from Kepler's third law by comparing it with the period of a small Earth satellite, we should obtain the value given above but with $m_2$ in place of $M$. This would yield a period about 4 hours too long.

In the CM frame of the Earth–Moon system, we see from (7.12) that both bodies move in ellipses around the centre of mass, with semi-major axes

$$a_1 = \frac{m_2}{M} a, \qquad a_2 = \frac{m_1}{M} a.$$

For the Earth–Moon system, $a = 3.84 \times 10^5$ km. The Earth therefore moves around the centre of mass in a small ellipse with semi-major axis $a_2 = a/82.3 = 4670$ km. (Note that the centre of mass is *inside* the Earth.) This leads to a small oscillation in the direction of the Sun (see Fig. 7.2 — which is not drawn to scale), of angular amplitude $\alpha \approx a_2/A$, where $A$ is the distance to the Sun; since $A = 1.5 \times 10^8$ km, $\alpha = 6.4$ seconds of arc.

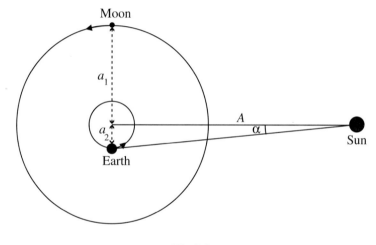

Fig. 7.2

A larger effect can be obtained by looking at the direction of some object closer to us than the Sun, such as an asteroid. This effect provides a method of determining the lunar mass.

In the frame in which the Sun is at rest, the centre of mass of the Earth–Moon system moves in an ellipse around the Sun, and the motion of the two bodies around the centre of mass is superimposed on this larger orbital motion. In the approximation in which the Sun's gravitational field is uniform over the dimension of the Earth–Moon system, the two types of motion are completely separate, and no transfer of energy or angular momentum from one to the other can occur. However, as we shall see later, there are small effects due to the non-uniformity of the Sun's field.

## 7.3 Elastic Collisions

A collision between two particles is called *elastic* if there is no loss of kinetic energy in the collision — that is, if the total kinetic energy after the collision is the same as that before it. Such collisions are typical of very hard bodies, like billiard balls. (They are also important in atomic and nuclear collision problems.)

It is easy to describe such a collision in the CM frame. The particles must approach each other with equal and opposite momenta, $\boldsymbol{p}^*$ and $-\boldsymbol{p}^*$, and recede after the collision again with equal and opposite momenta, $\boldsymbol{q}^*$ and $-\boldsymbol{q}^*$. Thus each particle is scattered through the same angle $\theta^*$ (see Fig. 7.3). Since the collision is elastic, we have, from (7.15),

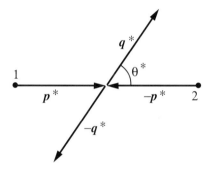

Fig. 7.3

$$T^* = \frac{\boldsymbol{p}^{*2}}{2\mu} = \frac{\boldsymbol{q}^{*2}}{2\mu}. \tag{7.17}$$

Thus the magnitudes of the momenta before and after the collision are the same,

$$p^* = q^*. \tag{7.18}$$

In practice, most experiments are performed with one particle initially at rest (or nearly so) in the laboratory. To interpret such an experiment, we therefore need to use the *laboratory (Lab) frame*, in which the momentum of particle 2, the *target*, before the collision is zero, so that $\boldsymbol{p}_2 = \boldsymbol{0}$.

We shall denote the Lab momentum of the incoming particle by $\boldsymbol{p}_1$, and the momenta after the collision by $\boldsymbol{q}_1$ and $\boldsymbol{q}_2$, and the angles of scattering and recoil by $\theta$ and $\alpha$ (see Fig. 7.4). We could work out the relations between these quantities by using the conservation laws of energy and mo-

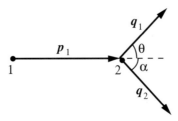

Fig. 7.4

mentum directly in the Lab frame, but it is actually simpler to relate them to the CM quantities, using the formulae obtained in §7.2.

Since $\boldsymbol{p}_2 = \boldsymbol{0}$, we have from (7.14),

$$\dot{\boldsymbol{R}} = \frac{1}{m_2}\boldsymbol{p}^*, \qquad (7.19)$$

and also

$$\boldsymbol{p}_1 = \frac{m_1}{m_2}\boldsymbol{p}^* + \boldsymbol{p}^* = \frac{M}{m_2}\boldsymbol{p}^*. \qquad (7.20)$$

The momenta after the collision are again given by (7.14), but with $\boldsymbol{q}^*$ in place of $\boldsymbol{p}^*$. They are then, using also (7.19),

$$\boldsymbol{q}_1 = \frac{m_1}{m_2}\boldsymbol{p}^* + \boldsymbol{q}^*, \qquad \boldsymbol{q}_2 = \boldsymbol{p}^* - \boldsymbol{q}^*.$$

All these relations may be conveniently summarized in a vector diagram, drawn in Fig. 7.5, which also incorporates the momentum conservation

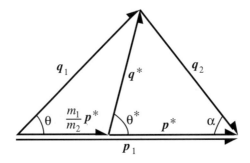

Fig. 7.5

equation in the Lab, $\boldsymbol{p}_1 = \boldsymbol{q}_1 + \boldsymbol{q}_2$. Any desired relation between momenta, energies or angles may be extracted from this diagram without difficulty. We consider explicitly only a few important relations.

By (7.18), the vectors $\boldsymbol{p}^*, \boldsymbol{q}^*, \boldsymbol{q}_2$ form an isosceles triangle. Thus the recoil angle and the magnitude of the recoil momentum $q_2$ are given in terms of CM quantities by

$$\alpha = \tfrac{1}{2}(\pi - \theta^*), \qquad q_2 = 2p^* \sin \tfrac{1}{2}\theta^*. \tag{7.21}$$

The Lab kinetic energy transferred to the target particle is therefore

$$T_2 = \frac{q_2^2}{2m_2} = \frac{2p^{*2}}{m_2} \sin^2 \tfrac{1}{2}\theta^*.$$

On the other hand, the total kinetic energy in the Lab is just the kinetic energy of the incoming particle,

$$T = \frac{p_1^2}{2m_1} = \frac{M^2 p^{*2}}{2m_1 m_2^2},$$

by (7.20). The interesting quantity is the fraction of the total kinetic energy which is transferred. This is

$$\frac{T_2}{T} = \frac{4m_1 m_2}{M^2} \sin^2 \tfrac{1}{2}\theta^*. \tag{7.22}$$

The maximum possible kinetic energy transfer occurs for a head-on collision ($\theta^* = \pi$), and is $T_2/T = 4m_1 m_2/(m_1 + m_2)^2$. Clearly this can be close to unity only if $m_1$ and $m_2$ are comparable in magnitude. If the incoming particle is very light, it bounces off the target with little loss of energy; if it is very heavy, it is hardly deflected at all from its original trajectory, and again loses little of its energy. For example, in a proton–$\alpha$-particle collision ($m_1/m_2 = 4$ or $\tfrac{1}{4}$), the maximum fractional energy transfer is 64 per cent. In an electron–proton collision ($m_1/m_2 = 1/1836$) it is about 0.2 per cent. (These conclusions are changed by relativistic effects for speeds approaching that of light.)

Another important relation is that between the Lab and CM scattering angles. It is easy to prove by elementary trigonometry (most simply by dropping a perpendicular from the upper vertex in Fig. 7.5) that

$$\tan \theta = \frac{\sin \theta^*}{(m_1/m_2) + \cos \theta^*}. \tag{7.23}$$

Note that this relation is independent of the momenta of the two particles, and depends only on their mass ratio.

As $\theta^*$ varies from 0 to $\pi$, the vector $\boldsymbol{q}^*$ sweeps out a semicircle of radius $p^*$. If the target is the heavier particle ($m_1/m_2 < 1$), the left-hand vertex of Fig. 7.5 lies inside this semicircle, and $\theta$ also varies from 0 to $\pi$. However, if the target is the lighter particle ($m_1/m_2 > 1$), the semicircle excludes this vertex. In that case, $\theta = 0$ both when $\theta^* = 0$ and when $\theta^* = \pi$, and there is an intermediate value of $\theta^*$ for which $\theta$ is a maximum. The maximum scattering angle occurs when $\boldsymbol{q}_1$ is tangent to the circle (see Fig. 7.6). It is given by

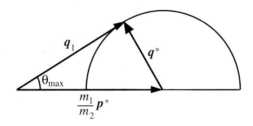

Fig. 7.6

$$\sin \theta_{\max} = m_2/m_1. \tag{7.24}$$

For instance, an $\alpha$-particle can only be scattered by a proton through an angle of almost $14.5°$, and a proton can only be scattered by an electron through an angle less than $0.031°$.

In the special case of equal-mass particles, the right side of (7.23) simplifies to $\tan \frac{1}{2}\theta^*$, so that we obtain the simple relation

$$\theta = \tfrac{1}{2}\theta^*, \qquad (m_1 = m_2). \tag{7.25}$$

In this case, the maximum scattering angle is $\pi/2$.

## 7.4   CM and Lab Cross-sections

Let us now consider the scattering of a beam of particles by a target. We shall suppose that the target contains a large number $N$ of identical target particles, but, as in §4.5, is still small enough so that the particles do not screen each other, and so that the geometry of the scattering event is not seriously affected by which target particle is struck. We also suppose that the beam is a parallel beam of particles, each with the same mass and velocity, with a particle flux of $f$ particles crossing unit area per unit time.

The incident particles may or may not be the same as those in the target, but for the moment we shall assume that they can be distinguished, so that it is possible to set up a detector to count the number of scattered particles emerging in some direction, without including also the recoiling target particles.

We can now introduce the concept of differential cross-section, just as we did for the case of a fixed target in §4.5. Indeed the fact that the particles that are struck recoil out of the target makes no essential difference. In any one collision, the scattering angle $\theta$ will be determined in some way (depending on the shape of the particles) by the impact parameter $b$. Thus the particles scattered through angles between $\theta$ and $\theta + d\theta$ will be those of the incident particles which strike any one of the target particles with impact parameters between the corresponding values $b$ and $b + db$. To find the number emerging within a solid angle in some specified direction, we have to calculate the corresponding cross-sectional area of the incident beam,

$$d\sigma = b\,|db|\,d\varphi. \tag{7.26}$$

(compare (4.39) and Fig. 4.9), and multiply by the number of target particles, $N$, and by the flux, $f$.

If we set up a detector of cross-sectional area $dA$ at a large distance $L$ from the target, the rate of detection will be

$$dw = Nf\frac{d\sigma}{d\Omega}\frac{dA}{L^2}, \tag{7.27}$$

exactly as in (4.44). The ratio $d\sigma/d\Omega$ is the *Lab differential cross-section.*

We shall return to the question of calculating the Lab differential cross-section later. First, however, we wish to discuss a slightly different type of experiment.

Let us imagine two beams of particles approaching from opposite directions — a 'colliding-beam' experiment. In particular, we shall be interested in the case where the momenta of the particles in the two beams are equal and opposite, so that we are directly concerned with the CM frame. To be specific, let us suppose that the particles in one beam are hard spheres of radius $a_1$ and those in the other beam are hard spheres of radius $a_2$. Evidently, a particular pair of particles will collide if the distance $b$ between the lines of motion of their centres is less than $a = a_1 + a_2$ (see Fig. 7.7).

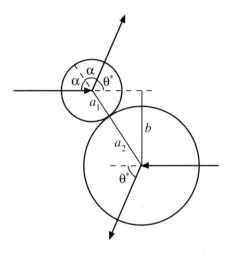

Fig. 7.7

Let us select one of the particles in the second beam. It will collide with any particles in the first beam whose centres cross an area $\sigma = \pi a^2$. This is the *total cross-section* for the collision, and is the same whether the target is at rest or moving towards the beam. Note that it is the *sum* of the radii which enters here. This serves to emphasize the fact that the cross-section is a property of the pair of particles involved in the collision, not of either particle individually.

To compute the number of collisions occurring in a short time interval $dt$, we may imagine the cross-sectional area $\sigma$ to be attached to our selected particle and moving with it. Then the probability that the centre of one of the particles in the first beam crosses this area in the time $dt$ will be $n_1 v \sigma \, dt$, where $n_1$ is the number of particles per unit volume in the beam, and $v$ is the relative velocity ($= v_1 + v_2$). If the number of particles in the second beam per unit volume is $n_2$, then the total number of collisions occurring in volume $V$ is $n_1 n_2 v \sigma V \, dt$.

Now, if we are interested in the number of particles emerging in a given solid angle, we have to find the relation between $b$ and the CM scattering angle $\theta^*$, and evaluate the corresponding cross-sectional area (7.26). In our case, it is clear from Fig. 7.7 that

$$b = a \sin \alpha = a \cos \tfrac{1}{2}\theta^*. \tag{7.28}$$

Hence, introducing the solid angle

$$d\Omega^* = \sin\theta^* \, d\theta^* \, d\varphi,$$

as in (4.42), we find that the *CM differential cross-section* is

$$\frac{d\sigma}{d\Omega^*} = \tfrac{1}{4}a^2. \tag{7.29}$$

If we set up a detector arranged to record particles scattered in a particular direction from a small volume $V$, then the detection rate of particles will be

$$w = n_1 n_2 v V \frac{d\sigma}{d\Omega^*} \frac{dA}{L^2}. \tag{7.30}$$

Note that $N$ and $f$ in (7.27) correspond to $n_2V$ and $n_1v$ in this formula.

In our particular case, of hard-sphere scattering, the differential cross-section (7.29) is independent of $\theta^*$, and the scattering is therefore isotropic in the CM frame. In other words, the number of particles detected is independent of the direction in which we place the detector (so long as it is outside the unscattered beam). This is a very special property of hard spheres, and is not true in general. Nor is it true in the Lab frame. Indeed, the reason for calculating the CM differential cross-section first is that the relation between $b$ and $\theta^*$ is much simpler than that between $b$ and $\theta$, so that the cross-section takes a simpler form in the CM frame than in the Lab frame.

The easiest way to find the Lab differential cross-section is often to compute the CM differential cross-section first, and then use the relation

$$\frac{d\Omega^*}{d\Omega} = \frac{\sin\theta^* \, d\theta^*}{\sin\theta \, d\theta} = \frac{dz^*}{dz}, \qquad \text{with} \qquad z = \cos\theta, \quad z^* = \cos\theta^*.$$

Let us assume (for a reason to be given later) that $m_1 < m_2$. Then it follows that

$$\frac{d\sigma}{d\Omega} = \frac{d\sigma}{d\Omega^*} \frac{dz^*}{dz}. \tag{7.31}$$

The value of $dz^*/dz$ may be found from (7.23), which may be written

$$z^2 = \frac{[(m_1/m_2) + z^*]^2}{1 + 2(m_1/m_2)z^* + (m_1/m_2)^2}.$$

It is straightforward, if tedious, to solve for $z^*$ and differentiate, but the resulting expression for the general case is not particularly illuminating, and we shall omit it.

If $m_1 > m_2$, there is a complication which arises from the fact that for each value of $\theta$ less than the maximum scattering angle $\theta_{max}$ there are *two* possible values for $\theta^*$, as is clear from Fig. 7.6. If we measure only the direction of the scattered particles, we cannot distinguish these two, and must add the contributions to $d\sigma/d\Omega$ from $d\sigma/d\Omega^*$ at these two separate values of $\theta^*$. (Moreover, for one of these two, $dz^*/dz$ is negative, and must be replaced by its absolute value.) The two values of $\theta^*$ can be distinguished if we can measure also the energy or momentum of the scattered particles.

We shall consider explicitly only the case of equal mass particles, $m_1 = m_2$.

## *Example:* Scattering of equal hard spheres

Find the Lab-frame differential cross-section for the scattering of equal-mass hard spheres.

Here the relation between the scattering angles, (7.23), simplifies to $\theta^* = 2\theta$, or $z^* = 2z^2 - 1$. Consequently, $dz^*/dz = 4z$, and the Lab differential cross-section corresponding to (7.29) is

$$\frac{d\sigma}{d\Omega} = a^2 \cos\theta, \qquad (\theta < \tfrac{1}{2}\pi). \tag{7.32}$$

Of course, no particles are scattered through angles greater than $\pi/2$, so

$$\frac{d\sigma}{d\Omega} = 0, \qquad (\theta > \tfrac{1}{2}\pi). \tag{7.33}$$

In this case, the cross-section is peaked towards the forward direction, and more particles will enter a detector placed at a small angle to the beam direction than one placed nearly at right angles.

As a check of this result, we may evaluate the total cross-section $\sigma$, which must agree with the value $\sigma = 4\pi \times \tfrac{1}{4}a^2 = \pi a^2$ obtained in the CM frame. We find

$$\sigma = \iint \frac{d\sigma}{d\Omega}\, d\Omega = \int_0^{2\pi} d\varphi \int_0^{\pi/2} a^2 \cos\theta \sin\theta\, d\theta = \pi a^2,$$

as expected. (Note the upper limit of $\pi/2$ on the $\theta$ integration, which arises because of (7.33).)

We assumed at the beginning of this section that the target particles and beam particles were distinguishable, and that the aim of the experiment was to count scattered beam particles only. However, it is easy to relax this condition. We shall now calculate the rate at which recoiling target particles enter the detector. In a collision in which the incoming particle has impact parameter $b$, and is moving in a plane specified by the angle $\varphi$, the target particle will emerge in a direction specified by the polar angles $\alpha, \psi$, where $\alpha$ is related to $\theta^*$ or $b$ by (7.21) and $\psi = \pi + \varphi$. The number of target particles emerging within the angular range $\mathrm{d}\alpha, \mathrm{d}\psi$ is therefore equal to the number of incident particles in the corresponding range $\mathrm{d}b, \mathrm{d}\varphi$. To determine the number of recoiling particles entering the detector, we have to relate $\mathrm{d}\sigma$ to the solid angle

$$\mathrm{d}\Omega_2 = \sin\alpha\,\mathrm{d}\alpha\,\mathrm{d}\psi.$$

By (7.21), we have $\cos\theta^* = 1 - 2\cos^2\alpha$, whence

$$\frac{\mathrm{d}\sigma}{\mathrm{d}\Omega_2} = \frac{\mathrm{d}\sigma}{\mathrm{d}\Omega^*}\,4\cos\alpha \qquad (\alpha < \tfrac{1}{2}\pi). \qquad (7.34)$$

In the particular case of equal-mass hard spheres, this differential cross-section has exactly the same form as (7.32). This shows that in the scattering of identical hard spheres the numbers of scattered particles and recoiling target particles entering the detector are precisely equal, and the total detection rate is obtained by doubling that found earlier.

## 7.5 Summary

For our later work, the most important result of this chapter is that the total momentum, angular momentum and kinetic energy in an arbitrary frame differ from those in the CM frame by an amount equal to the contribution of a particle of mass $M$ moving with the centre of mass.

We have seen that for a two-particle system the use of the CM frame is often a considerable simplification, and this is true also for more complicated systems. When we need results in some other frame, it is often best to solve the problem first in the CM frame, and then transform to the required frame. This is true for example in the calculation of scattering cross-sections in two-particle collisions.

## Problems

1. A double star is formed of two components, each with mass equal to that of the Sun. The distance between them is $1\,\mathrm{AU}$ (see Chapter 4, Problem 2). What is the orbital period?

2. Where is the centre of mass of the Sun–Jupiter system? (The mass ratio is $M_S/M_J = 1047$. See Chapter 4, Problems 2 and 3.) Through what angle does the Sun's position as seen from the Earth oscillate because of the gravitational attraction of Jupiter?

3. The parallax of a star (the angle subtended at the star by the radius of the Earth's orbit) is $\varpi$. The star's position is observed to oscillate with angular amplitude $\alpha$ and period $\tau$. If the oscillation is interpreted as being due to the existence of a planet moving in a circular orbit around the star, show that its mass $m_1$ is given by

$$\frac{m_1}{M_S} = \frac{\alpha}{\varpi}\left(\frac{M\tau_E}{M_S\tau}\right)^{2/3},$$

where $M$ is the total mass of star plus planet, $M_S$ is the Sun's mass, and $\tau_E = 1\,\mathrm{year}$. Evaluate the mass $m_1$ if $M = 0.25 M_S$, $\tau = 16\,\mathrm{years}$, $\varpi = 0.5''$ and $\alpha = 0.01''$. What conclusion can be drawn without making the assumption that the orbit is circular?

4. Two particles of masses $m_1$ and $m_2$ are attached to the ends of a light spring. The natural length of the spring is $l$, and its tension is $k$ times its extension. Initially, the particles are at rest, with $m_1$ at a height $l$ above $m_2$. At $t = 0$, $m_1$ is projected vertically upward with velocity $v$. Find the positions of the particles at any subsequent time (assuming that $v$ is not so large that the spring is expanded or compressed beyond its elastic limit).

5. *Prove that in an elastic scattering process the angle $\theta + \alpha$ between the emerging particles is related to the recoil angle $\alpha$ by

$$\frac{\tan(\theta + \alpha)}{\tan\alpha} = \frac{m_1 + m_2}{m_1 - m_2}.$$

(*Hint*: Express both tangents in terms of $\tan\frac{1}{2}\theta^*$.) What is the mass ratio if the particles emerge at right angles to each other?

6. A proton is elastically scattered through an angle of $56°$ by a nucleus, which recoils at an angle $\alpha = 60°$. Find the atomic mass of the nucleus, and the fraction of the kinetic energy transferred to it.

7. An experiment is to be designed to measure the differential cross-section for elastic pion–proton scattering at a CM scattering angle of 70° and a pion CM kinetic energy of 490 keV. (The electron-volt (eV) is the atomic unit of energy.) Find the angles in the Lab at which the scattered pions, and the recoiling protons, should be detected, and the required Lab kinetic energy of the pion beam. (The ratio of pion to proton mass is $1/7$.)

8. *An unstable particle of mass $M = m_1 + m_2$ decays into two particles of masses $m_1$ and $m_2$, releasing an amount of energy $Q$. Determine the kinetic energies of the two particles in the CM frame. Given that $m_1/m_2 = 4$, $Q = 1$ MeV, and that the unstable particle is moving in the Lab with kinetic energy 2.25 MeV, find the maximum and minimum Lab kinetic energies of the particle of mass $m_1$.

9. The molecules in a gas may be treated as identical hard spheres. Find the average loss of kinetic energy of a molecule with kinetic energy $T$ in a collision with a stationary molecule. (*Hint*: Use the fact that the collisions are isotropic in the CM frame, so that all values of $\cos \theta^*$ between $\pm 1$ are equally probable.) How many collisions are required, on average, to reduce the velocity of an exceptionally fast molecule by a factor of 1000?

10. Two identical charged particles, each of mass $m$ and charge $e$, are initially far apart. One of the particles is at rest at the origin, and the other is approaching it with velocity $v$ along the line $x = b, y = 0$, where $b = e^2/2\pi\epsilon_0 m v^2$. Find the scattering angle in the CM frame, and the directions in which the two particles emerge in the Lab. (See §4.7.)

11. *Find the distance of closest approach for the particles in Problem 10, and the velocity of each at the moment of closest approach.

12. Obtain the relation between the total kinetic energy in the CM and Lab frames. Discuss the limiting cases of very large and very small mass for the target.

13. *Suppose that the asteroid of Chapter 4, Problem 17, has a mass of $6 \times 10^{20}$ kg. Find the proportional change in the kinetic energy of the Earth in this encounter. What is the resulting change in the semi-major axis of the Earth's orbit? By how much is its orbital period lengthened? (*Note* that the postulated event is exceedingly improbable.)

14. Calculate the differential cross-section for the scattering of identical hard spheres directly in the Lab frame.

15. Find the Lab differential cross-section for the scattering of identical particles of charge $e$ and mass $m$, if the incident velocity is $v$. (See (4.50).)

16. At low energies, protons and neutrons behave roughly like hard spheres of equal mass and radius about $1.3 \times 10^{-14}$ m. A parallel beam of neutrons, with a flux of $3 \times 10^{10}$ neutrons m$^{-2}$ s$^{-1}$, strikes a target containing $4 \times 10^{22}$ protons. A circular detector of radius 20 mm is placed 0.7 m from the target, in a direction making an angle of 30° to the beam direction. Calculate the rate of detection of neutrons, and of protons.

17. *Write down the equations of motion for a pair of charged particles of equal masses $m$, and of charges $q$ and $-q$, in a uniform electric field $\boldsymbol{E}$. Show that the field does not affect the motion of the centre of mass. Suppose that the particles are moving in circular orbits with angular velocity $\omega$ in planes parallel to the $xy$-plane, with $\boldsymbol{E}$ in the $z$-direction. Write the equations in a frame rotating with angular velocity $\omega$, and hence find the separation of the planes.

# Chapter 8

# Many-Body Systems

Any material object may be regarded as composed of a large number of small particles, small enough to be treated as essentially point-like, but still large enough to obey the laws of classical rather than quantum mechanics. These particles interact in complicated ways with each other and with the environment. However, as we shall see, if we are interested only in the motion of the object as a whole, many of these details are irrelevant.

We shall consider in this chapter a general system of $N$ particles labelled by an index $i = 1, 2, \ldots, N$, interacting through two-body forces, and also subjected to external forces due to bodies outside the system. We denote the force on the $i$th particle due to the $j$th particle by $\boldsymbol{F}_{ij}$, and the external force on the $i$th particle by $\boldsymbol{F}_i$. Then the equations of motion are

$$m_i \ddot{\boldsymbol{r}}_i = \boldsymbol{F}_{i1} + \boldsymbol{F}_{i2} + \cdots + \boldsymbol{F}_{iN} + \boldsymbol{F}_i \equiv \sum_j \boldsymbol{F}_{ij} + \boldsymbol{F}_i. \qquad (8.1)$$

Here, and throughout this chapter, the sum is over all particles of the system, $j = 1, 2, \ldots, N$. Of course, there is no force on the $i$th particle due to itself, and so $\boldsymbol{F}_{ii} = \boldsymbol{0}$. The sum in this case is really over the other $N - 1$ particles.

## 8.1 Momentum; Centre-of-mass Motion

The position $\boldsymbol{R}$ of the centre of mass is defined, as in (7.2), by

$$\boldsymbol{R} = \frac{1}{M} \sum_i m_i \boldsymbol{r}_i, \qquad \text{where} \qquad M = \sum_i m_i. \qquad (8.2)$$

The total momentum is

$$\boldsymbol{P} = \sum_i m_i \dot{\boldsymbol{r}}_i = M\dot{\boldsymbol{R}}. \qquad (8.3)$$

It is equal to the momentum of a particle of mass $M$ located at the centre of mass.

The rate of change of momentum, by (8.1), is

$$\dot{\boldsymbol{P}} = \sum_i \sum_j \boldsymbol{F}_{ij} + \sum_i \boldsymbol{F}_i. \qquad (8.4)$$

Now the two-body forces must satisfy Newton's third law

$$\boldsymbol{F}_{ji} = -\boldsymbol{F}_{ij}. \qquad (8.5)$$

Thus, for every term $\boldsymbol{F}_{ij}$ in the double sum in (8.4), there is an equal and opposite term $\boldsymbol{F}_{ji}$. The terms therefore cancel in pairs and the double sum is zero. (Compare the discussion at the end of §1.3 for the three-particle case.)

Hence we obtain the important result that the rate of change of momentum is equal to the sum of the *external* forces alone:

$$\dot{\boldsymbol{P}} = M\ddot{\boldsymbol{R}} = \sum_i \boldsymbol{F}_i. \qquad (8.6)$$

In the special case of an *isolated* system of particles, acted on by no external forces, this yields the law of *conservation of momentum*,

$$\boldsymbol{P} = M\dot{\boldsymbol{R}} = \text{constant}. \qquad (8.7)$$

Let us now regard our system of particles as forming a composite body. If the body is isolated, then according to Newton's first law it moves with uniform velocity. Thus we see that to maintain this law for composite bodies, we should define the *position* of such a body to mean the position of its *centre of mass*. Moreover, with this definition, Newton's second law is just (8.6), provided that we interpret the *force* on the body in the obvious way to mean the sum of the external forces on all of its constituent particles, and the *mass* as the sum of their masses. It is also clear that if Newton's third law applies to each pair of particles from two composite bodies, then it will apply to the bodies as a whole.

Thus, with a suitable (and very natural) interpretation of the concepts involved, Newton's three basic laws may be applied to composite bodies as well as to point particles. (Though we have phrased the discussion in terms of collections of particles, we could also include the case of a continuous distribution of matter, by dividing it up into infinitesimal particles.) It follows that, so long as we are interested only in the motion of a body as a whole, we may replace it by a particle of mass $M$ located at the centre of mass. This is a result of the greatest importance, for it allows us to apply our earlier discussion of particle motion to real physical bodies. We have of course implicitly assumed it in many of our applications; for example, we treated the planets as point particles in discussing their orbital motion.

There is one point about which one must be careful. We may still need some information about the actual shape of the body to calculate the total force acting on it. This force is not necessarily equal to that on a particle of mass $M$ at the centre of mass. In the gravitational case, it happens to be so if the body is spherical, or if the gravitational field is uniform, but not in general otherwise.

As an example of the use of the momentum conservation equation for an isolated system, we consider the motion of a rocket.

### *Example:* **Rockets**

A rocket, initially at rest, emits matter with a constant velocity $u$ relative to the rocket (though not necessarily at a uniform rate). Gravity is negligible. If the initial total mass of the rocket is $M_0$, find its velocity $v$ when the mass has fallen to $M$.

Consider the emission of a small mass $\mathrm{d}m$. After the emission, the mass of the rocket will be reduced by this amount,·

$$\mathrm{d}M = -\mathrm{d}m, \tag{8.8}$$

and its velocity will be increased to some value $v + \mathrm{d}v$. (See Fig. 8.1.) This velocity can be found from the momentum conservation equation

$$(M - \mathrm{d}m)(v + \mathrm{d}v) + \mathrm{d}m(v - u) = Mv.$$

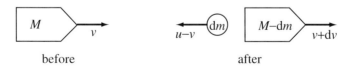

| before | after |

Fig. 8.1

Neglecting second-order infinitesimals, we find

$$M \, dv = u \, dm. \tag{8.9}$$

By (8.8), we can write this in the form

$$\frac{dv}{u} = -\frac{dM}{M},$$

whence, integrating, and using the initial condition $M = M_0$ when $v = 0$, we have

$$\frac{v}{u} = -\ln \frac{M}{M_0}.$$

This relation may be solved to give the mass as a function of the velocity attained:

$$M = M_0 e^{-v/u}. \tag{8.10}$$

This shows that to accelerate a rocket to a velocity equal to its ejection velocity $u$, we must eject all but a fraction $1/e$ of its original mass.

Note that the velocity of the rocket depends only on the ejection velocity and the fraction of the initial mass which has been ejected, and not on the rate of ejection. It makes no difference whether the acceleration is brief and intense, or prolonged and gentle. (This assumes of course that the rocket is not subjected to other forces, such as gravity, during its acceleration. It would clearly be useless to try to escape from the Earth with a rocket providing a long, slow acceleration, because it would be constantly retarded by gravity.)

For interplanetary flights, rockets are normally used only for brief spells, between which the spacecraft moves in a free orbit — as, for example, in Chapter 4, Problem 16, where these brief spells are at or near planetary departures and arrivals. If the duration of the rocket bursts is sufficiently short for the change in position of the rocket during each burst to be negligible, then it is a good approximation to assume that on each occasion the velocity is changed instantaneously, by an amount known as the *velocity impulse*. The relevant quantity for determining the mass of the rocket required to deliver a given payload, using a given ejection velocity, via a given orbital manoeuvre, is then the sum of the velocity impulses (in the ordinary, not the vector, sense). For example, the minimum velocity impulse needed to escape from the Earth is $11 \, \text{km s}^{-1}$. For the return trip, it is $22\,\mathrm{km\,s^{-1}}$, if the deceleration on re-entry into the atmosphere is to be produced by the rockets rather than atmospheric friction.

## 8.2 Angular Momentum; Central Internal Forces

The total angular momentum of our system of particles is

$$\boldsymbol{J} = \sum_i m_i \boldsymbol{r}_i \wedge \dot{\boldsymbol{r}}_i. \tag{8.11}$$

The rate of change of $\boldsymbol{J}$ is

$$\dot{\boldsymbol{J}} = \sum_i m_i \boldsymbol{r}_i \wedge \ddot{\boldsymbol{r}}_i = \sum_i \sum_j \boldsymbol{r}_i \wedge \boldsymbol{F}_{ij} + \sum_i \boldsymbol{r}_i \wedge \boldsymbol{F}_i. \tag{8.12}$$

Now let us examine the contribution to (8.12) from the internal force between a particular pair of particles, say 1 and 2. (See Fig. 8.2.) For simplicity, let us write $\boldsymbol{r} = \boldsymbol{r}_1 - \boldsymbol{r}_2$, and $\boldsymbol{F} = \boldsymbol{F}_{12}$, so that also $\boldsymbol{F}_{21} = -\boldsymbol{F}$.

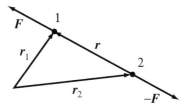

Fig. 8.2

The contribution consists of two terms,

$$\boldsymbol{r}_1 \wedge \boldsymbol{F}_{12} + \boldsymbol{r}_2 \wedge \boldsymbol{F}_{21} = \boldsymbol{r}_1 \wedge \boldsymbol{F} - \boldsymbol{r}_2 \wedge \boldsymbol{F} = \boldsymbol{r} \wedge \boldsymbol{F}. \tag{8.13}$$

This contribution will be zero if $\boldsymbol{F}$ is a *central* internal force, parallel to $\boldsymbol{r}$.

Let us for the moment simply *assume* that all the internal forces are central. (We shall discuss the possible justification for this assumption shortly.) Then all the terms in the double sum in (8.12) will cancel in pairs, just as they did in the evaluation of the rate of change of total linear momentum. The total moment of all the internal forces will be zero. So, when the internal forces are central, the rate of change of angular momentum is equal to the sum of the moments of the *external* forces only:

$$\dot{\boldsymbol{J}} = \sum_i \boldsymbol{r}_i \wedge \boldsymbol{F}_i. \tag{8.14}$$

In particular, for an isolated system, with all $\boldsymbol{F}_i = \boldsymbol{0}$, we have the law of *conservation of angular momentum*,

$$\boldsymbol{J} = \text{constant}. \tag{8.15}$$

More generally, this is true if all the external forces are directed towards, or away from, the origin (which may be true for example if they are due to the gravitational attraction of a spherical body).

Although we have only shown that (8.14) and (8.15) hold in the special case of central internal forces, they are actually of much more general validity, for reasons that should become clearer in Chapter 12. It is certainly *not* true that all the internal forces in real solids, liquids or gases are central. Many are, but there are obvious exceptions, notably the electromagnetic force between moving charges. It is better to regard (8.14) as a basic postulate of the dynamics of composite bodies, justified by the fact that the predictions derived from it agree with observations.

It is often convenient, as we did in Chapter 7, to separate the contributions to $\boldsymbol{J}$ from the centre-of-mass motion and the relative motion. We define the positions $\boldsymbol{r}_i^*$ of the particles relative to the centre of mass by

$$\boldsymbol{r}_i = \boldsymbol{R} + \boldsymbol{r}_i^*. \tag{8.16}$$

Clearly, the position of the centre of mass relative to itself is zero, so that

$$M\boldsymbol{R}^* = \sum_i m_i \boldsymbol{r}_i^* = \boldsymbol{0}. \tag{8.17}$$

Now, substituting (8.16) into (8.11), we obtain

$$\boldsymbol{J} = \left(\sum_i m_i\right) \boldsymbol{R} \wedge \dot{\boldsymbol{R}} + \left(\sum_i m_i \boldsymbol{r}_i^*\right) \wedge \dot{\boldsymbol{R}} + \boldsymbol{R} \wedge \left(\sum_i m_i \dot{\boldsymbol{r}}_i^*\right)$$
$$+ \sum_i m_i \boldsymbol{r}_i^* \wedge \dot{\boldsymbol{r}}_i^*.$$

The second and third terms vanish by virtue of (8.17). Thus we can write

$$\boldsymbol{J} = M\boldsymbol{R} \wedge \dot{\boldsymbol{R}} + \boldsymbol{J}^*, \tag{8.18}$$

where $\boldsymbol{J}^*$ is the angular momentum about the centre of mass, namely

$$\boldsymbol{J}^* = \sum_i m_i \boldsymbol{r}_i^* \wedge \dot{\boldsymbol{r}}_i^*. \tag{8.19}$$

It is easy to find the rate of change of $\boldsymbol{J}^*$. For, by (8.6),

$$\frac{\mathrm{d}}{\mathrm{d}t}(M\boldsymbol{R} \wedge \dot{\boldsymbol{R}}) = M\boldsymbol{R} \wedge \ddot{\boldsymbol{R}} = \boldsymbol{R} \wedge \sum_i \boldsymbol{F}_i. \tag{8.20}$$

Hence, subtracting from (8.14), and using (8.16), we obtain

$$\dot{\boldsymbol{J}}^* = \sum_i \boldsymbol{r}_i^* \wedge \boldsymbol{F}_i. \tag{8.21}$$

Thus the rate of change of the angular momentum $\boldsymbol{J}^*$ about the centre of mass is equal to the sum of the moments of the external forces about the centre of mass. This is a remarkable result. For, it must be remembered that the centre of mass is not in general moving uniformly. In general, we may take moments about the origin of any *inertial* frame, but it would be quite wrong to take moments about an *accelerated* point. Only in very special cases, such as where the point is the centre of mass, are we allowed to do this.

This result means that in discussing the rotational motion of a body we can ignore the motion of the centre of mass, and treat it as though it were fixed. It is particularly important in the case of rigid bodies, which we discuss in Chapter 9.

For an isolated system, it follows from (8.21) that $\boldsymbol{J}^*$ as well as $\boldsymbol{J}$ is a constant. More generally, $\boldsymbol{J}^*$ is a constant if the external forces have zero total moment about the centre of mass. For example, for a system of particles in a uniform gravitational field, the resultant force acts at the centre of mass, and $\boldsymbol{J}^*$ is therefore a constant.

## 8.3 The Earth–Moon System

As an interesting example of the use of the angular momentum conservation law, we consider the system comprising the Earth and the Moon. (We ignore the other planets, and treat the Sun as fixed.)

The angular momentum of this system is

$$\boldsymbol{J} = M\boldsymbol{R} \wedge \dot{\boldsymbol{R}} + \boldsymbol{J}^*, \tag{8.22}$$

where $M = m + m'$ is the total mass of Earth plus Moon, and $J^*$ is the angular momentum about the centre of mass. This angular momentum $J^*$ can again be separated into an orbital term, due to the motion of the Earth and Moon around their common centre of mass, and two rotational terms, $J^*_E$ and $J^*_M$, due to the rotation of each body about its own centre. The orbital angular momentum, according to (7.15), is $\mu r \wedge \dot{r}$, where $\mu = mm'/M$. Thus

$$J^* = \mu r \wedge \dot{r} + J^*_E + J^*_M. \tag{8.23}$$

Let us first neglect the non-uniformity of the Sun's gravitational field over the dimensions of the Earth–Moon system. Then $J^*$ as well as $J$ will be constant. Indeed, to a good approximation, the individual terms of (8.23) are separately constant. However, over long periods of time they do change, for at least two reasons. One is the non-spherical shape of the Earth, which, as we saw in §6.5, leads to a precession of the Moon's orbital plane. This is too small to be of much importance in the present context (the non-uniformity of the Sun's field produces a much bigger effect), but because of angular momentum conservation there is a corresponding precession of the Earth's angular momentum $J^*_E$ which *is* important. We shall discuss this effect further in §10.1.

The second reason is the effect of tidal friction. The dissipation of energy by the tides has the effect of gradually slowing the Earth's rotation. We can picture this effect as follows: as the Earth rotates, it tries to carry with it the tidal 'bulges', while the Moon's attraction is pulling them back into line. (See Fig. 8.3.) Thus there is a couple acting to slow the Earth's

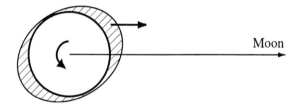

Fig. 8.3

rotation, and a corresponding equal and opposite couple tending to increase the orbital angular momentum.

To investigate this effect, we shall make the simplest possible approximations. We neglect the rotational angular momentum $J^*_M$ of the Moon (which is in fact very small compared to the other terms in (8.23)), and assume that the Earth's axis is perpendicular to the Moon's orbital plane, so that the remaining terms of (8.23) have the same direction. The Earth's rotational angular momentum, as we shall see in the next chapter, has the form $J_{\mathrm{E}}^* = I\omega$, where $I$ is the so-called *moment of inertia*, and $\omega$ the rotational angular velocity. For a sphere of uniform density, mass $m$ and radius $r$, $I$ has the value $I = \frac{2}{5}mr^2$. In fact, the Earth's density increases towards its centre, and $I$ is therefore somewhat less than this, approximately $I = 0.33mr^2$. Thus the total angular momentum about the centre of mass is

$$J^* = \mu a^2 \Omega + 0.33mr^2\omega, \tag{8.24}$$

where $\Omega$ is the Moon's orbital angular velocity. It is convenient to express the angular velocities in terms of the present value $\omega_0$ of $\omega$, given by (5.1). Using $m/\mu = 82.3$, we may write the conservation law (8.24) (dividing by $\mu r^2 \omega_0$) as

$$\left(\frac{a}{r}\right)^2 \frac{\Omega}{\omega_0} + 27.2\frac{\omega}{\omega_0} = 160, \tag{8.25}$$

where the value of the constant on the right is obtained by using the present values $a/r = 60.3$ and $\Omega/\omega_0 = 0.0365$. Note that at present the first term of (8.25) is almost five times as large as the second.

Now the orbital angular velocity $\Omega$ is related to $a$ by (4.32), *i.e.*,

$$\Omega^2 a^3 = GM,$$

or, equivalently

$$\frac{\Omega}{\omega_0} = 17.1 \left(\frac{r}{a}\right)^{3/2}. \tag{8.26}$$

Inserting this into (8.25) yields for the Earth's rotational angular velocity

$$\frac{\omega}{\omega_0} = 5.88 - 0.629 \left(\frac{a}{r}\right)^{1/2}. \tag{8.27}$$

As $\omega$ decreases due to tidal friction, $a$ must increase, *i.e.*, the Moon moves further away from the Earth. Indeed this effect can now be measured: the Moon is currently receding from the Earth at about $3.8\,\mathrm{cm\,yr^{-1}}$. Hence $\Omega$ decreases too. If we use (8.26) and (8.27) to plot $\Omega$ and $\omega$ as functions of $a$ we obtain curves like those of Fig. 8.4 (which are *not* drawn to scale).

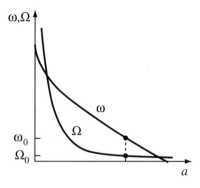

Fig. 8.4

It is important to note that if at any time $\Omega$ were actually larger than $\omega$, so that the Moon revolved around the Earth in less than a day, then the tidal forces would act to *increase* the angular velocities and *decrease* the separation. On this simple picture, the smallest distance there can have been between the bodies corresponds to the first point where the curves cross (about 2.3 Earth radii, with a common rotation period of under 5 hours). However, it is in fact impossible that the Moon should ever have been so close, because it would then have been inside the *Roche limit* (about 2.9 Earth radii) where the tidal forces become so strong that they would break up the Moon (see Chapter 6, Problem 20). It is now generally believed that the material that later formed the Moon was ejected from a proto-Earth in a violent collision with another planet early in its history, and then gradually condensed into the present Moon in an orbit with a radius of a few Earth radii.

The final state of the system in the very distant future would correspond to the second point where the curves cross. Then $\omega$ and $\Omega$ would again be equal, and tidal forces would play no further role. In this final state, the Earth would always present the same face to the Moon, as the Moon already does to the Earth (except for a small wobble). The distance between Earth and Moon would then be about half as large again as it now is, and the common rotation period (the 'day' or the 'month') would be nearly 48 days.

There are, however, several factors which might modify this picture. It was at one time believed that Mercury and the Sun had reached the final state described above, and that Mercury always presents the same face to the Sun. However, it is now known that the ratio of the orbital and rotation periods is not unity but 3/2. There are several other cases in the solar system of such rational ratios. The stability of this configuration depends on the unusually large eccentricity of Mercury's orbit. It is unlikely that such a state could be reached by the Earth–Moon system.

The 'final' state would in any case not be the end of the evolution of the system, because the *solar* tides would still be active. They would tend to slow the Earth's rotation even more, and would therefore gradually pull the Moon back towards the Earth. In the very remote future, the two bodies would collide, although it is probable that the Sun will have reached the end of its life, and engulfed the whole system, long before that happens.

A significant over-simplification in our argument is the assumption that the Moon's orbit is a circle in a plane normal to the Earth's axis. In reality, both its eccentricity and its inclination can change with time, and of course the Earth's axis also precesses is a complicated way. The effect of tidal dissipation is not only to make the Moon's orbit expand, but also to make it more eccentric and less inclined. In the distant past, it may have been considerably more steeply inclined than it now is.

The Sun has another effect which has been omitted from our discussion. The solar ocean tides are indeed small compared to the lunar ones. However, there are quite substantial semi-diurnal tides in the Earth's atmosphere, noticeable as fluctuations in the barometric pressure at sea level. They are predominantly of thermal, rather than gravitational, origin, due to heating during the hours of daylight, and are very different in character from the ocean tides. The most important component has a period of half a *solar*, rather than lunar, day. It is a little surprising that the dominant period is 12 hours, rather than 24. This may be an example of the phenomenon of resonance (see §2.6): the amplitude of the 12-hour oscillation may be enhanced by being closer to a natural oscillation period of the atmosphere than is 24 hours. As on the oceans, the Sun exerts a gravitational couple on the resulting bulges in the atmosphere. (So does the Moon of course, but its effect averages out to zero over a lunar month.) It happens that the maxima in this case occur *before*, not after, 12 noon and 12 midnight. Thus the couple acts to speed the Earth's rotation rather than to slow it. This process has the effect of transferring angular momentum from the Earth's orbital motion to its rotation. It therefore tends to decrease the Earth's orbital radius, pulling it (immeasurably slowly) closer to the Sun. A purely gravitational process could not have this effect, for the energy (kinetic plus potential) is actually increasing. This energy comes from solar heating, with the Earth acting as a kind of giant heat engine converting heat to mechanical work. However, the net effect of the atmospheric tides in speeding the Earth's rotation is not nearly large enough to counterbalance that of the ocean tides in slowing it.

## 8.4   Energy; Conservative Forces

The kinetic energy of our system of particles is

$$T = \sum_i \tfrac{1}{2} m_i \dot{\boldsymbol{r}}_i^2. \tag{8.28}$$

As in the case of angular momentum, $T$ can be separated into centre-of-mass and relative terms. Substituting (8.16) into (8.28), we again find that the cross terms drop out because of (8.17). Thus we obtain

$$T = \tfrac{1}{2} M \dot{\boldsymbol{R}}^2 + T^*, \tag{8.29}$$

where $T^*$, the kinetic energy relative to the centre of mass, is

$$T^* = \sum_i \tfrac{1}{2} m_i \dot{\boldsymbol{r}}_i^{*2}. \tag{8.30}$$

The relations (8.3), (8.18) and (8.29) show that, exactly as in the two-particle case (see (7.16)), the momentum, angular momentum and kinetic energy in any frame are obtained from those in the CM frame by adding the contribution of a particle of mass $M$ located at the centre of mass $\boldsymbol{R}$.

The rate of change of kinetic energy is

$$\dot{T} = \sum_i m_i \dot{\boldsymbol{r}}_i \cdot \ddot{\boldsymbol{r}}_i = \sum_i \sum_j \dot{\boldsymbol{r}}_i \cdot \boldsymbol{F}_{ij} + \sum_i \dot{\boldsymbol{r}}_i \cdot \boldsymbol{F}_i. \tag{8.31}$$

This equation may be compared with (8.12). Let us again examine the contribution from the internal force $\boldsymbol{F}_{12} = -\boldsymbol{F}_{21} = \boldsymbol{F}$ between particles 1 and 2 (see Fig. 8.2). As in (8.13), it is

$$\dot{\boldsymbol{r}}_1 \cdot \boldsymbol{F} - \dot{\boldsymbol{r}}_2 \cdot \boldsymbol{F} = \dot{\boldsymbol{r}} \cdot \boldsymbol{F}.$$

This is of course the rate at which the force $\boldsymbol{F}$ does work. However, in contrast to the previous situation, there is in general no reason for this to vanish, and $\dot{T}$ cannot therefore be written in terms of the external forces alone. This is not at all surprising, because in general we must expect there to be some change in the internal potential energy of the system.

We may note that there are certain special cases in which a force does no work, and may be omitted from the energy change equation. This will be the case, for example, if $\dot{\boldsymbol{r}} = \boldsymbol{0}$ (*e.g.*, the reaction at a fixed pivot), or if $\boldsymbol{F}$ is always perpendicular to $\dot{\boldsymbol{r}}$ (*e.g.*, the magnetic force on a charged particle). An important special case is that of a rigid body, in which the distances between all the particles are fixed. Then $r$ is constant, so that $\boldsymbol{r} \cdot \dot{\boldsymbol{r}} = 0$. If the force is central, then it has the same direction as $\boldsymbol{r}$, and is always perpendicular to $\dot{\boldsymbol{r}}$. Thus a *central* force between particles whose distance apart is fixed does no work.

Now, however, let us make the less restrictive assumption that all the internal forces are conservative. Then the force $\boldsymbol{F}$ must correspond to a potential energy function of particles 1 and 2, and the rate of working $\dot{\boldsymbol{r}} \cdot \boldsymbol{F}$ will be equal to minus the rate of change of this potential energy. Let us denote the internal potential energy — the sum of the potential energies for all the $\frac{1}{2}N(N-1)$ pairs of particles — by $V_{\text{int}}$. Then clearly the rate of working of all the internal forces will be minus the rate of change of $V_{\text{int}}$. Thus we obtain

$$\frac{\mathrm{d}}{\mathrm{d}t}(T + V_{\text{int}}) = \sum_i \dot{\boldsymbol{r}}_i \cdot \boldsymbol{F}_i. \tag{8.32}$$

The rate of change of the kinetic plus internal potential energy is equal to the rate at which the external forces do work.

We can also find the rate of change of energy relative to the centre of mass. From (8.6), we have

$$\frac{\mathrm{d}}{\mathrm{d}t}(\tfrac{1}{2}M\dot{\boldsymbol{R}}^2) = M\dot{\boldsymbol{R}} \cdot \ddot{\boldsymbol{R}} = \dot{\boldsymbol{R}} \cdot \sum_i \boldsymbol{F}_i.$$

Hence, subtracting from (8.32) and using (8.29) and (8.16), we find

$$\frac{\mathrm{d}}{\mathrm{d}t}(T^* + V_{\text{int}}) = \sum_i \dot{\boldsymbol{r}}_i^* \cdot \boldsymbol{F}_i. \tag{8.33}$$

This equation is the analogue of (8.21). Note that $V_{\text{int}}$ is a function only of the differences $\boldsymbol{r}_i - \boldsymbol{r}_j$, and is therefore unaffected by the choice of reference frame. Thus $T^* + V_{\text{int}}$ is the total energy in the CM frame.

Now let us suppose that the external forces are also conservative. Then there must exist a corresponding external potential energy function $V_{\text{ext}}$, whose rate of change is minus the rate of working of the *external* forces. From (8.32) we then find the law of *conservation of energy*

$$T + V = E = \text{constant}, \qquad \text{where} \qquad V = V_{\text{ext}} + V_{\text{int}}. \tag{8.34}$$

It is important to notice, however, that $V_{\text{ext}}$ depends on the co-ordinates themselves, not merely the co-ordinate differences, and is therefore dependent on the choice of frame, unlike $V_{\text{int}}$. Thus in general it is *not* possible to write down a similar equation in which only the kinetic energy $T^*$ relative to the centre of mass appears.

## 8.5   Lagrange's Equations

We close this chapter with a brief discussion of the extension of the Lagrangian method of §3.7 to a many-particle system. We consider only the case where all the forces are conservative, so that there exists a total potential energy function $V$. We saw in §7.1 (see the discussion following (7.9)) that the derivatives of the potential energy function $V_{\text{int}}(\boldsymbol{r}_1 - \boldsymbol{r}_2)$ corresponding to a two-body force with respect to the components of $\boldsymbol{r}_1$ or $\boldsymbol{r}_2$ give correctly minus the forces on the corresponding particles. Thus the components of the total force on the $i$th particle will be minus the derivatives of $V$ with respect to the components of $\boldsymbol{r}_i$. The equations of motion may thus be written

$$m_i \ddot{x}_i = -\frac{\partial V}{\partial x_i}, \qquad (i = 1, 2, \ldots, N), \tag{8.35}$$

together with similar equations for the $y$ and $z$ components.

Now, clearly,

$$\frac{\partial T}{\partial \dot{x}_i} = m_i \dot{x}_i,$$

while $\partial T/\partial x_i$ and $\partial V/\partial \dot{x}_i$ both vanish. Thus (8.35) are identical with Lagrange's equations

$$\frac{\mathrm{d}}{\mathrm{d}t}\left(\frac{\partial L}{\partial \dot{x}_i}\right) = \frac{\partial L}{\partial x_i}, \qquad (i = 1, 2, \ldots, N), \tag{8.36}$$

with $L = T - V$.

Exactly as in §3.7, it follows that the action integral,

$$I = \int_{t_0}^{t_1} L\,\mathrm{d}t,$$

is stationary under arbitrary small variations of the $3N$ co-ordinates $\boldsymbol{r}_i = (x_i, y_i, z_i)$ which vanish at the limits of integration. Then, again as in §3.7, we may re-express $L$ in terms of any other set of $3N$ co-ordinates,

$q_1, q_2, \ldots, q_{3N}$. The corresponding equations of motion will be Lagrange's equations

$$\frac{\mathrm{d}}{\mathrm{d}t}\left(\frac{\partial L}{\partial \dot{q}_\alpha}\right) = \frac{\partial L}{\partial q_\alpha}, \qquad (\alpha = 1, 2, \ldots, 3N). \qquad (8.37)$$

As an example, let us suppose that external forces are provided by a uniform gravitational field $\boldsymbol{g}$.

## Example: Uniform gravitational field

Show that the centre-of-mass and internal motions can be completely separated in the case of a uniform gravitational field, with two separate energy conservation laws.

The potential energy function for the uniform field is

$$V_{\mathrm{ext}} = -\sum_i m_i \boldsymbol{g} \cdot \boldsymbol{r}_i = -M\boldsymbol{g} \cdot \boldsymbol{R}.$$

Now, let us write the Lagrangian in terms of the three co-ordinates $\boldsymbol{R}$ and the co-ordinates $\boldsymbol{r}_i^*$ (of which only $3N - 3$ are independent, because of (8.17)). Then, using (8.29), we find

$$L = \tfrac{1}{2}M\dot{\boldsymbol{R}}^2 + M\boldsymbol{g} \cdot \boldsymbol{R} + \sum_i m_i \dot{\boldsymbol{r}}_i^{*2} - V_{\mathrm{int}}. \qquad (8.38)$$

Since $V_{\mathrm{int}}$ is a function only of the co-ordinate differences $\boldsymbol{r}_i - \boldsymbol{r}_j = \boldsymbol{r}_i^* - \boldsymbol{r}_j^*$, this Lagrangian function exhibits a complete separation between the terms involving $\boldsymbol{R}$ and those involving the $\boldsymbol{r}_i^*$, just as in the two-particle case of §7.1. For the case of a *uniform* gravitational field, the motion of the centre of mass and the relative motion are uncoupled. In particular, there are two separate conservation laws for energy,

$$\tfrac{1}{2}M\dot{\boldsymbol{R}}^2 - M\boldsymbol{g} \cdot \boldsymbol{R} = \mathrm{constant},$$

and

$$T^* + V_{\mathrm{int}} = \mathrm{constant}.$$

As we saw in §8.2, there is also in this case a separate conservation law for angular momentum relative to the centre of mass.

There is an important physical consequence of this separation: no observation of the *internal* motion of a system can ever reveal the presence of a uniform gravitational field, since such a field in no way affects the equations of motion for the relative co-ordinates $r_i^*$. Note that a laboratory is subjected to *other* external forces besides the Earth's gravitational field, since it is supported by the ground. Indeed if the Earth were removed, but the supporting forces somehow retained, there would be no observable difference inside the laboratory, which would of course be accelerated upwards with acceleration $g$! This was an important consideration in the argument which led to Einstein's general theory of relativity.

## 8.6   Summary

The centre of mass of any system moves like a particle of mass $M$ acted on by a force equal to the total force on the system. The contribution of this motion to the angular momentum or kinetic energy may be completely separated from the contributions of the relative motion, and $J$ or $T$ may be written as a sum of two corresponding terms. (Of course, the *only* contribution to $P$ comes from the centre-of-mass motion.)

When the internal forces are central, the rate of change of angular momentum is equal to the sum of the moments of the external forces. When they are conservative, the rate of change of the kinetic energy plus the internal potential energy is equal to the rate of working of the external forces. In both cases, the same thing is true for the motion relative to the centre of mass.

If the external forces are also central, or conservative, then the total angular momentum, or total energy (including external potential energy), respectively, are conserved. In particular, for an isolated system, $P$, $J$ and $T + V_{\text{int}}$ are all constants.

---

## Problems

1. A rocket is launched from the surface of the Earth, to reach a height of 50 km. Find the required velocity impulse, neglecting the variation of $g$ with height (and the Earth's rotation). Given that the mass of the payload and rocket without fuel is 100 kg, and the ejection velocity is $2\,\text{km}\,\text{s}^{-1}$, find the required initial mass.

2. A satellite is orbiting the Earth in a circular orbit 230 km above the equator. Calculate the total velocity impulse needed to place it in a

synchronous orbit (see Chapter 4, Problem 1), using an intermediate semi-elliptical transfer orbit which just touches both circles. (*Hint:* From the orbit parameters $a$ and $ae$, find $l$, and hence the velocities.) Given that the final mass to be placed in orbit is 30 kg, and the ejection velocity of the rocket is $2.5 \, \mathrm{km \, s^{-1}}$, find the necessary initial mass.

3. Assume that the residual mass of a rocket, without payload or fuel, is a given fraction $\lambda$ of the initial mass including fuel (but still without payload). Show that the total take-off mass required to accelerate a payload $m$ to velocity $v$ is

$$M_0 = m \, \frac{1 - \lambda}{e^{-v/u} - \lambda}.$$

If $\lambda = 0.15$, what is the upper limit to the velocity attainable with an ejection velocity of $2.5 \, \mathrm{km \, s^{-1}}$?

4. Find a formula analogous to that of Problem 3 for a two-stage rocket, in which each stage produces the same velocity impulse. (The first-stage rocket is discarded when its fuel is burnt out.) With the figures of Problem 3, what is the minimum number of stages required to reach escape velocity $(11.2 \, \mathrm{km \, s^{-1}})$? With that number, what take-off mass is required, if the payload mass is 100 kg?

5. Find the velocity impulse needed to launch the spacecraft on its trip to Jupiter, described in Chapter 4, Problem 16. (*Hint:* Use energy conservation to find the velocity at the surface of the Earth needed to give the appropriate relative velocity for the spacecraft once it has escaped.) If a three-stage rocket is used, and the parameter $\lambda$ of Problem 3 is 0.1, what is the minimum required ejection velocity? Given that $u = 2.5 \, \mathrm{km \, s^{-1}}$, and that the mass of the payload is 500 kg, find the total take-off mass.

6. Find the gain in kinetic energy when a rocket emits a small amount of matter. Hence calculate the total energy which must be supplied from chemical or other sources to accelerate the rocket to a given velocity. Show that this is equal to the energy required if an equal amount of matter is ejected while the rocket is held fixed on a test-bed.

7. *A rocket with take-off mass $M_0$ is launched vertically upward, as in Problem 1. Consider the effect of a finite burn-up time. Show that, if the rocket ejects matter at a constant rate $a$, then its height at time $t$ is

$$z = ut - \frac{uM}{a} \ln \frac{M_0}{M} - \tfrac{1}{2}gt^2, \qquad \text{with} \qquad M = M_0 - at.$$

Hence, show that if it burns out after a time $t_1$, leaving a final mass $M_1$, then [provided that $t_1 < (u/g)\ln(M_0/M_1)$], the maximum height reached is

$$z = \frac{u^2}{2g}\left(\ln\frac{M_0}{M_1}\right)^2 - ut_1\left(\frac{M_0}{M_0 - M_1}\ln\frac{M_0}{M_1} - 1\right).$$

Using the same values of $u$, $M_0$ and $M_1$ as in Problem 1, find the maximum height reached if the burn-up time is (a) 10 s, and (b) 30 s.

8. *A satellite is to be launched into a synchronous orbit directly from the surface of the Earth, using a rocket launched vertically from a point on the equator. Find the required launch speed to achieve the desired apogee, and the required velocity increment. With the same figures as in Problem 2, find the required take-off mass.

9. Two billiard balls are resting on a smooth table, and just touching. A third identical ball moving along the table with velocity $v$ perpendicular to their line of centres strikes both balls simultaneously. Find the velocities of the three balls immediately after impact, assuming that the collision is elastic.

10. A spherical satellite of radius $r$ is moving with velocity $v$ through a uniform tenuous atmosphere of density $\rho$. Find the retarding force on the satellite if each particle which strikes it (a) adheres to the surface, and (b) bounces off it elastically. Can you explain why the two answers are equal, in terms of the scattering cross-section of a hard sphere?

11. *If the orbit of the satellite of Problem 10 is highly elliptical, the retarding force is concentrated almost entirely in the lowest part of the orbit. Replace it by an impulsive force of impulse $I$ delivered once every orbit, at perigee. By considering changes in energy and angular momentum, find the changes in the parameters $a$ and $l$. Show that $\delta l = \delta a(1 - e)^2$, and hence that the effect is to decrease the period and apogee distance, while leaving the perigee distance unaffected. (The orbit therefore becomes more and more circular with time.) Show that the velocity at apogee increases, while that at perigee decreases.

12. *Suppose that the satellite of Problems 10 and 11 has achieved a circular orbit of radius $a$. Find the rates of change of energy and angular momentum, and hence show that the rates of change of $a$ and $l$ are equal, so that the orbit remains approximately circular. Show also that the velocity of the satellite must *increase*.

13. If the satellite orbit of Problem 12 is 500 km above the Earth's surface, the mass and radius of the satellite are 30 kg and 0.7 m, and

$\rho = 10^{-13}\,\text{kg}\,\text{m}^{-3}$, find the changes in orbital period and height in a year. If the height is $200\,\text{km}$ and $\rho = 10^{-10}\,\text{kg}\,\text{m}^{-3}$, find the changes in a single orbit.

14. Find the lengths of the 'day' and the 'month' (a) when the Moon was 10 Earth radii away, and (b) when the solar and lunar tides become equal in magnitude. (See Chapter 6, Problem 16.)

15. Show that in a conservative $N$-body system, a state of minimal total energy for a given total $z$-component of angular momentum is necessarily one in which the system is rotating as a rigid body about the $z$-axis. [Use the method of Lagrange multipliers (see Appendix A, Problem 11), and treat the components of the positions $r_i$ and velocities $\dot{r}_i$ as independent variables.]

16. *A planet of mass $M$ is surrounded by a cloud of small particles in orbits around it. Their mutual gravitational attraction is negligible. Due to collisions between the particles, the energy will gradually decrease from its initial value, but the angular momentum will remain fixed, $J = J_0$, say. The system will thus evolve towards a state of minimum energy, subject to this constraint. Show that the particles will tend to form a ring around the planet. [As in Problem 15, the constraint may be imposed by the method of Lagrange multipliers. In this case, because there are three components of the constraint equation, we need three Lagrange multipliers, say $\omega_x, \omega_y, \omega_z$. We have to minimize the function $E - \boldsymbol{\omega} \cdot (\boldsymbol{J} - \boldsymbol{J}_0)$ with respect to variations of the positions $r_i$ and velocities $\dot{r}_i$, and with respect to $\boldsymbol{\omega}$. Show by minimizing with respect to $\dot{r}_i$ that once equilibrium has been reached the cloud rotates as a rigid body, and by minimizing with respect to $r_i$ that all particles occupy the same orbit.] What happens to the energy lost? Why does the argument not necessarily apply to a cloud of particles around a hot star?

17. *An $N$-body system is interacting only through the gravitational forces between the bodies. Show that the potential energy function $V$ satisfies the equation

$$\sum_i r_i \cdot \nabla_i V = -V,$$

where $\boldsymbol{\nabla}_i = (\partial/\partial x_i, \partial/\partial y_i, \partial/\partial z_i)$. (*Hint*: Show that each two-body term $V_{ij}$ satisfies this equation. This condition expresses the fact that $V$ is a homogeneous function of the co-ordinates of degree $-1$.)

18. *Under the conditions of Problem 17, show that the total kinetic and potential energies $T$ and $V$ satisfy the *virial equation*,

$$2T + V = \frac{\mathrm{d}^2 K}{\mathrm{d}t^2}, \qquad \text{where} \qquad K = \tfrac{1}{2}\sum_i m_i r_i^2.$$

(Note that $K$ relates to the overall scale of the system. We may define a root-mean-square radius $r$ by $K = \tfrac{1}{2}Mr^2$.) Deduce that, if the scale of the system as measured by $K$ is, on average, neither growing nor shrinking, then the time-averaged value of the total energy is equal to minus the time-averaged value of the kinetic energy — the *virial theorem*. (Compare Chapter 4, Problem 19.)

# Chapter 9

# Rigid Bodies

The principal characteristic of a solid body is its rigidity. Under normal circumstances, its size and shape vary only slightly under stress, changes in temperature, and the like. Thus it is natural to consider the idealization of a perfectly rigid body, whose size and shape are permanently fixed. Such a body may be characterized by the requirement that the distance between any two points of the body remains fixed. In this chapter, we shall be concerned with the mechanics of rigid bodies.

## 9.1 Basic Principles

It will be convenient to simplify the notation of the previous chapter by omitting the particle label $i$ from sums over all particles in the rigid body. Thus, for example, we shall write

$$\boldsymbol{P} = \sum m\dot{\boldsymbol{r}}, \qquad \text{and} \qquad \boldsymbol{J} = \sum m\boldsymbol{r} \wedge \dot{\boldsymbol{r}},$$

in place of (8.3) and (8.11).

The motion of the centre of mass of the body is completely specified by (8.6):

$$\dot{\boldsymbol{P}} = M\ddot{\boldsymbol{R}} = \sum \boldsymbol{F}. \tag{9.1}$$

Our main interest in this chapter will be centred on the rotational motion of the rigid body. For the moment, let us assume, as we originally did in §8.3, that the internal forces are central. Then, according to (8.14),

$$\dot{\boldsymbol{J}} = \sum \boldsymbol{r} \wedge \boldsymbol{F}. \tag{9.2}$$

We shall see later that these two equations are sufficient to determine the motion completely.

One very important application of (9.1) and (9.2) should be noted. For a rigid body at rest, $\dot{\boldsymbol{r}} = \boldsymbol{0}$ for every particle, and thus both $\boldsymbol{P}$ and $\boldsymbol{J}$ vanish. Clearly, the body can remain at rest only if the right hand sides of both (9.1) and (9.2) vanish, *i.e.*, if the sum of the forces and the sum of their moments are both zero. In fact, as we shall see, this is not only a necessary, but also a sufficient, condition for equilibrium.

Under the same assumption of central internal forces, we saw in §8.5 that the internal forces do no work, so that

$$\dot{T} = \sum \dot{\boldsymbol{r}} \cdot \boldsymbol{F}. \tag{9.3}$$

This might at first sight appear to be a third independent equation. However, we shall see later that it is actually a consequence of the other two. It is of course particularly useful in the case when the external forces are conservative, since it then leads to the conservation law

$$T + V = E = \text{constant}, \tag{9.4}$$

where $V$ is the *external* potential energy, previously denoted by $V_{\text{ext}}$. (Note that there is no mention here of internal potential energy; in a rigid body, this does not change.)

The assumption that the internal forces are central is much stronger than it need be. Indeed, as we discussed in §8.3, it could not be justified from our knowledge of the internal forces in real solids, which are certainly not exclusively central, and in any case cannot be adequately described by classical mechanics. All we actually require is the validity of the basic equations (9.1) and (9.2), and it is better to regard these as basic assumptions of rigid body dynamics, whose justification lies in the fact that their consequences agree with observation. To some extent, the success of these postulates, particularly (9.2), may seem rather fortuitous. However, we shall see in Chapter 12 that it is closely related to the relativity principle discussed in Chapter 1.

## 9.2   Rotation about an Axis

Let us now apply these basic equations to a rigid body which is free to rotate only about a fixed axis, which for simplicity we take to be the $z$-axis.

We also choose the origin on this axis, positioned so that the $z$ co-ordinate of the centre of mass is zero.

In cylindrical polars, the $z$ and $\rho$ co-ordinates of every point are fixed, while the $\varphi$ co-ordinate varies according to $\dot{\varphi} = \omega$, the angular velocity of the body (not necessarily constant). We examine first the component of angular momentum about the axis of rotation. By (3.29), it is

$$J_z = \sum m\rho v_\varphi = \sum m\rho^2\omega = I\omega, \tag{9.5}$$

where

$$I = \sum m\rho^2 \tag{9.6}$$

is the *moment of inertia* about the axis. Since $I$ is obviously constant, the $z$ component of (9.2) yields

$$\dot{J}_z = I\dot{\omega} = \sum \rho F_\varphi. \tag{9.7}$$

This equation determines the rate of change of angular velocity, and may be called the *equation of motion* of the rotating body.

The condition for equilibrium here is simply that the right hand side of (9.7) vanishes, *i.e.*, that the forces have no net moment about the $z$-axis. For example, consider a rectangular *lamina* (*i.e.*, a plane, two-dimensional object) of size $a \times b$ and negligible mass, pivoted at one corner, carrying a weight $Mg$ and supported by a horizontal force $F$, as shown in Fig. 9.1. The total moment here is $bF - aMg$, so for equilibrium we require $F = (a/b)Mg$.

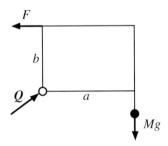

Fig. 9.1

The kinetic energy may also be expressed in terms of $I$. For

$$T = \sum \tfrac{1}{2}m(\rho\dot{\varphi})^2 = \tfrac{1}{2}I\omega^2. \tag{9.8}$$

The equation (9.3) is, therefore,

$$\dot{T} = I\omega\dot{\omega} = \sum(\rho\dot{\varphi})F_\varphi = \omega\sum\rho F_\varphi.$$

This is clearly a consequence of (9.7), and gives us no additional information.

The momentum equation (9.1) serves to determine the reaction at the
axis, which has of course no moment about the axis, and so does not appear
in (9.7). Let us denote the force on the body at the axis by $\boldsymbol{Q}$, and separate
this from all the other forces on the body. Then from (9.1)

$$\dot{\boldsymbol{P}} = M\ddot{\boldsymbol{R}} = \boldsymbol{Q} + \sum\boldsymbol{F}. \tag{9.9}$$

For example, in the case illustrated in Fig. 9.1, the equilibrium condition
$\dot{\boldsymbol{P}} = \boldsymbol{0}$ immediately yields $\boldsymbol{Q} = (F, Mg, 0)$. (Note that the $z$-axis here is
horizontal.)

The centre of mass is of course fixed in the body, so that, by (5.2),
$\dot{\boldsymbol{R}} = \boldsymbol{\omega} \wedge \boldsymbol{R}$. Differentiating again to find the acceleration of the centre of
mass, we obtain

$$\ddot{\boldsymbol{R}} = \dot{\boldsymbol{\omega}} \wedge \boldsymbol{R} + \boldsymbol{\omega} \wedge \dot{\boldsymbol{R}} = \dot{\boldsymbol{\omega}} \wedge \boldsymbol{R} + \boldsymbol{\omega} \wedge (\boldsymbol{\omega} \wedge \boldsymbol{R}). \tag{9.10}$$

The first term is the tangential acceleration, $R\dot{\omega}$ in the $\varphi$ direction, and the
second the radial acceleration, $-\omega^2 R$ in the $\rho$ direction. (Compare (3.47).)
Together, equations (9.9) and (9.10) serve to determine $\boldsymbol{Q}$.

As a simple example, let us consider a *compound pendulum* — a rigid
body pivoted about a horizontal axis, and moving under gravity.

## *Example:* Compound pendulum

Find the period of small oscillations of a compound pendulum,
and the reaction at the pivot as a function of angle.

To be consistent, we shall still take the $z$-axis to be the axis of
rotation, and choose the $x$-axis vertically downward. (See Fig. 9.2.)
Then the force (the weight) acts at the centre of mass, and has the
components $\boldsymbol{F} = (Mg, 0, 0)$. Thus the equation of motion (9.7) is

$$I\ddot{\varphi} = -MgR\sin\varphi. \tag{9.11}$$

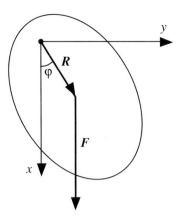

Fig. 9.2

This is identical with the equation of motion of a simple pendulum of length $l = I/MR$. (This is the length of the *equivalent simple pendulum*) Thus the period of small oscillations is $2\pi\sqrt{l/g} = 2\pi\sqrt{I/MgR}$. As we shall see, $l$ is always larger than the distance $R$ from the pivot to the centre of mass.

The energy conservation equation is

$$T + V = \tfrac{1}{2}I\dot\varphi^2 - MgR\cos\varphi = E = \text{constant}. \tag{9.12}$$

It may be obtained from (9.11) by multiplying by $\dot\varphi$ and integrating. As usual, $E$ is determined by the initial conditions, and (9.12) then serves to fix the angular velocity $\dot\varphi$ for any given inclination $\varphi$.

The components of the reaction $\boldsymbol{Q}$, obtained from (9.9), are

$$Q_z = 0$$
$$Q_\rho = -Mg\cos\varphi - MR\dot\varphi^2,$$
$$Q_\varphi = Mg\sin\varphi + MR\ddot\varphi.$$

Thus, substituting for $\ddot\varphi$ in terms of $\varphi$ using (9.11), and for $\dot\varphi$ in terms of $\varphi$ by (9.12), we find

$$Q_z = 0$$
$$Q_\rho = -Mg(1 + 2R/l)\cos\varphi - 2E/l,$$
$$Q_\varphi = Mg(1 - R/l)\sin\varphi.$$

Note, however, that this result is *not* the same as the reaction at the pivot of a simple pendulum of length $l$.

As a rather different illustration of the use of these equations, let us consider the effect of a sudden blow.

### *Example:* Effect of sudden blow

Suppose that the pendulum is initially in equilibrium and is given a sharp horizontal blow of impulse $K$ at a distance $d$ below the axis (see Fig. 9.3). For what value of $d$ does the impulsive reaction at the pivot vanish?

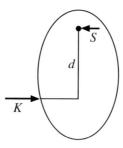

Fig. 9.3

From (9.11) integrated over a short time interval, we learn that the angular velocity $\omega$ just after the blow is given by $I\omega = dK$. As before, (9.9) serves to fix the reaction at the axis, in this case the impulsive reaction, $S$ say. Since the velocity of the centre of mass immediately after the blow is $\omega R$, the integrated form of (9.9) is

$$M\omega R = -S + K,$$

whence

$$S = \left(1 - \frac{MdR}{I}\right)K.$$

The interesting thing about this result is that it is possible to make $S$ vanish by choosing $d$ equal to the length of the equivalent simple pendulum, $d = I/MR = l$. This is the ideal point at which to hit a ball with a bat (sometimes called the 'sweet spot' in sporting applications — cricket, tennis, baseball, etc.).

## 9.3 Perpendicular Components of Angular Momentum

We now consider the remaining components of the angular momentum vector. We shall see that they give us further information about the reaction on the axis.

In Cartesian co-ordinates, the velocity of a point $r$ of the body is given by

$$\dot{x} = -\omega y, \qquad \dot{y} = \omega x, \qquad \dot{z} = 0. \tag{9.13}$$

Thus we find

$$J_x = \sum m(-z\dot{y}) = -\sum mxz\omega,$$
$$J_y = \sum m(z\dot{x}) \ \ = -\sum myz\omega.$$

We can write all three components of $\boldsymbol{J}$ in the form

$$J_x = I_{xz}\omega, \qquad J_y = I_{yz}\omega, \qquad J_z = I_{zz}\omega, \tag{9.14}$$

where

$$I_{xz} = -\sum mxz, \quad I_{yz} = -\sum myz, \quad I_{zz} = \sum m(x^2 + y^2). \tag{9.15}$$

Here $I_{zz}$ is the moment of inertia about the $z$-axis, previously denoted by $I$. The quantities $I_{xz}$ and $I_{yz}$ are called *products of inertia*.

At first sight it may seem surprising that $\boldsymbol{J}$ has components in directions perpendicular to $\omega$. A simple example may help to clarify the reason for this. Consider a light rigid rod with equal masses $m$ at its two ends, rigidly fixed at an angle $\theta$ to an axis through its mid-point. (See Fig. 9.4.) If the positions of the masses are $r$ and $-r$, the total angular momentum is

$$\boldsymbol{J} = m\boldsymbol{r} \wedge \dot{\boldsymbol{r}} + m(-\boldsymbol{r}) \wedge (-\dot{\boldsymbol{r}}) = 2m\boldsymbol{r} \wedge (\omega \wedge \boldsymbol{r}).$$

Clearly, $\boldsymbol{J}$ is perpendicular to $r$, as shown in the figure. When the rod is in the $xz$-plane, the masses are moving in the $\pm y$ directions, and there is a component of angular momentum about the $x$-axis as well as one about the $z$-axis.

In this example, the centre of mass lies on the axis. Thus if no external force is applied, the total reaction on the axis is zero. There is, however, a resultant *couple* on the axis (here represented by the forces $\boldsymbol{F}'$), which is required to balance the couple produced by the centrifugal forces. The magnitude of this couple may be determined from the remaining pair of the

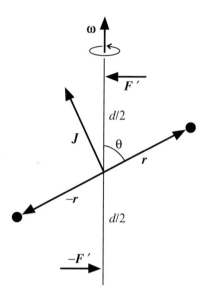

Fig. 9.4

equations (9.2). From (9.15), we see that, unlike $I_{zz}$ the products of inertia are not constants, in general. Using (9.13), we find

$$\dot{I}_{xz} = -\omega I_{yz}, \qquad \dot{I}_{yz} = \omega I_{xz}, \qquad \dot{I}_{zz} = 0.$$

Thus if $G$ is the couple on the axis, we have from (9.2)

$$\dot{J}_x = I_{xz}\dot{\omega} - I_{yz}\omega^2 = G_x + \sum(yF_z - zF_y),$$
$$\dot{J}_y = I_{yz}\dot{\omega} + I_{xz}\omega^2 = G_y + \sum(zF_x - xF_z).$$

If there are no other external forces, then $\omega$ is a constant, and $G$ precisely balances the centrifugal couple. When the rod is in the $xz$-plane, the only non-vanishing component of $G$ is the moment about the $y$-axis,

$$G_y = dF'_x = I_{xz}\omega^2 = -2mr^2\omega^2 \sin\theta \cos\theta.$$

This determines $F'$.

## 9.4 Principal Axes of Inertia

We have seen that, in general, the angular momentum vector $\boldsymbol{J}$ is in a different direction from the angular velocity vector $\boldsymbol{\omega}$. There are special cases, however, in which the products of inertia $I_{xz}$ and $I_{yz}$ vanish. Then $\boldsymbol{J}$ is also in the $z$ direction. In that case, the $z$-axis is called a *principal axis of inertia*. When a body is rotating freely about a principal axis through its centre of mass, there is no resultant force or couple on the axis.

The $z$-axis is a principal axis, in particular, if the $xy$-plane is a plane of reflection symmetry; for then the contributions to the products of inertia $I_{xz}$ and $I_{yz}$ from any point $(x, y, z)$ is exactly cancelled by that from $(x, y, -z)$. Similarly, it is a principal axis if it is an axis of rotational symmetry, for then the contribution from $(x, y, z)$ is cancelled by that from $(-x, -y, z)$.

For bodies with three symmetry axes — such as a rectangular parallelepiped or an ellipsoid — there are obviously three perpendicular principal axes, and we shall see later that this is true more generally. In this case, it is clearly an advantage to choose these as our co-ordinate axes. So we shall no longer assume that the axis of rotation is the $z$-axis, but take it to have an arbitrary inclination. Then $\boldsymbol{\omega}$ has three components $(\omega_x, \omega_y, \omega_z)$.

The angular momentum $\boldsymbol{J}$ may be written, using (A.16), in the form

$$\boldsymbol{J} = \sum m\boldsymbol{r} \wedge (\boldsymbol{\omega} \wedge \boldsymbol{r}) = \sum m[r^2\boldsymbol{\omega} - (\boldsymbol{r} \cdot \boldsymbol{\omega})\boldsymbol{r}]. \qquad (9.16)$$

Its components are therefore linear functions of the components of $\boldsymbol{\omega}$, which we may write in matrix notation as

$$\begin{bmatrix} J_x \\ J_y \\ J_z \end{bmatrix} = \begin{bmatrix} I_{xx} & I_{xy} & I_{xz} \\ I_{yx} & I_{yy} & I_{yz} \\ I_{zx} & I_{zy} & I_{zz} \end{bmatrix} \begin{bmatrix} \omega_x \\ \omega_y \\ \omega_z \end{bmatrix}, \qquad (9.17)$$

that is, $J_x = I_{xx}\omega_x + I_{xy}\omega_y + I_{xz}\omega_z$, and two similar equations. The nine elements $I_{xx}, I_{xy}, \ldots, I_{zz}$ of the $3 \times 3$ matrix may be regarded as the components of a single entity $\mathsf{I}$, in much the same way as the three quantities $J_x, J_y, J_z$ are regarded as the components of the vector $\boldsymbol{J}$. The entity $\mathsf{I}$ is called a *tensor*, in this case the *inertia tensor*. The elementary properties of tensors are described in §A.9. However, we shall need to use only one result from this general theory.

It is obvious from the definition (9.15) that the products of inertia satisfy symmetry relations like $I_{xy} = I_{yx}$. The $3 \times 3$ matrix in (9.17) is therefore unchanged by reflection in the leading diagonal. A tensor $\mathsf{I}$ with this property is called *symmetric*.

If the three co-ordinate axes are all axes of symmetry, then all the products of inertia vanish, and $\mathbf{I}$ has the diagonal form

$$\begin{bmatrix} I_{xx} & 0 & 0 \\ 0 & I_{yy} & 0 \\ 0 & 0 & I_{zz} \end{bmatrix}.$$

In this case, the relations (9.17) simplify to

$$J_x = I_{xx}\omega_x, \qquad J_y = I_{yy}\omega_y, \qquad J_z = I_{zz}\omega_z. \tag{9.18}$$

Thus $\mathbf{J}$ is parallel to $\boldsymbol{\omega}$ if $\boldsymbol{\omega}$ is along any one of the three symmetry axes, but not in general otherwise.

It is shown in §A.10 that for *any* given symmetric tensor one can always find a set of axes with respect to which it is diagonal. Thus for any rigid body, we can find three perpendicular axes through any given point which are principal axes of inertia. It will be convenient to introduce three unit vectors $\mathbf{e}_1, \mathbf{e}_2, \mathbf{e}_3$ along these axes. Then, if we write

$$\boldsymbol{\omega} = \omega_1 \mathbf{e}_1 + \omega_2 \mathbf{e}_2 + \omega_3 \mathbf{e}_3, \tag{9.19}$$

the components of $\mathbf{J}$ in these three directions will be obtained by multiplying the components of $\boldsymbol{\omega}$ by the appropriate moments of inertia, as in (9.18). Thus we obtain

$$\mathbf{J} = I_1\omega_1 \mathbf{e}_1 + I_2\omega_2 \mathbf{e}_2 + I_3\omega_3 \mathbf{e}_3. \tag{9.20}$$

The three diagonal elements of the inertia tensor, $I_1, I_2, I_3$ are called *principal moments of inertia*. We shall always use a single subscript for the principal moments, to distinguish them from moments of inertia about arbitrary axes.

It is important to realize that the principal axes are fixed in the body, not in space, and therefore rotate with it. It is often convenient to use these axes to define our frame of reference, particularly since the principal moments $I_1, I_2, I_3$ are constants. This is, however, a rotating frame, not an inertial one.

The kinetic energy may also be expressed in terms of the angular velocity and the inertia tensor. We have

$$T = \sum \tfrac{1}{2} m \dot{\mathbf{r}}^2 = \sum \tfrac{1}{2} m (\boldsymbol{\omega} \wedge \mathbf{r})^2 = \sum \tfrac{1}{2} m [\omega^2 r^2 - (\boldsymbol{\omega} \cdot \mathbf{r})^2],$$

by a standard formula of vector algebra (see Appendix A, Problem 7). Comparing with (9.16), we see that

$$T = \tfrac{1}{2}\boldsymbol{\omega} \cdot \boldsymbol{J}. \tag{9.21}$$

(Equations (9.5) and (9.8) provide a special case of this general result.) Thus from (9.19) and (9.20), we find

$$T = \tfrac{1}{2}I_1\omega_1^2 + \tfrac{1}{2}I_2\omega_2^2 + \tfrac{1}{2}I_3\omega_3^2. \tag{9.22}$$

These equations for rotational motion may be compared with the corresponding ones for translational motion with velocity $\boldsymbol{v}$, namely $T = \tfrac{1}{2}Mv^2 = \tfrac{1}{2}\boldsymbol{v}\cdot\boldsymbol{P}$. The principal difference is that the mass, unlike the inertia tensor, has no directional properties, so that the coefficients of $v_x^2, v_y^2$ and $v_z^2$ are all equal.

### Symmetric bodies

The term *symmetric* applied to a rigid body has a special, technical significance. It means that two of the principal moments of inertia coincide, say $I_1 = I_2$. This is true in particular if the third axis $\boldsymbol{e}_3$ is a symmetry axis of appropriate type. It may be an axis of cylindrical symmetry — for example, the axis of a spheroid or a circular cylinder or cone. More generally, $I_1 = I_2$ if $\boldsymbol{e}_3$ is an axis of more than two-fold rotational symmetry — for example the axis of an equilateral triangular prism, or a square pyramid. (In the case of the prism, this is not entirely obvious. It follows from the fact that one cannot find a pair of perpendicular axes $\boldsymbol{e}_1$ and $\boldsymbol{e}_2$ which are in any way distinguished from other possible pairs.)

For a symmetric body, (9.20) becomes

$$\boldsymbol{J} = I_1(\omega_1\boldsymbol{e}_1 + \omega_2\boldsymbol{e}_2) + I_3\omega_3\boldsymbol{e}_3. \tag{9.23}$$

Note that if $\omega_3 = 0$, then $\boldsymbol{J} = I_1\boldsymbol{\omega}$, so $\boldsymbol{J}$ is parallel to $\boldsymbol{\omega}$. In other words, *any* axis in the plane of $\boldsymbol{e}_1$ and $\boldsymbol{e}_2$ is a principal axis. Assuming that $I_3 \neq I_1$, the symmetry axis $\boldsymbol{e}_3$ is of course uniquely determined, but the axes $\boldsymbol{e}_1$ and $\boldsymbol{e}_2$ are not. We may choose them to be any pair of orthogonal axes in the plane normal to $\boldsymbol{e}_3$. Indeed, they need not even be fixed in the body, so long as they always remain perpendicular to $\boldsymbol{e}_3$ and to each other. Equation (9.23) will still hold so long as this condition is satisfied. This freedom of choice will prove to be very useful later.

It can happen that all three principal moments of inertia are equal, $I_1 = I_2 = I_3$. This is the case for a sphere (with origin at the centre), a cube, or a regular tetrahedron (or, indeed, any of the five regular solids). It may of course also happen by accidental coincidence of the three values. We may call such bodies *totally symmetric*. They have the property that $\boldsymbol{J} = I_1\boldsymbol{\omega}$ whatever direction $\boldsymbol{\omega}$ may take, so that every axis is a principal axis, and the choice of the orthonormal triad $\boldsymbol{e}_1, \boldsymbol{e}_2, \boldsymbol{e}_3$ is completely arbitrary.

## 9.5   Calculation of Moments of Inertia

The moments and products of inertia of any body with respect to a given origin may be calculated from the definitions (9.15). For continuous distributions of matter, we must replace sums by integrals:

$$I_{xx} = \iiint \rho(\boldsymbol{r})(y^2 + z^2)\, \mathrm{d}^3\boldsymbol{r}, \qquad I_{xy} = \iiint \rho(\boldsymbol{r})(-xy)\, \mathrm{d}^3\boldsymbol{r}, \qquad (9.24)$$

*etc.*, where $\rho(\boldsymbol{r})$ is the density.

### Shift of origin

When the rigid body is pivoted so that one point is fixed, it is convenient to choose that point to be the origin. If there is no fixed point, we generally choose the origin to be the centre of mass. Thus it is useful to be able to relate the moments and products of inertia about an arbitrary point to those about the centre of mass.

As usual, we shall distinguish quantities referred to the centre of mass as origin by an asterisk. To find the desired relations, we substitute in (9.15) the relation (8.16), $\boldsymbol{r} = \boldsymbol{R} + \boldsymbol{r}^*$, and use the conditions (8.17), namely

$$\sum mx^* = \sum my^* = \sum mz^* = 0.$$

Because of these relations, the cross terms between $\boldsymbol{R}$ and $\boldsymbol{r}^*$ drop out, exactly as they did in (8.18) or (8.29). For example,

$$I_{xy} = -\sum m(X + x^*)(Y + y^*) = -MXY - \sum mx^* y^*.$$

The last term here is just the product of inertia $I_{xy}^*$ referred to the centre of mass as origin. Thus we obtain relations of the form

$$I_{xx} = M(Y^2 + Z^2) + I_{xx}^*, \qquad I_{xy} = -MXY + I_{xy}^*, \qquad (9.25)$$

*etc.* Note that, as in previous cases, the components of the inertia tensor with respect to an arbitrary origin are obtained from those with respect to the centre of mass by adding the contribution of a particle of mass $M$ at $\mathbf{R}$. (For the moments of inertia, this is known as the *parallel axes theorem*.) Because of this result, it is only necessary, for any given body, to compute the moments and products of inertia with respect to the centre of mass; those with respect to any other origin are then given by (9.25).

It is important to realize that the principal axes at a given origin are not necessarily parallel to those at the centre of mass. If we choose the axes at the centre of mass to be principal axes, then the products of inertia $I^*_{xy}, I^*_{xz}, I^*_{yz}$ will all be zero, but it is clear from (9.25) that this does not necessarily imply that $I_{xy}, I_{xz}, I_{yz}$ are zero. In fact, this will be true if, and only if, the chosen origin lies on one of the principal axes through the centre of mass, so that two of the three co-ordinates $X, Y, Z$ are zero.

### Routh's rule

There is a simple and convenient formula that allows us to write down at once the moments of inertia of a large class of bodies. We consider bodies with uniform density $\rho$ and three perpendicular symmetry planes. The principal axes are then obvious, and we shall take them to be the co-ordinate axes.

From (9.24), we see that the principal moments of inertia may be written in the form

$$I^*_1 = K_y + K_z, \qquad I^*_2 = K_x + K_z, \qquad I^*_3 = K_x + K_y,$$

where, for example,

$$K_z = \iiint_V \rho z^2 \, \mathrm{d}x \, \mathrm{d}y \, \mathrm{d}z,$$

integrated over the volume $V$ of the body. The mass of the body is of course

$$M = \iiint_V \rho \, \mathrm{d}x \, \mathrm{d}y \, \mathrm{d}z.$$

Now let us denote the lengths of the three co-ordinate axes by $2a, 2b, 2c$, respectively, and consider together all bodies of the same type (*e.g.*, ellipsoids, or rectangular parallelepipeds) but with different values of $a, b, c$. It is then easy to find how $M$ and $K_z$ depend on these three lengths. Making the substitution $x = a\xi, y = b\eta, z = c\zeta$, we see that

$$M = \rho abc \iiint_{V_0} \mathrm{d}\xi \, \mathrm{d}\eta \, \mathrm{d}\zeta,$$

where $V_0$ is a standard body of this type with $\rho = 1$ and $a = b = c = 1$. Hence $M \propto \rho abc$. In a similar way, $K_z \propto \rho abc^3$. It follows that $K_z = \lambda_z M c^2$, where $\lambda_z$ is a dimensionless number, the same for all bodies of this type. Hence we obtain *Routh's rule*, which asserts that

$$
\begin{aligned}
I_1^* &= M(\lambda_y b^2 + \lambda_z c^2), \\
I_2^* &= M(\lambda_x a^2 + \lambda_z c^2), \\
I_3^* &= M(\lambda_x a^2 + \lambda_y b^2).
\end{aligned}
\tag{9.26}
$$

The values of the constants $\lambda_x, \lambda_y, \lambda_z$ may be found by examining the standard volume $V_0$. For ellipsoids, for example, $V_0$ is a sphere of unit radius, for which

$$
K_{z0} = \iiint_{V_0} \zeta^2 \, d\xi \, d\eta \, d\zeta = \int_{-1}^{1} \zeta^2 \pi (1 - \zeta^2) \, d\zeta = \frac{4\pi}{15}.
$$

Since also $M_0 = 4\pi/3$, we have $\lambda_z = K_{z0}/M_0 = \frac{1}{5}$. Thus for an *ellipsoid*

$$
\lambda_x = \lambda_y = \lambda_z = \frac{1}{5}. \qquad \text{(ellipsoid)}
$$

A similar calculation shows that for a *rectangular parallelepiped*

$$
\lambda_x = \lambda_y = \lambda_z = \frac{1}{3}. \qquad \text{(parallelepiped)}
$$

As an example of a body for which the three constants are not all equal, we may quote the *elliptic cylinder* of length $2c$, for which

$$
\lambda_x = \lambda_y = \frac{1}{4} \qquad \lambda_z = \frac{1}{3}. \qquad \text{(cylinder)}
$$

All three examples may be summarized by saying that $\lambda$ is $1/3$ for 'rectangular' axes, $1/4$ for 'elliptic' ones, and $1/5$ for 'ellipsoidal' ones.

This formula covers several special cases of particular interest. A *sphere* is of course an ellipsoid with $a = b = c$, and each of its three principal moments of inertia is $\frac{2}{5} M a^2$. Similarly a *cube* is a parallelepiped with $a = b = c$, and each of its principal moments is $\frac{2}{3} M a^2$. A parallelepiped with $c = 0$ is a flat rectangular plate. Its principal moments are

$$
I_1^* = \frac{1}{3} M b^2, \qquad I_2^* = \frac{1}{3} M a^2, \qquad I_3^* = \frac{1}{3} M (a^2 + b^2).
$$

Similarly, a flat circular plate is a cylinder with $a = b$ and $c = 0$. In that case,

$$I_1^* = I_2^* = \frac{1}{4}Ma^2, \qquad I_3^* = \frac{1}{2}Ma^2.$$

Finally, a thin rod is a limiting case of either a parallelepiped or a cylinder with $a = b = 0$. Its principal moments are

$$I_1^* = I_2^* = \frac{1}{3}Mc^2, \qquad I_3^* = 0.$$

## 9.6    Effect of a Small Force on the Axis

So far we have assumed that the axis about which the rigid body is rotating is fixed. We now go on to consider how it moves if only one point on the axis is fixed. As we shall see, rapidly rotating bodies have great stability, which forms the basis, for example, of the gyroscope.

We suppose that the rigid body is free to rotate about a fixed smooth pivot, and that initially it is rotating freely about a principal axis, say $e_3$. Then, if the angular velocity is $\boldsymbol{\omega} = \omega e_3$, the angular momentum will be $\boldsymbol{J} = I_3 \omega e_3$. So long as no external force acts on the body the angular momentum equation (9.2) tells us that

$$\dot{\boldsymbol{J}} = I_3 \dot{\boldsymbol{\omega}} = \boldsymbol{0},$$

so the axis will remain fixed in space and the angular velocity will be constant. (This would not be true if the axis of rotation were not a principal axis. We shall see what happens in that case later.)

Now suppose that a small force $\boldsymbol{F}$ is applied to the axis at a point $\boldsymbol{r}$ (see Fig. 9.5). Then the equation of motion becomes

$$\dot{\boldsymbol{J}} = \boldsymbol{r} \wedge \boldsymbol{F}. \qquad (9.27)$$

This force will cause the axis to change direction, and the body will therefore acquire a small component of angular velocity perpendicular to its axis $e_3$. However, provided the force is small enough, the angular velocity with which the axis moves will be small in comparison to the angular velocity of rotation about the axis. In that case, we may neglect the angular

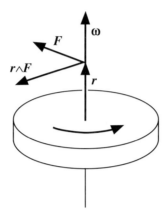

Fig. 9.5

momentum components normal to the axis, and again write

$$\dot{\boldsymbol{J}} = I_3\dot{\boldsymbol{\omega}} = \boldsymbol{r} \wedge \boldsymbol{F}. \tag{9.28}$$

Since the moment $\boldsymbol{r} \wedge \boldsymbol{F}$ is perpendicular to $\boldsymbol{\omega}$, the magnitude of the angular velocity does not change. (Recall that $\mathrm{d}(\omega^2)/\mathrm{d}t = 2\boldsymbol{\omega} \cdot \dot{\boldsymbol{\omega}}$.) However, its direction does change, and the axis will move in the direction of $\boldsymbol{r} \wedge \boldsymbol{F}$, that is *perpendicular* to the applied force.

The effect may be rather surprising at first sight. However, it is easy to see it in action in, for example, a freely rolling bicycle wheel (see Fig. 9.6). If the wheel leans to the right, it will not necessarily fall over, as it would if it were not rotating, but will turn to the right. The force $\boldsymbol{F}$ in this case is the force of gravity, which produces a moment about a horizontal axis, as shown in Fig. 9.6. Thus, in a small time interval $\mathrm{d}t$, the angular velocity $\boldsymbol{\omega}$ of a wheel acquires a small additional horizontal component, $\mathrm{d}\boldsymbol{\omega}$. This shows that the axis of the wheel must change direction.

When turning a corner on a bicycle, this effect plays a role, but there are other more important effects, related to the way the handlebars and front fork are pivoted. (See Jones, *Physics Today*, **23**, 34–40, 1970 and Fajans, *American Journal of Physics*, **68**, 654–659, 2000.)

Another familiar example is the spinning top, or toy gyroscope (which we shall discuss in more detail in Chapters 10 and 12). The force in this case is again the force of gravity, $\boldsymbol{F} = -Mg\boldsymbol{k}$, acting at the centre of mass position, $\boldsymbol{R} = R\boldsymbol{e}_3$ (see Fig. 9.7). Thus (9.28) yields

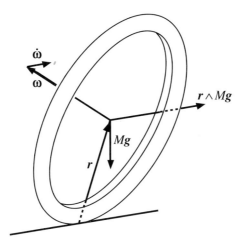

Fig. 9.6

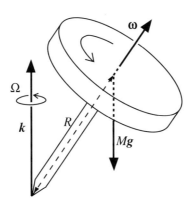

Fig. 9.7

$$I_3 \omega \dot{e}_3 = -MgRe_3 \wedge k.$$

This equation is of a form we have encountered several times already. It may be written

$$\dot{e}_3 = \Omega \wedge e_3,$$

where

$$\Omega = \frac{MgR}{I_3\omega}\mathbf{k}. \qquad (9.29)$$

Thus (see §5.1) it describes a vector $\mathbf{e}_3$ rotating with constant angular velocity $\Omega$ about the vertical. The effect is similar to the Larmor precession (see §5.6) or the precession of a satellite orbit (see §6.5). Note that the angular velocity $\Omega$ is independent of the inclination of the axis.

Our treatment here is valid only if $\Omega \ll \omega$, or, equivalently, if $MgR \ll I_3\omega^2$. In other words, we require the rotational kinetic energy $(T = \frac{1}{2}I_3\omega^2)$ to be much greater than the possible change in gravitational potential energy (of order $MgR$). We shall discuss what happens when this condition is not satisfied later (see §10.3 and §12.4).

Note that the precessional angular velocity (9.29) is inversely proportional both to the moment of inertia and the rotational angular velocity $\omega$. Thus, to minimize the effect of a given force, we should use a fat, rapidly spinning body.

The great stability of rapidly rotating bodies is the basis of the *gyroscope*. Essentially, this consists of a spinning body suspended in such a way that its axis is free to rotate relative to its support. The bearings are made as nearly frictionless as possible, to minimize the torques on the gyroscope. Then, no matter how we turn the support, the axis of the gyroscope will remain pointing very closely to the same direction in space.

The gyroscope is particularly useful for navigational instruments in aircraft. It may be employed, for example, to provide an 'artificial horizon' which allows the pilot to fly a level course even in cloud (where our normal sense of balance is notoriously unreliable). A similar instrument, designed to show the orientation in a horizontal plane, serves as a direction indicator.

## Precession of the equinoxes

The non-spherical shape of the Earth leads to another example of the phenomenon of precession. The effect is closely related to the precession of a satellite orbit discussed in §6.5. Since the Earth exerts a moment on the satellite, the satellite exerts an equal and opposite moment on the Earth. For a small satellite, this is of course quite negligible, but a similar moment is exerted by both the Sun and the Moon, and, although the effect is still quite small, it leads to observable changes over long periods of time.

Let us consider for example the effect of the Sun. Because it attracts the nearer equatorial bulge more strongly that the farther one, it exerts a couple on the Earth's axis, in a direction which would tend (if the Earth were not rotating) to align it with the normal to the plane of the Earth's orbit (called the *ecliptic* plane, because it is the plane in which eclipses can occur.) (See Fig. 9.8.) However, because of the Earth's rotation, the

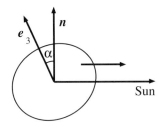

Fig. 9.8

actual effect is to make the axis precess around the normal to the ecliptic plane. Since the equator is inclined to the ecliptic at about 23.45°, the axis therefore describes a cone in space with semi-vertical angle 23.45°.

It turns out that this effect depends on the same parameter, (6.41) or (6.42), that appeared in the discussion of the tides. Thus, as in that case, the effect of the Moon is rather more than twice as large as that of the Sun (though somewhat complicated by the fact that the Moon's orbital plane is not fixed in space). The two bodies together lead to a precessional angular velocity of about 50″ per year, which corresponds to a precessional period of 26 000 years. The effect means that the pole of the Earth's axis moves relative to the fixed stars, so that 13 000 years ago our present pole star was some 47° away from the pole. (The irregularity of the Moon's motion, which is due to the moment exerted by the Sun on the Earth–Moon system, leads to a superimposed wobble of the axis with an amplitude of 9″ and a period of 18.6 years.)

The equinoxes occur when the apparent direction of the Sun crosses the equator. The points on the ecliptic where this happens also precess with the 26 000-year period, which is why this phenomenon is known as 'precession of the equinoxes'.

## 9.7   Instantaneous Angular Velocity

We shall now turn to the problem of the general motion of a rigid body, without assuming that it is spinning about a principal axis, or that the forces acting are small. In this section, we shall discuss the concept of angular velocity, and show that, no matter how the body is moving, it is always possible to define an *instantaneous* angular velocity vector $\boldsymbol{\omega}$ (which in general is constant in neither direction nor magnitude).

We consider first a rigid body free to rotate about a fixed pivot. Then the position of every point of the body is fixed if we specify the directions of the three principal axes $\boldsymbol{e}_1, \boldsymbol{e}_2, \boldsymbol{e}_3$. (This would be true for any three axes fixed in the body, but it will be convenient later if we choose them to be principal axes.)

Since the position of any point of the body is fixed relative to these axes, its velocity is determined by the three velocities $\dot{\boldsymbol{e}}_1, \dot{\boldsymbol{e}}_2, \dot{\boldsymbol{e}}_3$. In fact, if the point in question is

$$\boldsymbol{r} = r_1 \boldsymbol{e}_1 + r_2 \boldsymbol{e}_2 + r_3 \boldsymbol{e}_3, \tag{9.30}$$

then $r_1, r_2, r_3$ are constants, and

$$\dot{\boldsymbol{r}} = r_1 \dot{\boldsymbol{e}}_1 + r_2 \dot{\boldsymbol{e}}_2 + r_3 \dot{\boldsymbol{e}}_3. \tag{9.31}$$

We now wish to show that, no matter how the body is moving, we can always define an instantaneous angular velocity $\boldsymbol{\omega}$. Let us define the nine quantities

$$a_{ij} = \boldsymbol{e}_i \cdot \dot{\boldsymbol{e}}_j.$$

Now, since $\boldsymbol{e}_1$ is a unit vector, $\boldsymbol{e}_1^2 = 1$. Hence, differentiating, we obtain

$$0 = \frac{\mathrm{d}}{\mathrm{d}t}(\boldsymbol{e}_1^2) = 2\boldsymbol{e}_1 \cdot \dot{\boldsymbol{e}}_1 = 2a_{11}.$$

It follows that $\dot{\boldsymbol{e}}_1$ has the form

$$\dot{\boldsymbol{e}}_1 = a_{21}\boldsymbol{e}_2 + a_{31}\boldsymbol{e}_3. \tag{9.32}$$

Moreover, since $\boldsymbol{e}_1$ and $\boldsymbol{e}_2$ are always perpendicular, we have

$$0 = \frac{\mathrm{d}}{\mathrm{d}t}(\boldsymbol{e}_1 \cdot \boldsymbol{e}_2) = \dot{\boldsymbol{e}}_1 \cdot \boldsymbol{e}_2 + \boldsymbol{e}_1 \cdot \dot{\boldsymbol{e}}_2 = a_{21} + a_{12}.$$

Thus we can define a vector $\boldsymbol{\omega}$ by setting

$$\omega_1 = -a_{23} = a_{32},$$
$$\omega_2 = -a_{31} = a_{13}, \qquad (9.33)$$
$$\omega_3 = -a_{12} = a_{21}.$$

With this definition, (9.32) becomes simply

$$\dot{\boldsymbol{e}}_1 = \omega_3 \boldsymbol{e}_2 - \omega_2 \boldsymbol{e}_3 = \boldsymbol{\omega} \wedge \boldsymbol{e}_1, \qquad (9.34)$$

where of course

$$\boldsymbol{\omega} = \omega_1 \boldsymbol{e}_1 + \omega_2 \boldsymbol{e}_2 + \omega_3 \boldsymbol{e}_3$$

is the instantaneous angular velocity. Substituting this, and two similar equations, into (9.31), we find that the velocity of any point in the body is given by

$$\dot{\boldsymbol{r}} = \boldsymbol{\omega} \wedge \boldsymbol{r}. \qquad (9.35)$$

Thus we have shown that an instantaneous angular velocity vector $\boldsymbol{\omega}$ always exists. We note that the direction of $\boldsymbol{\omega}$ is the instantaneous axis of rotation: for points $\boldsymbol{r}$ lying on this axis, the instantaneous velocity $\dot{\boldsymbol{r}}$ is zero.

Now let us consider the general case, in which no point of the body is permanently fixed. Then we may specify the position of the body by the position $\boldsymbol{R}$ of its centre of mass (or of any other designated point), and by the orientations of the principal axes at that point. We can now define, as before, an angular velocity vector $\boldsymbol{\omega}$ relative to the centre of mass. If $\dot{\boldsymbol{R}}$ is the velocity of the centre of mass, then the instantaneous velocity of a point whose position relative to the centre of mass is $\boldsymbol{r}^*$ is

$$\dot{\boldsymbol{r}} = \dot{\boldsymbol{R}} + \dot{\boldsymbol{r}}^*.$$

But, since $\boldsymbol{r}^*$ is fixed in the body, we can apply the same argument as above to define an instantaneous angular velocity vector $\boldsymbol{\omega}$ and write

$$\dot{\boldsymbol{r}} = \dot{\boldsymbol{R}} + \boldsymbol{\omega} \wedge \boldsymbol{r}^*. \qquad (9.36)$$

We have not attached an asterisk to $\boldsymbol{\omega}$, because the angular velocity is in fact independent of the choice of origin. For example, if the body is rotating with angular velocity $\boldsymbol{\omega}$ about a fixed pivot, then

$$\dot{\boldsymbol{r}} = \boldsymbol{\omega} \wedge \boldsymbol{r} = \boldsymbol{\omega} \wedge \boldsymbol{R} + \boldsymbol{\omega} \wedge \boldsymbol{r}^*.$$

But the instantaneous velocity of the centre of mass is of course $\dot{\boldsymbol{R}} = \boldsymbol{\omega} \wedge \boldsymbol{R}$, whence subtracting from (9.36), we find

$$\dot{\boldsymbol{r}}^* = \boldsymbol{\omega} \wedge \boldsymbol{r}^*,$$

which shows that $\boldsymbol{\omega}$ is also the angular velocity relative to the centre of mass.

The equation of motion of a rigid body rotating about a fixed pivot is the angular momentum equation (9.2),

$$\dot{\boldsymbol{J}} = \sum \boldsymbol{r} \wedge \boldsymbol{F}, \tag{9.37}$$

where $\boldsymbol{J}$ is related to $\boldsymbol{\omega}$ by (9.20):

$$\boldsymbol{J} = I_1 \omega_1 \boldsymbol{e}_1 + I_2 \omega_2 \boldsymbol{e}_2 + I_3 \omega_3 \boldsymbol{e}_3.$$

As in §9.2, the momentum equation (9.1) serves to determine the reaction at the pivot.

In the general case, we have to determine both the velocity of the centre of mass and the angular velocity. The motion of the centre of mass is described by the momentum equation (9.1):

$$\dot{\boldsymbol{P}} = M\ddot{\boldsymbol{R}} = \sum \boldsymbol{F}, \tag{9.38}$$

and the rotational motion by

$$\dot{\boldsymbol{J}}^* = \sum \boldsymbol{r}^* \wedge \boldsymbol{F}. \tag{9.39}$$

The angular momentum $\boldsymbol{J}^*$ about the centre of mass is of course related to the angular velocity by

$$\boldsymbol{J}^* = I_1^* \omega_1 \boldsymbol{e}_1^* + I_2^* \omega_2 \boldsymbol{e}_2^* + I_3^* \omega_3 \boldsymbol{e}_3^*,$$

where $\boldsymbol{e}_1^*, \boldsymbol{e}_2^*, \boldsymbol{e}_3^*$ are the principal axes at the centre of mass, and $I_1^*, I_2^*, I_3^*$ are the corresponding principal moments.

## 9.8   Rotation about a Principal Axis

The principal axes $\boldsymbol{e}_1, \boldsymbol{e}_2, \boldsymbol{e}_3$ rotate with the body. Thus if we wish to use directly the expression for $\boldsymbol{J}$ in terms of its components with respect to these axes, we have to remember that they constitute a rotating frame. In this section, we shall return to the notation of Chapter 5, and distinguish the absolute rate of change $\mathrm{d}\boldsymbol{J}/\mathrm{d}t$ from the relative rate of change $\dot{\boldsymbol{J}}$. The equation of motion (9.37) refers of course to the absolute rate of change, so we must write it as

$$\frac{\mathrm{d}\boldsymbol{J}}{\mathrm{d}t} = \sum \boldsymbol{r} \wedge \boldsymbol{F} = \boldsymbol{G}, \tag{9.40}$$

say. (The argument here may be applied equally to a rigid body rotating about a fixed pivot, or to the rotation about the centre of mass.)

The relative rate of change is

$$\dot{\boldsymbol{J}} = I_1 \dot{\omega}_1 \boldsymbol{e}_1 + I_2 \dot{\omega}_2 \boldsymbol{e}_2 + I_3 \dot{\omega}_3 \boldsymbol{e}_3,$$

since the principal moments of inertia are constants. The two rates of change are related by

$$\frac{\mathrm{d}\boldsymbol{J}}{\mathrm{d}t} = \dot{\boldsymbol{J}} + \boldsymbol{\omega} \wedge \boldsymbol{J}.$$

Substituting in (9.40), we obtain

$$\dot{\boldsymbol{J}} + \boldsymbol{\omega} \wedge \boldsymbol{J} = \boldsymbol{G},$$

or, in terms of components,

$$\begin{aligned}
I_1 \dot{\omega}_1 + (I_3 - I_2)\omega_2\omega_3 &= G_1, \\
I_2 \dot{\omega}_2 + (I_1 - I_3)\omega_3\omega_1 &= G_2, \\
I_3 \dot{\omega}_3 + (I_2 - I_1)\omega_1\omega_2 &= G_3.
\end{aligned} \tag{9.41}$$

These equations may be solved, in principle, to give the angular velocity components as functions of time. However, when there are external forces, they are not often particularly useful, because these forces are usually specified in terms of their components with respect to a given fixed set of axes. Even if the external force $\boldsymbol{F}$ is a constant, its components $F_1, F_2, F_3$ are variable, and depend on the unknown orientation of the body. For this kind of problem, an alternative method of solution, using Lagrange's equations (to be described in the next chapter), is usually preferable.

We shall confine our discussion here to the case where there are *no* external forces, so that the right hand sides of (9.41) vanish. If the body is initially rotating about the principal axis $\boldsymbol{e}_3$, so that $\omega_1 = \omega_2 = 0$, then we see from (9.41) that they remain zero, and $\omega_3$ is a constant. Thus we have verified our earlier assertion that a body rotating freely about a principal axis will continue to rotate with constant angular velocity. Now, however, we wish to investigate the stability of this motion.

### *Example:* **Stability of free rotation about a principal axis**

A rigid body is rotating freely about a principal axis, say $e_3$, and is given a small displacement. Will its rotation axis remain close to $e_3$?

If the displacement is small enough, we may treat $\omega_1$ and $\omega_2$ (at least initially) as small, and neglect the product $\omega_1\omega_2$. Thus the third equation, with $G_3 = 0$, tells us that $\omega_3$ remains almost constant. We can solve the remaining two equations (with $G_1 = G_2 = 0$) by looking for solutions of the form

$$\omega_1 = a_1 e^{pt}, \qquad \omega_2 = a_2 e^{pt},$$

where $a_1$, $a_2$ and $p$ are constants. (This is essentially the method we used to treat the the problem of stability of equilibrium in one dimension in §§2.2, 2.3. We shall discuss the general problem of stability by a similar method in the next chapter.) Substituting, and dividing by $e^{pt}$, we obtain

$$I_1 p a_1 + (I_3 - I_2)\omega_3 a_2 = 0,$$
$$I_2 p a_2 + (I_1 - I_3)\omega_3 a_1 = 0.$$

Eliminating the ratio $a_1/a_2$, we find

$$p^2 = \frac{(I_3 - I_2)(I_1 - I_3)}{I_1 I_2}\omega_3^2.$$

The denominator is obviously positive, as is $\omega_3^2$. Thus if $I_3 > I_1$ and $I_3 > I_2$, or if $I_3 < I_1$ and $I_3 < I_2$, the two roots for $p$ are pure imaginary, and we have an oscillatory solution. On the other hand, if $I_1 > I_3 > I_2$ or $I_1 < I_3 < I_2$, then the roots are real, and the values of $\omega_1$ and $\omega_2$ will in general increase exponentially with time, until the approximation of treating them as small is no longer valid.

Therefore, we may conclude that the rotation about the axis $e_3$ is stable if $I_3$ is either the largest or the smallest of the three principal moments, but not if it is the middle one.

This interesting result is easy to verify. For example, if one throws a matchbox in the air, it is not hard to get it to spin about its longest or its shortest axis, but it will not spin stably about the other one. The result applies to any rigid body. It is often called the *tennis racquet theorem*, and we will examine this problem from a different perspective in §13.6.

## 9.9   Euler's Angles

The orientation of a rigid body about a fixed point — or about its centre of mass — must be specified by three angles. These may be chosen in various ways, but one convenient choice is the set of angles known as *Euler's angles*. Two of these are required to fix the direction of one of the axes, say $e_3$ — they are just the polar angles $\theta, \varphi$ — and the third specifies the angle through which the body has been rotated from a standard position about this axis.

We can reach an arbitrary orientation by starting with the body in a standard position, and making three successive rotations. Let us suppose that initially the axes $e_1, e_2, e_3$ of the body coincide with the fixed axes $i, j, k$. (See Fig. 9.9.) The first two rotations are designed to bring the

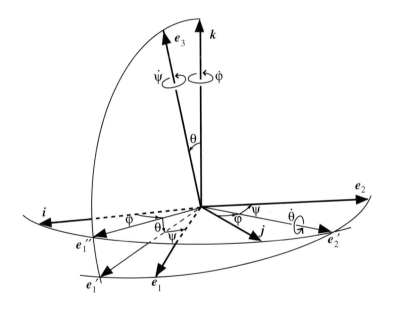

Fig. 9.9

axis $e_3$ to its required position, specified by the polar angles $\theta, \varphi$. First, we make a rotation through an angle $\varphi$ about the axis $k$. This brings the three axes into the positions $e_1'', e_2', k$. Next, we make a rotation through an angle $\theta$ about the axis $e_2'$, bringing the axes to the positions $e_1', e_2', e_3$.

Finally, we make a rotation through an angle $\psi$ about $e_3$. This brings all three axes to their required positions $e_1, e_2, e_3$.

Since the three Euler angles, $\varphi, \theta, \psi$, fix the orientations of the three axes, they specify completely the orientation of the rigid body. The angular velocity of the body is clearly determined by the rates of change of these three angles. A small change in $\varphi$ corresponds to a rotation about $k$. Hence if $\varphi$ is changing at a rate $\dot{\varphi}$, and the other angles are constant, the angular velocity is $\dot{\varphi}k$. Similarly, if only $\theta$ or $\psi$ is changing, the angular velocity is $\dot{\theta}e_2'$ or $\dot{\psi}e_3$. When all three angles are changing, the angular velocity is the sum of these three contributions,

$$\boldsymbol{\omega} = \dot{\varphi}\boldsymbol{k} + \dot{\theta}\boldsymbol{e}_2' + \dot{\psi}\boldsymbol{e}_3. \qquad (9.42)$$

In order to find the angular momentum or kinetic energy, using (9.20) or (9.22), we should have to find the components of $\boldsymbol{\omega}$ in the directions of the three principal axes, $e_1, e_2, e_3$. This is not difficult, but we shall not go through the derivation in the general case. Instead, we limit the discussion to the case of a *symmetric body*, with symmetry axis $e_3$, and $I_1 = I_2$. Then, according to the discussion of §9.4, any orthogonal pair of axes in the plane of $e_1$ and $e_2$ will serve as principal axes, together with $e_3$. It is not necessary that the axes we choose be fixed in the body, so long as they are always principal axes. In particular, we may use the axes $e_1', e_2'$. This choice has the advantage that two of the three terms in (9.44) are already expressed in terms of these axes $e_1', e_2', e_3$.

The components of $k$ in the directions of $e_1', e_2', e_3$ are clearly $(-\sin\theta, 0, \cos\theta)$, so that

$$\boldsymbol{k} = -\sin\theta\,\boldsymbol{e}_1' + \cos\theta\,\boldsymbol{e}_3. \qquad (9.43)$$

Thus from (9.42), we find

$$\boldsymbol{\omega} = -\dot{\varphi}\sin\theta\,\boldsymbol{e}_1' + \dot{\theta}\boldsymbol{e}_2' + (\dot{\psi} + \dot{\varphi}\cos\theta)\boldsymbol{e}_3. \qquad (9.44)$$

We can now find the angular momentum from (9.20). All we have to do is to multiply the three components of $\boldsymbol{\omega}$ by the principal moments $I_1, I_1, I_3$. We obtain

$$\boldsymbol{J} = -I_1\dot{\varphi}\sin\theta\,\boldsymbol{e}_1' + I_1\dot{\theta}\boldsymbol{e}_2' + I_3(\dot{\psi} + \dot{\varphi}\cos\theta)\boldsymbol{e}_3. \qquad (9.45)$$

To find the equations of motion, we could either re-express $\boldsymbol{J}$ in terms of the fixed axes $\boldsymbol{i}, \boldsymbol{j}, \boldsymbol{k}$, or write down equations directly in the rotating frame defined by $e_1', e_2', e_3$. We shall, however, find a simpler method in the next chapter.

The kinetic energy is obtained similarly from (9.22). It is

$$T = \tfrac{1}{2}I_1\dot{\varphi}^2 \sin^2\theta + \tfrac{1}{2}I_1\dot{\theta}^2 + \tfrac{1}{2}I_3(\dot{\psi} + \dot{\varphi}\cos\theta)^2. \tag{9.46}$$

### Free motion of a symmetric rigid body

As an example of the use of the Euler angles, let us consider a symmetric body moving under no forces. Its centre of mass moves of course with uniform velocity. The interesting part of the motion is the rotation about the centre of mass. The same formalism applies to a body rotating freely about a fixed pivot.

Since there are no external forces, the angular momentum equation is simply $\mathrm{d}\mathbf{J}/\mathrm{d}t = \mathbf{0}$. The angular momentum vector therefore points in a fixed direction in space. Though we could solve the problem in terms of arbitrary axes, it will be convenient to choose the axis $\mathbf{k}$ to be in the direction of the constant vector $\mathbf{J}$. Then, by (9.43),

$$\mathbf{J} = J\mathbf{k} = -J\sin\theta\, e_1' + J\cos\theta\, e_3.$$

For consistency, this must be identically equal to the expression (9.45). Hence, equating the three components, we obtain

$$I_1\dot{\varphi}\sin\theta = J\sin\theta,$$
$$I_1\dot{\theta} = 0, \tag{9.47}$$
$$I_3(\dot{\psi} + \dot{\varphi}\cos\theta) = J\cos\theta.$$

From the second equation, we learn that $\theta$ is a constant, and from the other two that (provided $\theta \neq 0$ or $\pi$) $\dot{\varphi}$ and $\dot{\psi}$ are constants. Hence the axis of the body, $e_3$ rotates around the direction of $\mathbf{J}$ at a constant rate $\dot{\varphi}$, maintaining a constant angle $\theta$ to it, and in addition the body spins about its axis with constant angular velocity $\dot{\psi}$ (see Fig 9.10).

The angular velocity vector

$$\boldsymbol{\omega} = -\dot{\varphi}\sin\theta\, e_1' + (\dot{\psi} + \dot{\varphi}\cos\theta)e_3 \tag{9.48}$$

has fixed components with respect to $e_1', e_2', e_3$. It too therefore maintains a constant angle with $e_3$, and also with $\mathbf{J}$. In space, the angular velocity vector $\boldsymbol{\omega}$ thus describes a cone around the direction of $\mathbf{J}$, precessing at the

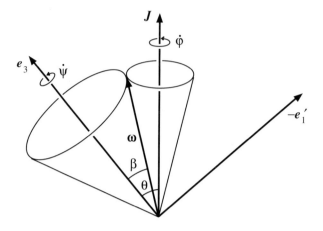

Fig. 9.10

rate $\dot{\varphi}$. This is known as the *space cone*. In the body, since $\boldsymbol{\omega}$ maintains a constant angle to $\boldsymbol{e}_3$, it must describe a cone around this axis — the *body cone*. The rate at which it describes this body cone is $\dot{\psi}$. (For, $\boldsymbol{\omega}$ is fixed with respect to the axes $\boldsymbol{e}_1', \boldsymbol{e}_2', \boldsymbol{e}_3$, and the body rotates relative to them with angular velocity $\dot{\psi}$.)

Since the direction of $\boldsymbol{\omega}$ is the instantaneous axis of rotation, the body cone is instantaneously at rest along the line in the direction of $\boldsymbol{\omega}$, which is where it touches the space cone. (See Fig. 9.10.) Thus the motion may be described by saying that the body cone rolls round the fixed space cone.

From (9.48) we see that the semi-vertical angle $\beta$ of the body cone is given by

$$\tan\beta = -\frac{\omega_1}{\omega_3} = \frac{\dot{\varphi}\sin\theta}{\dot{\psi} + \dot{\varphi}\cos\theta} = \frac{I_3}{I_1}\tan\theta, \qquad (9.49)$$

by (9.47). In particular, if the body is nearly totally symmetric, so that $|I_1 - I_3| \ll I_3$, then the angles $\beta$ and $\theta$ are nearly equal, and the space cone is much narrower than the body cone. In this case, it is easy to see that $\dot{\psi} \ll \dot{\varphi}$, so that the period taken to describe the body cone is much greater than that for the space cone.

The Earth's axis has a very small oscillation of this kind, called the *Chandler wobble*. It precesses around a fixed direction in space roughly once every day, and moves round a cone in the Earth with semi-vertical angle of order $0.1''$ (a circle of about $3\,\mathrm{m}$ radius at the pole) in a period which should be about 300 days for a perfectly rigid Earth, but is actually around 14 months, with considerable variation. In this case $I_3 > I_1$, and therefore $\beta > \theta$, so the body cone actually rolls around with the space cone inside it, and $\dot\psi < 0$ (see Fig. 9.11). (Note that this precession is quite independent

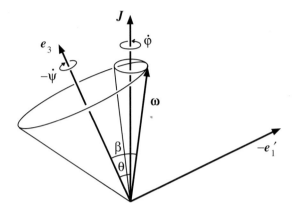

Fig. 9.11

of the much larger effect, the precession of the equinoxes, discussed at the end of §9.6).

## 9.10   Summary

Determining the motion of a rigid body under known forces can be divided into two quite separate problems. The motion of the centre of mass is determined, via the momentum equation (9.1), by the total force acting on the body — or, if one point is fixed, the force is determined by the motion of the centre of mass. The rotational motion about the centre of mass is determined, via the angular momentum equation (9.2), by the total moment of the forces acting.

   The relation between angular momentum and angular velocity is described by the inertia tensor. It is always possible to find a set of principal axes (fixed in the body) for which all the products of inertia vanish. Then $\boldsymbol{J}$ is obtained from $\boldsymbol{\omega}$ simply by multiplying each component by the appropriate principal moment of inertia. Similarly the kinetic energy $T$ is the sum of the three terms corresponding to rotation about each of the three principal axes.

In general, it is convenient to describe the rotational motion in terms of a set of principal axes. Normally these rotate with the body, though in the special case of a symmetric rigid body (with $I_1 = I_2$) they may be chosen as in §9.9. In that case, the orientation of the body is conveniently described by Euler's angles. We shall see in the following chapter that the Lagrangian method is very useful for obtaining equations of motion in terms of Euler's angles.

---

## Problems

1. A uniform solid cube of edge length $2a$ is suspended from a horizontal axis along one edge. Find the length of the equivalent simple pendulum. Given that the cube is released from rest with its centre of mass level with the axis, find its angular velocity when it reaches the lowest point.

2. An insect of mass 100 mg is resting on the edge of a flat uniform disc of mass 3 g and radius 50 mm, which is rotating at 60 r.p.m. about a smooth pivot. The insect crawls in towards the centre of the disc. Find the angular velocity when it reaches it, and the gain in kinetic energy. Where does this kinetic energy come from, and what happens to it when the insect crawls back out to the edge?

3. A uniform solid cube of edge $2a$ is sliding with velocity $v$ on a smooth horizontal table when its leading edge is suddenly brought to rest by a small ridge on the table. Which dynamical variables are conserved (a) before impact, (b) during impact, and (c) after impact? Find the angular velocity immediately after impact, and the fractional loss of kinetic energy. Determine the minimum value of $v$ for which the cube topples over rather than falling back.

4. A pendulum consists of a light rigid rod of length 250 mm, with two identical uniform solid spheres of radius 50 mm attached one on either side of its lower end. Find the period of small oscillations (a) perpendicular to the line of centres, and (b) along it.

5. A uniform rod of mass $M$ and length $2a$ hangs from a smooth hinge at one end. Find the length of the equivalent simple pendulum. It is struck sharply, with impulse $X$, at a point a distance $b$ below the hinge. Use the angular momentum equation to find the initial value of the angular velocity. Find also the initial momentum. Determine the point at which the rod may be struck without producing any impulsive reaction at the hinge. Show that, if the rod is struck elsewhere, the

direction of the impulsive reaction depends on whether the point of impact is above or below this point.

6. *(a) A simple pendulum supported by a light rigid rod of length $l$ is released from rest with the rod horizontal. Find the reaction at the pivot as a function of the angle of inclination.

   (b) For the cube of Problem 1, find the horizontal and vertical components of the reaction on the axis as a function of its angular position. Compare your answer with the corresponding expressions for the equivalent simple pendulum.

7. *Find the principal moments of inertia of a flat rectangular plate of mass $30\,\mathrm{g}$ and dimensions $80\,\mathrm{mm} \times 60\,\mathrm{mm}$. Given that the plate is rotating about a diagonal with angular velocity $15\,\mathrm{rad\,s^{-1}}$, find the components of the angular momentum parallel to the edges. Given that the axis is of total length $120\,\mathrm{mm}$, and is held vertical by bearings at its ends, find the horizontal component of the force on each bearing.

8. Find the principal moments of inertia of a uniform solid cube of mass $m$ and edge length $2a$ (a) with respect to the mid-point of an edge, and (b) with respect to a vertex.

9. Find the moment of inertia about an axis through its centre of a uniform hollow sphere of mass $M$ and outer and inner radii $a$ and $b$. (*Hint*: Think of it as a sphere of density $\rho$ and radius $a$, with a sphere of density $\rho$ and radius $b$ removed.)

10. A spaceship of mass $3\,\mathrm{t}$ has the form of a hollow sphere, with inner radius $2.5\,\mathrm{m}$ and outer radius $3\,\mathrm{m}$. Its orientation in space is controlled by a uniform circular flywheel of mass $10\,\mathrm{kg}$ and radius $0.1\,\mathrm{m}$. Given that the flywheel is set spinning at $2000$ r.p.m., find how long it takes the spaceship to rotate through $1°$. Find also the energy dissipated in this manoeuvre.

11. A long, thin hollow cylinder of radius $a$ is balanced on a horizontal knife edge, with its axis parallel to it. It is given a small displacement. Calculate the angular displacement at the moment when the cylinder ceases to touch the knife edge. (*Hint*: This is the moment when the radial component of the reaction falls to zero.)

12. *Calculate the principal moments of inertia of a uniform, solid cone of vertical height $h$, and base radius $a$ about its vertex. For what value of the ratio $h/a$ is every axis through the vertex a principal axis? For this case, find the position of the centre of mass and the principal moments of inertia about it.

13. A top consists of a uniform, solid cone of height $50\,\text{mm}$ and base radius $20\,\text{mm}$. It is spinning with its vertex fixed at 7200 r.p.m. Find the precessional period of the axis about the vertical.

14. A gyroscope consisting of a uniform solid sphere of radius $0.1\,\text{m}$ is spinning at 3000 r.p.m. about a horizontal axis. Due to faulty construction, the fixed point is not precisely at the centre, but $20\,\mu\text{m}$ away from it along the axis. Find the time taken for the axis to move through $1°$.

15. A gyroscope consisting of a uniform circular disc of mass $100\,\text{g}$ and radius $40\,\text{mm}$ is pivoted so that its centre of mass is fixed, and is spinning about its axis at 2400 r.p.m. A $5\,\text{g}$ mass is attached to the axis at a distance of $100\,\text{mm}$ from the centre. Find the angular velocity of precession of the axis.

16. *A uniformly charged sphere is spinning freely with angular velocity $\boldsymbol{\omega}$ in a uniform magnetic field $\boldsymbol{B}$. Taking the $z$ axis in the direction of $\boldsymbol{\omega}$, and $\boldsymbol{B}$ in the $xz$-plane, write down the moment about the centre of the magnetic force on a particle at $\boldsymbol{r}$. Evaluate the total moment of the magnetic force on the sphere, and show that it is equal to $(q/2M)\boldsymbol{J}\wedge\boldsymbol{B}$, where $q$ and $M$ are the total charge and mass, respectively. Hence show that the axis will precess around the direction of the magnetic field with precessional angular velocity equal to the Larmor frequency of §5.5. What difference would it make if the charge distribution were spherically symmetric, but non-uniform?

17. *A wheel of radius $a$, with its mass concentrated on the rim, is rolling with velocity $v$ round a circle of radius $R$ $(\gg a)$, maintaining a constant inclination $\alpha$ to the vertical. Show that $v = a\omega = R\Omega$, where $\omega$ is the angular velocity of the wheel about its axis, and $\Omega$ $(\ll \omega)$ is the precessional angular velocity of the axis. Use the momentum equation to find the horizontal and vertical components of the force at the point of contact. Then show from the angular momentum equation about the centre of mass that $R = 2v^2/g\tan\alpha$. Evaluate $R$ for $v = 5\,\text{m s}^{-1}$ and $\alpha = 30°$.

18. A solid rectangular box, of dimensions $100\,\text{mm}\times60\,\text{mm}\times20\,\text{mm}$, is spinning freely with angular velocity 240 r.p.m. Determine the frequency of small oscillations of the axis, if the axis of rotation is (a) the longest, and (b) the shortest, axis.

19. *A rigid body of spheroidal shape, spinning rapidly about its axis of symmetry, is placed on a smooth flat table. Show by considering the moment of the force at the point of contact that its axis will precess in one direction if it is oblate $(c < a = b$, e.g. a discus$)$ and in the

opposite direction if it is prolate ($c > a = b$, *e.g.* a rugby ball). Show also that if there is a small frictional force, the axis will become more nearly vertical, so that if the body is oblate its centre of mass will fall, but if it is prolate it will *rise*.

20. *The average moment exerted by the Sun on the Earth is, except for sign, identical with the expression found in Chapter 6, Problem 26, provided we interpret $m$ as the mass of the Sun, and $r$ as the distance to the Sun. Show that $Q = -2(I_3 - I_1)$ and hence that the precessional angular velocity produced by this moment is

$$\mathbf{\Omega} = -\frac{3}{2}\frac{I_3 - I_1}{I_3}\frac{\varpi^2}{\omega}\cos\alpha\,\mathbf{n},$$

where $\varpi$ is the Earth's orbital angular velocity, and $\alpha = 23.45°$ is the tilt between the Earth's axis and the normal to the orbital plane (the ecliptic). Show also that $(I_3 - I_1)/I_3 \approx \epsilon$, the oblateness of the Earth, and hence evaluate $\mathbf{\Omega}$. Why is this effect less sensitive to the distribution of density within the Earth than the complementary one discussed in §6.5?

21. *The axis of a gyroscope is free to rotate within a smooth horizontal circle in colatitude $\lambda$. Due to the Coriolis force, there is a couple on the gyroscope. To find the effect of this couple, use the equation for the rate of change of angular momentum in a frame rotating with the Earth (*e.g.*, that of Fig. 5.7), $\dot{\mathbf{J}} + \mathbf{\Omega} \wedge \mathbf{J} = \mathbf{G}$, where $\mathbf{G}$ is the couple restraining the axis from leaving the horizontal plane, and $\mathbf{\Omega}$ is the Earth's angular velocity. (Neglect terms of order $\Omega^2$, in particular the contribution of $\mathbf{\Omega}$ to $\mathbf{J}$.) From the component along the axis, show that the angular velocity $\omega$ about the axis is constant; from the vertical component show that the angle $\varphi$ between the axis and east obeys the equation

$$I_1\ddot{\varphi} - I_3\omega\Omega\sin\lambda\cos\varphi = 0.$$

Show that the stable position is with the axis pointing north. Determine the period of small oscillations about this direction if the gyroscope is a flat circular disc spinning at 6000 r.p.m. in latitude 30° N. Explain why this system is sensitive to the horizontal component of $\mathbf{\Omega}$, and describe the effect qualitatively from the point of view of an inertial observer.

# Chapter 10

# Lagrangian Mechanics

We have already seen that the equations of motion for a system of $N$ particles moving under conservative forces may be obtained from the Lagrangian function in terms of any set of $3N$ independent co-ordinates. (See §8.5.) In this chapter, we shall give a more systematic account of the Lagrangian method, and apply it in particular to the case of rigid bodies.

## 10.1 Generalized Co-ordinates; Holonomic Systems

Let us consider a rigid body, composed of a large number $N$ of particles. The positions of all the particles may be specified by $3N$ co-ordinates. However, these $3N$ co-ordinates cannot all vary independently, but are subject to constraints — the rigidity conditions. In fact, the position of every particle may be fixed by specifying the values of just six quantities — for instance, the three co-ordinates $X, Y, Z$ of the centre of mass and the three Euler angles $\varphi, \theta, \psi$ which determine the orientation. These six constitute a set of *generalized co-ordinates* for the rigid body.

In particular problems, these co-ordinates may be subject to further constraints, which may be of two kinds. First, we might for example fix the position of one point of the body, say the centre of mass. Such constraints are represented by algebraic conditions on the co-ordinates (*e.g.*, $X = Y = Z = 0$), which may be used to eliminate some of the co-ordinates. In this particular case, the three Euler angles alone suffice to fix the position of every particle.

The second type of constraint is represented by conditions on the *velocities* rather than the co-ordinates. For example, we might constrain the centre of mass to move with uniform velocity, or to move round a circle with uniform angular velocity. Then, in place of algebraic equations, we have differential equations (*e.g.*, $\dot{X} = u$). In simple cases, these equations can be solved to find some of the co-ordinates as explicit functions of time (*e.g.*, $X = X_0 + ut$). Then the position of every particle will be determined by the values of the remaining generalized co-ordinates *and* the time $t$.

In general, we say that $q_1, q_2, \ldots, q_n$ is a set of *generalized co-ordinates* for a given system if the position of every particle in the system is a function of these variables, and perhaps also explicitly of time:

$$\boldsymbol{r}_i = \boldsymbol{r}_i(q_1, q_2, \ldots, q_n, t). \tag{10.1}$$

The number of co-ordinates which can vary independently is called the number of *degrees of freedom* of the system. If it is possible to solve the constraint equations, and eliminate some of the co-ordinates, leaving a set equal in number to the number of degrees of freedom, the system is called *holonomic*. If this elimination introduces explicit functions of time, the system is said to be *forced*; on the other hand, if all the constraints are purely algebraic, so that $t$ does not appear explicitly in (10.1), the system is *natural*.

There do exist non-holonomic systems, for which the constraint equations cannot be solved to eliminate some of the co-ordinates. Consider, for example, a sphere rolling on a rough plane. Its position may be specified by five generalized co-ordinates — $X, Y, \varphi, \theta, \psi$. ($Z$ is a constant and may be omitted.) However, the sphere can roll in only two directions, and the number of degrees of freedom is two (three if we allow it also to spin about the vertical). The constraint equations — the rolling conditions — serve to determine the angular velocity in terms of the velocity of the centre of mass. Using the fact that the instantaneous axis of rotation must be a horizontal axis through the point of contact, it is not hard to show that $\boldsymbol{\omega} = \boldsymbol{k} \wedge \dot{\boldsymbol{R}}/a$, where $a$ is the radius of the sphere. (This follows from the fact that $\dot{\boldsymbol{R}} = \boldsymbol{\omega} \wedge a\boldsymbol{k}$, together with $\boldsymbol{k} \cdot \boldsymbol{\omega} = 0$.) However, these equations cannot be integrated to find the orientation in terms of the position of the centre of mass; for, we could roll the sphere round a circle so that it returns to its starting point but with a different orientation.

For the present, we shall assume that the constraint equations *can* be solved, and the number of generalized coordinates taken equal to the number of degrees of freedom.

The distinction between a natural and a forced system can be expressed in another way, which will be useful later. Differentiating (10.1), we find that the velocity of the $i$th particle is a linear function of $\dot{q}_1, \ldots, \dot{q}_n$, though

in general it depends in a more complicated way on the co-ordinates themselves:

$$\dot{\boldsymbol{r}}_i = \sum_{\alpha=1}^{n} \frac{\partial \boldsymbol{r}_i}{\partial q_\alpha} \dot{q}_\alpha + \frac{\partial \boldsymbol{r}_i}{\partial t}.$$

The last term arises from the explicit dependence on $t$ in (10.1), and is
absent for a natural system. When we substitute in $T = \sum \frac{1}{2} m \dot{r}^2$, we
obtain a quadratic function of the time derivatives $\dot{q}_1, \ldots, \dot{q}_n$. For a natural
system, it is a *homogeneous* quadratic function; but for a forced system
there are also linear terms and terms independent of the velocities.

For example, according to (8.29) and (9.46), the kinetic energy of a
symmetric rigid body is

$$T = \tfrac{1}{2} M (\dot{X}^2 + \dot{Y}^2 + \dot{Z}^2) + \tfrac{1}{2} I_1^* (\dot{\varphi}^2 \sin^2 \theta + \dot{\theta}^2) + \tfrac{1}{2} I_3^* (\dot{\psi} + \dot{\varphi} \cos \theta)^2.$$

If we impose further algebraic constraints, such as $X = 0$, the corresponding
terms drop out, and we are still left with a homogeneous quadratic function
of the remaining time derivatives. On the other hand, if we impose a
differential constraint, such as $\dot{X} = u$ or $\dot{\varphi} = \omega$, we obtain a function with
constant or linear terms.

## 10.2    Lagrange's Equations

To obtain a more general form of Lagrange's equations than the one found
in §3.7 — in particular, one that is not restricted to conservative forces —
let us return for the moment to the case of a single particle, moving under
an arbitrary force $\boldsymbol{F}$. We consider variations of the integral

$$I = \int_{t_0}^{t_1} T \, \mathrm{d}t, \qquad \text{where} \qquad T = \tfrac{1}{2} m \dot{r}^2. \tag{10.2}$$

Now let us make a small variation $\delta x(t)$ in the $x$ co-ordinate, subject to
the boundary conditions $\delta x(t_0) = \delta x(t_1) = 0$. Then clearly $\delta T = m \dot{x} \, \delta \dot{x}$.
Substituting in (10.2), and performing an integration by parts, in which
the integrated term vanishes by virtue of the boundary conditions, we find

$$\delta I = - \int_{t_0}^{t_1} m \ddot{x} \, \delta x \, \mathrm{d}t.$$

Now, the integrand is $-F_x \, \delta x = -\delta W$, where $\delta W$ is the work done by the
force $\boldsymbol{F}$ in the displacement $\delta x$. Similarly, if we make variations of all three

coordinates, we find

$$\delta I = -\int_{t_0}^{t_1} \delta W \, \mathrm{d}t, \qquad (10.3)$$

where $\delta W = \boldsymbol{F} \cdot \delta \boldsymbol{r}$ is the work done by the force in the displacement $\delta \boldsymbol{r}$.

The advantage of (10.3) is that, like Hamilton's principle (see §3.7), it makes no explicit reference to any particular set of co-ordinates. Hence we can use it to find equations of motion in terms of an arbitrary set of co-ordinates $q_1, q_2, q_3$. Let us define the *generalized forces* $F_1, F_2, F_3$ corresponding to $\boldsymbol{F}$ by

$$\delta W = F_1 \delta q_1 + F_2 \delta q_2 + F_3 \delta q_3. \qquad (10.4)$$

Then we may equate $\delta I$ as given by (10.3) to the general expression (3.37). Consider, for example, a variation $\delta q_1$ of $q_1$. Then, according to (3.37),

$$\delta I = \int_{t_0}^{t_1} \left[ \frac{\partial T}{\partial q_1} - \frac{\mathrm{d}}{\mathrm{d}t} \left( \frac{\partial T}{\partial \dot{q}_1} \right) \right] \delta q_1(t) \, \mathrm{d}t.$$

On the other hand, by (10.3) and (10.4),

$$\delta I = -\int_{t_0}^{t_1} F_1 \, \delta q_1(t) \, \mathrm{d}t.$$

These two expressions must be equal for arbitrary variations subject only to the boundary conditions that $\delta q_1(t_0) = \delta q_1(t_1) = 0$. Hence the integrands must be equal, and we obtain *Lagrange's equations* in the form

$$\frac{\mathrm{d}}{\mathrm{d}t} \left( \frac{\partial T}{\partial \dot{q}_\alpha} \right) = \frac{\partial T}{\partial q_\alpha} + F_\alpha. \qquad (10.5)$$

For example, for a particle moving in a plane, and described by polar co-ordinates, $T = \frac{1}{2}m(\dot{r}^2 + r^2\dot{\theta}^2)$. Hence Lagrange's equations are

$$\frac{\mathrm{d}}{\mathrm{d}t}(m\dot{r}) = mr\dot{\theta}^2 + F_r,$$
$$\frac{\mathrm{d}}{\mathrm{d}t}(mr^2\dot{\theta}) = F_\theta. \qquad (10.6)$$

It is important to remember that $F_\theta$ here does *not* mean the component of the vector $\boldsymbol{F}$ in the direction of increasing $\theta$. (Compare the discussion following (3.47).) It is defined by (10.4), or

$$\delta W = F_r \delta r + F_\theta \delta\theta.$$

The work done in producing a small *angular* displacement $\delta\theta$ is equal to $\delta\theta$ multiplied by the *moment* of the force $\boldsymbol{F}$ about the origin. Thus $F_\theta$ here is in fact the moment of $\boldsymbol{F}$ about the origin. (Recall that the generalized momentum $p_\theta = mr^2\dot\theta$ is the angular momentum. Thus the second of Eqs. (10.6) expresses the fact that the rate of change of angular momentum is equal to the moment of the force.)

If the force is conservative, then $\delta W = -\delta V$, where $V$ is the potential energy. It then follows from (10.4) that

$$F_\alpha = -\frac{\partial V}{\partial q_\alpha}. \tag{10.7}$$

Thus, defining the *Lagrangian function* $L = T - V$, we can write (10.5) in the form (3.44) obtained earlier,

$$\frac{\mathrm{d}}{\mathrm{d}t}\left(\frac{\partial L}{\partial \dot q_\alpha}\right) = \frac{\partial L}{\partial q_\alpha}. \tag{10.8}$$

Note that the 'generalized force' $\partial L/\partial q_\alpha$ on the right hand side of this equation includes not only the force $F_\alpha$ determined by $\boldsymbol{F}$ but also the 'apparent' force $\partial T/\partial q_\alpha$ arising from the curvilinear nature of the co-ordinates (for example, the centrifugal term $mr\dot\theta^2$ in (10.6)).

More generally, even if $\boldsymbol{F}$ is non-conservative, it may be possible to find an effective 'potential energy' function $V$ depending on the velocities as well as the co-ordinates, and such that

$$F_\alpha = -\frac{\partial V}{\partial q_\alpha} + \frac{\mathrm{d}}{\mathrm{d}t}\left(\frac{\partial V}{\partial \dot q_\alpha}\right). \tag{10.9}$$

In this case too, we can define $L = T - V$, and write Lagrange's equations in the form (10.8). The most important example of a force of this type in the electromagnetic force on a charged particle, which we shall discuss in §10.5. Note that in this case the 'potential energy' $V$ contributes not only to the 'generalized force' on the right side of (10.8) but also to the 'generalized momentum' on the left.

There is nothing in the discussion leading to Lagrange's equations (10.5) or (10.8) that restricts their applicability to the case of a single particle. The general principle embodied in (10.3) applies to any system whatsoever, containing any number of particles. If the system is holonomic, with $n$ degrees of freedom, then we can make independent arbitrary variations of the generalized co-ordinates $q_1, q_2, \ldots, q_n$, and obtain Lagrange's equations

(10.5), in which the generalized forces are defined, as in (10.4), by

$$\delta W = F_1 \delta q_1 + F_2 \delta q_2 + \cdots + F_n \delta q_n = \sum_{\alpha=1}^{n} F_\alpha \delta q_\alpha. \qquad (10.10)$$

Finally, if the forces are conservative, or more generally if they can be written in the form (10.9), then Lagrange's equations may be expressed in the form (10.8).

## 10.3   Precession of a Symmetric Top

As a first example of the use of Lagrange's equations, we consider in more detail the problem of the symmetric top, discussed in the previous chapter. This is a symmetric rigid body, pivoted at a point on its axis of symmetry, and moving under gravity. (Compare Fig. 9.7.)

> ### *Example:* Steady precession of a top
>
> Find the conditions under which steady precession at a given constant angle of inclination of the axis, $\theta$, can occur. Show that for any $\theta < \pi/2$, there is a minimum value of the rotational angular velocity about the axis for which this is possible.

This system has three degrees of freedom, and we use the three Euler angles as generalized co-ordinates. The kinetic energy is given by (9.46), and the potential energy is $V = MgR\cos\theta$. Thus the Lagrangian function is

$$L = \tfrac{1}{2} I_1 \dot\varphi^2 \sin^2\theta + \tfrac{1}{2} I_1 \dot\theta^2 + \tfrac{1}{2} I_3 (\dot\psi + \dot\varphi \cos\theta)^2 - MgR\cos\theta. \qquad (10.11)$$

Lagrange's equation (10.8) for $\theta$ is therefore

$$\frac{\mathrm{d}}{\mathrm{d}t}(I_1 \dot\theta) = I_1 \dot\varphi^2 \sin\theta \cos\theta - I_3(\dot\psi + \dot\varphi \cos\theta)\dot\varphi \sin\theta + MgR\sin\theta. \qquad (10.12)$$

The Lagrangian function does not involve the other two Euler angles, but only their time derivatives, so the corresponding equations express the constancy of the generalized momenta $p_\varphi$ and $p_\psi$:

$$\frac{\mathrm{d}}{\mathrm{d}t}[I_1 \dot\varphi \sin^2\theta + I_3(\dot\psi + \dot\varphi \cos\theta)\cos\theta] = 0,$$
$$\frac{\mathrm{d}}{\mathrm{d}t}[I_3(\dot\psi + \dot\varphi \cos\theta)] = 0. \qquad (10.13)$$

Note that this last equation tells us that the component of angular velocity $\omega_3$ about the axis of symmetry is constant

$$\omega_3 = \dot{\psi} + \dot{\varphi}\cos\theta = \text{constant}. \qquad (10.14)$$

From the equations (10.13), given that $\theta$ remains constant, we learn that both $\dot{\varphi}$ and $\dot{\psi}$ must be constant. Thus the axis of the top precesses around the vertical with constant angular velocity $\dot{\varphi} = \Omega$, say. To find the relation between the angular velocities $\Omega$ and $\omega_3$, we examine the $\theta$ equation, (10.12). Since the left side must vanish, we obtain (for $\sin\theta \neq 0$, that is unless the axis is vertical),

$$I_1\Omega^2\cos\theta - I_3\omega_3\Omega + MgR = 0. \qquad (10.15)$$

The minimum value of $\omega_3$ for which real roots for $\Omega$ exist is given by

$$I_3^2\omega_3^2 = 4I_1 MgR\cos\theta.$$

If the top is spinning more slowly than this, it will start to wobble. For any larger value of $\omega_3$ there are two possible values of $\Omega$, the roots of (10.15).

It is interesting to examine the special case where $\omega_3$ is large. Then there is one root for $\Omega$ which is much less than $\omega_3$, given by $\Omega \approx MgR/I_3\omega_3$, and another which is comparable in magnitude to $\omega_3$, $\Omega \approx I_3\omega_3/I_1\cos\theta$. The first of these is just the approximate result we obtained in §9.6 on the assumption that $\Omega$ was much smaller than $\omega_3$ (see (9.29)). The second represents a rapid precessional motion in which the gravitational force is negligible. It corresponds precisely to the free precessional motion of a rigid body, discussed in §9.9.

Note that if we allow $\theta$ to be greater than $\pi/2$ — that is, if we consider a top suspended *below* its point of support — then steady precession is possible for any value of $\omega_3$. In particular, for $\omega_3 = 0$, we find the possible angular velocities of a compound pendulum swinging in a circle,

$$\Omega = \pm\sqrt{\frac{MgR}{I_1|\cos\theta|}}.$$

We shall discuss the more general motion of this system in Chapter 12, using the Hamiltonian methods developed there.

## 10.4    Pendulum Constrained to Rotate about an Axis

Next, let us consider the system illustrated in Fig. 10.1. It consists of a
light rigid rod, of length $l$, carrying a mass $m$ at one end, and hinged at

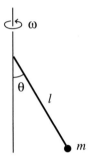

Fig. 10.1

the other to a vertical axis, so that it can swing freely in a vertical plane.
We suppose initially that, in addition to gravity, a known torque $G(t)$ is
applied to rotate it about the axis.

This system has two degrees of freedom. Its position may be described
by the two polar angles $\theta, \varphi$. (For convenience, we take the $z$-axis vertically
downward, so that $\theta = 0$ is the equilibrium position.)

The Lagrangian function is

$$L = \tfrac{1}{2}ml^2(\dot{\theta}^2 + \dot{\varphi}^2 \sin^2 \theta) - mgl(1 - \cos\theta). \qquad (10.16)$$

Since the work done by the torque $G$ is $\delta W = G\,\delta\varphi$, Lagrange's equations
are

$$\frac{\mathrm{d}}{\mathrm{d}t}\left(\frac{\partial L}{\partial \dot{\theta}}\right) = \frac{\partial L}{\partial \theta}, \qquad \frac{\mathrm{d}}{\mathrm{d}t}\left(\frac{\partial L}{\partial \dot{\varphi}}\right) = \frac{\partial L}{\partial \varphi} + G,$$

or, explicitly,

$$ml^2\ddot{\theta} = ml^2\dot{\varphi}^2 \sin\theta\cos\theta - mgl\sin\theta, \qquad (10.17)$$

$$\frac{\mathrm{d}}{\mathrm{d}t}(ml^2\dot{\varphi}\sin^2\theta) = G. \qquad (10.18)$$

Now let us suppose that the torque $G$ is adjusted to constrain the system
to rotate with constant angular velocity $\omega$ about the vertical. This imposes

the constraint $\dot{\varphi} = \omega$, and the system may be regarded as a system with one degree of freedom, described by the co-ordinate $\theta$. Substituting this constraint into the Lagrangian function (10.16), we find

$$L = \tfrac{1}{2}ml^2(\dot{\theta}^2 + \omega^2 \sin^2 \theta) - mgl(1 - \cos\theta). \tag{10.19}$$

Note the appearance in the kinetic energy part of $L$ of a term independent of $\dot{\theta}$, which is characteristic of a forced system. It is often useful in problems of this kind to separate the Lagrangian function into a 'kinetic' term $T'$, quadratic in the time derivatives, and a 'potential' term $-V'$, independent of them. In this case, we write $L = T' - V'$, where

$$T' = \tfrac{1}{2}ml^2\dot{\theta}^2, \qquad V' = mgl(1 - \cos\theta) - \tfrac{1}{2}ml^2\omega^2 \sin^2 \theta.$$

Physically, this corresponds to using a rotating frame, rotating with angular velocity $\omega$. Here $T'$ is the kinetic energy of the motion relative to this frame, and the extra term in $V'$ is the potential energy corresponding to the centrifugal force.

For a forced system, the constraining forces can do work on the system, so the total energy $T+V$ is not in general a constant. However, in our case, there is still a conservation law. Multiplying (10.17) by $\dot{\theta}$ and integrating, we find

$$T' + V' = E' = \text{constant}. \tag{10.20}$$

This is *not* the total energy $T + V$, because the centrifugal term appears with the opposite sign.

It is easy to verify the consistency of this by calculating the work done on the system. The rate at which the torque does work is $G\omega$. Thus, from (10.18),

$$\frac{\mathrm{d}}{\mathrm{d}t}(T + V) = G\omega = \frac{\mathrm{d}}{\mathrm{d}t}(ml^2\omega^2 \sin^2 \theta).$$

Since $T' + V' = T + V - ml^2\omega^2 \sin^2 \theta$, this is equivalent to (10.20).

We can use this conservation law, just as we did in §2.1, to find the qualitative features of the motion.

### *Example:* **Types of motion of the constrained pendulum**

Describe the possible types of motion of the constrained pendulum for a given value of $\omega$.

To do this, we draw the 'effective potential energy' diagram for the function $V'$. There are two distinct cases, depending on the value of $\omega$. If $\omega^2 < g/l$ (that is, if the rotation period is longer than the free oscillation period of the pendulum), then $V'$ has a minimum at $\theta = 0$, and a maximum at $\theta = \pi$, as shown in the upper curve in Fig. 10.2. The motion is qualitatively just like that of the ordinary pendulum, though the period is longer.

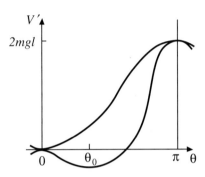

Fig. 10.2

If $\omega^2 > g/l$, then $V'$ has maxima at both $0$ and $\pi$, and an intermediate minimum at $\theta = \theta_0 = \arccos(g/l\omega^2)$, as shown in the lower curve in Fig. 10.2. The equilibrium position at $\theta = 0$ is then unstable. The stable equilibrium position at $\theta_0$ is the one where the transverse components of the gravitational and centrifugal forces are in equilibrium. For this case, three types of motion are possible. If $E' < 0$, the pendulum oscillates around this stable position, without ever reaching the vertical. For $0 < E' < 2mgl$, it swings from one side to the other as before, though $\theta = 0$ is no longer the position of maximum velocity. Finally, if $E' > 2mgl$, we have a continuous circular motion, passing through both the upward and downward verticals.

There is a more general way of handling constraints (holonomic or not), using the method of *Lagrange multipliers* (see Appendix A, Problem 10). To impose the constraint $\dot{\varphi} = \omega$ on our system, we introduce a new variable $\lambda$, the *Lagrange multiplier*, and subtract $\lambda$ times the constraint from the Lagrangian function. From (10.16), we obtain in this way

$$L = \tfrac{1}{2}ml^2(\dot{\theta}^2 + \dot{\varphi}^2 \sin^2 \theta) - mgl(1 - \cos \theta) - \lambda(\dot{\varphi} - \omega). \qquad (10.21)$$

The Euler–Lagrange equation for $\theta$, (10.17) is unchanged. That for $\varphi$ becomes

$$\frac{\mathrm{d}}{\mathrm{d}t}(ml^2\dot{\varphi}\sin^2\theta - \lambda) = 0. \tag{10.22}$$

In addition, we have the Euler–Lagrange equation for $\lambda$, which simply reproduces the constraint,

$$0 = \dot{\varphi} - \omega.$$

Comparing (10.22) with (10.18), we see that $\dot{\lambda}$ is physically the torque $G$ required to impose the constraint. Equivalently, we could have integrated the constraint, and subtracted $\lambda(\varphi - \omega t)$ from $L$. In that case, $\lambda$ itself would be the torque. In general, the physical significance of $\lambda$ may be found by considering a small virtual change in the constraint equation. This method is particularly useful when we want to find the constraining forces or torques.

## 10.5 Charged Particle in an Electromagnetic Field

We now wish to consider one of the most important examples of a non-conservative force. We consider a particle of charge $q$ moving in an electric field $\boldsymbol{E}$ and a magnetic field $\boldsymbol{B}$. The force on the particle is then

$$\boldsymbol{F} = q(\boldsymbol{E} + \boldsymbol{v} \wedge \boldsymbol{B}), \tag{10.23}$$

or, in terms of components,

$$F_x = qE_x + q(\dot{y}B_z - \dot{z}B_y), \tag{10.24}$$

together with two similar equations obtained by cyclic permutation of $x, y, z$.

Now, we wish to show that this force may be written in the form (10.9) with a suitably chosen function $V$. To do this, we have to make use of a standard result of electromagnetic theory (see §A.7), according to which it is always possible to find a *scalar potential* $\phi$ and a *vector potential* $\boldsymbol{A}$, functions of $\boldsymbol{r}$ and $t$, such that

$$\boldsymbol{E} = -\boldsymbol{\nabla}\phi - \frac{\partial\boldsymbol{A}}{\partial t}, \qquad \boldsymbol{B} = \boldsymbol{\nabla} \wedge \boldsymbol{A}. \tag{10.25}$$

For time-independent fields, $\phi$ is simply the electrostatic potential of Chapter 6.

Now, let us consider the function

$$V = q\phi(\mathbf{r}, t) - q\dot{\mathbf{r}} \cdot \mathbf{A}(\mathbf{r}, t)$$
$$= q\phi - q(\dot{x}A_x + \dot{y}A_y + \dot{z}A_z). \tag{10.26}$$

Clearly,

$$-\frac{\partial V}{\partial x} = -q\frac{\partial \phi}{\partial x} + q\left(\dot{x}\frac{\partial A_x}{\partial x} + \dot{y}\frac{\partial A_y}{\partial x} + \dot{z}\frac{\partial A_z}{\partial x}\right).$$

Also,

$$\frac{\mathrm{d}}{\mathrm{d}t}\left(\frac{\partial V}{\partial \dot{x}}\right) = -q\frac{\mathrm{d}A_x}{\mathrm{d}t},$$

so, bearing in mind that $A_x$ varies with time *both* because of its explicit time dependence, *and* because of its dependence on the particle position $\mathbf{r}(t)$, we find

$$\frac{\mathrm{d}}{\mathrm{d}t}\left(\frac{\partial V}{\partial \dot{x}}\right) = -q\left(\frac{\partial A_x}{\partial t} + \frac{\partial A_x}{\partial x}\dot{x} + \frac{\partial A_x}{\partial y}\dot{y} + \frac{\partial A_x}{\partial z}\dot{z}\right).$$

Adding, we find that the terms in $\dot{x}$ cancel, whence

$$-\frac{\partial V}{\partial x} + \frac{\mathrm{d}}{\mathrm{d}t}\left(\frac{\partial V}{\partial \dot{x}}\right) = q\left(-\frac{\partial \phi}{\partial x} - \frac{\partial A_x}{\partial t}\right) +$$
$$q\left[\dot{y}\left(\frac{\partial A_y}{\partial x} - \frac{\partial A_x}{\partial y}\right) + \dot{z}\left(\frac{\partial A_z}{\partial x} - \frac{\partial A_x}{\partial z}\right)\right]$$
$$= qE_x + q(\dot{y}B_z - \dot{z}B_y),$$

by (10.25). But this is just the expression for $F_x$ given by (10.24). Hence we have verified that $\mathbf{F}$ is given by (10.9) with $V$ equal to (10.26).

It follows that the equations of motion for a particle in an electromagnetic field may be obtained from the Lagrangian function

$$L = \tfrac{1}{2}m\dot{r}^2 + q\dot{\mathbf{r}} \cdot \mathbf{A}(\mathbf{r}, t) - q\phi(\mathbf{r}, t). \tag{10.27}$$

Note the appearance in $L$ of terms linear in the time derivatives. This function *cannot* be separated into two parts, $T' - V'$, one quadratic in $\dot{r}$ and one independent of it. Another consequence of the appearance of these terms is that the generalized momentum $p_x$ is no longer equal to the familiar mechanical momentum, $m\dot{x}$. Instead,

$$p_x = \frac{\partial L}{\partial \dot{x}} = m\dot{x} + qA_x,$$

or, more generally,

$$\boldsymbol{p} = m\dot{\boldsymbol{r}} + q\boldsymbol{A}. \tag{10.28}$$

We can now obtain the equations of motion in terms of arbitrary co-ordinates from the Lagrangian function (10.27). For example, in terms of cylindrical polars, it reads

$$L = \tfrac{1}{2}m(\dot{\rho}^2 + \rho^2\dot{\varphi}^2 + \dot{z}^2) + q(\dot{\rho}A_\rho + \rho\dot{\varphi}A_\varphi + \dot{z}A_z) - q\phi. \tag{10.29}$$

### *Example:* **Uniform magnetic field**

A charged particle moves in a uniform static magnetic field $\boldsymbol{B}$. Find the solutions of the equations of motion in which $\rho$ is constant.

For a uniform magnetic field, we may take

$$\phi = 0 \qquad \text{and} \qquad \boldsymbol{A} = \tfrac{1}{2}\boldsymbol{B}\wedge\boldsymbol{r},$$

or, if we choose the $z$-axis in the direction of $\boldsymbol{B}$,

$$\phi = 0, \qquad A_\rho = 0, \qquad A_\varphi = \tfrac{1}{2}B\rho, \qquad A_z = 0.$$

(It is easy to verify the relation $\boldsymbol{B} = \boldsymbol{\nabla}\wedge\boldsymbol{A}$ using (A.55).) Thus the Lagrangian function is

$$L = \tfrac{1}{2}m(\dot{\rho}^2 + \rho^2\dot{\varphi}^2 + \dot{z}^2) + \tfrac{1}{2}qB\rho^2\dot{\varphi}. \tag{10.30}$$

Lagrange's equations are therefore

$$m\ddot{\rho} = m\rho\dot{\varphi}^2 + qB\rho\dot{\varphi},$$

$$\frac{\mathrm{d}}{\mathrm{d}t}(m\rho^2\dot{\varphi} + \tfrac{1}{2}qB\rho^2) = 0, \tag{10.31}$$

$$m\ddot{z} = 0.$$

In particular, in the case where $\rho$ is a constant, we learn from the last two equations that $\dot{\varphi}$ and $\dot{z}$ are also constants, and from the first equation that either $\dot{\varphi} = 0$ (particle moving parallel to the $z$-axis), or

$$\dot{\varphi} = -\frac{qB}{m}.$$

This is of course precisely the solution we obtained in §5.2.

The second of the three equations (10.31) is particularly interesting. It shows that, although in general the $z$ component of the particle angular momentum is not a constant, there is still a corresponding conservation law for the quantity

$$p_\varphi = m\rho^2\dot\varphi + \tfrac{1}{2}qB\rho^2.$$

The reason for the existence of such a conservation law will be discussed in Chapter 12.

## 10.6   The Stretched String

As a final example of the use of the Lagrangian method, we consider a rather different kind of problem. This is an example of a system with an *infinite* number of degrees of freedom — a string of length $l$, and mass $\mu$ per unit length, with fixed ends, and stretched to a tension $F$.

We shall consider small transverse oscillations of the string. In place of a finite set of generalized co-ordinates, we now have a continuous function, the displacement $y(x,t)$ of the string from its equilibrium position. For the partial derivatives of $y$ we use the notations

$$\dot y = \frac{\partial y}{\partial t}, \qquad y' = \frac{\partial y}{\partial x}.$$

The kinetic energy of a small element of string of length $dx$ is $\tfrac{1}{2}(\mu\,dx)\dot y^2$. Thus the total kinetic energy is

$$T = \int_0^l \tfrac{1}{2}\mu\dot y^2\,dx.$$

When the string is in equilibrium, its length is $l$. However, when it is displaced, its length is increased, say to $l + \Delta l$, and is given by (3.34), namely

$$l + \Delta l = \int_0^l \sqrt{1 + y'^2}\,dx.$$

The work done against the tension in increasing the length by $\Delta l$ is $F\,\Delta l$. This is the potential energy of the string. For small displacements, we can replace $\sqrt{1 + y'^2}$ by $1 + \tfrac{1}{2}y'^2$. Thus we may take

$$V = \int_0^l \tfrac{1}{2}Fy'^2\,dx. \tag{10.32}$$

Our Lagrangian function is, therefore,

$$L = \int_0^l \left( \tfrac{1}{2}\mu\dot{y}^2 - \tfrac{1}{2}Fy'^2 \right) dx. \tag{10.33}$$

As before, we can use Hamilton's principle to derive the equation of motion. To do this, it will be useful to consider a more general problem, in which the Lagrangian function has the form

$$L = \int_0^l \mathcal{L}(y, \dot{y}, y') \, dx, \tag{10.34}$$

where $\mathcal{L}$ may be called the *Lagrangian density*. (In our particular case, $\mathcal{L}$, given by the integrand of (10.33), is actually independent of $y$, and a function only of its derivatives.)

The action integral is

$$I = \int_{t_0}^{t_1} \int_0^l \mathcal{L}(y, \dot{y}, y') \, dx \, dt.$$

Let us consider a small variation of the displacement $\delta y(x, t)$ which vanishes at $t_0$ and $t_1$, and also (because the ends of the string are fixed) at $x = 0$ and $x = l$. The variation of the action integral is

$$\delta I = \int_{t_0}^{t_1} \int_0^l \left[ \frac{\partial \mathcal{L}}{\partial y} \delta y + \frac{\partial \mathcal{L}}{\partial \dot{y}} \frac{\partial(\delta y)}{\partial t} + \frac{\partial \mathcal{L}}{\partial y'} \frac{\partial(\delta y)}{\partial x} \right] dx \, dt.$$

Now, as in §3.6, we integrate by parts, with respect to $t$ in the second term, and with respect to $x$ in the third. In both cases, the integrated terms vanish because $\delta y$ is zero at the limits of integration. Thus we obtain

$$\delta I = \int_{t_0}^{t_1} \int_0^l \left[ \frac{\partial \mathcal{L}}{\partial y} - \frac{\partial}{\partial t}\left(\frac{\partial \mathcal{L}}{\partial \dot{y}}\right) - \frac{\partial}{\partial x}\left(\frac{\partial \mathcal{L}}{\partial y'}\right) \right] \delta y(x, t) \, dx \, dt.$$

Hamilton's principle requires that this expression should be zero for arbitrary variations $\delta y(x, t)$ vanishing at the limits. This is possible only if the integrand vanishes identically. Hence we obtain Lagrange's equation

$$\frac{\partial \mathcal{L}}{\partial y} - \frac{\partial}{\partial t}\left(\frac{\partial \mathcal{L}}{\partial \dot{y}}\right) - \frac{\partial}{\partial x}\left(\frac{\partial \mathcal{L}}{\partial y'}\right) = 0. \tag{10.35}$$

In the particular case of the stretched string, the derivatives of the Lagrangian density are

$$\frac{\partial \mathcal{L}}{\partial y} = 0, \qquad \frac{\partial \mathcal{L}}{\partial \dot{y}} = \mu\dot{y}, \qquad \frac{\partial \mathcal{L}}{\partial y'} = -Fy'.$$

Thus Lagrange's equation becomes

$$\ddot{y} = c^2 y'', \qquad \text{where} \qquad c^2 = F/\mu. \tag{10.36}$$

In place of a discrete set of ordinary differential equations for the generalized co-ordinates $q_\alpha(t)$, we now have a *partial* differential equation for the function $y(x,t)$ of two independent variables. Note that $c^2$ has the dimensions of (velocity)$^2$. In fact, as we show below, $c$ is the velocity of propagation of waves along the string.

Equation (10.36) is the one-dimensional *wave equation*. Similar equations turn up in many branches of physics, whenever wave phenomena are encountered.

### *Example:* **Solution of the stretched-string equation**

Given the initial displacement $y(x,0)$ and velocity $\dot{y}(x,0)$ of the string find its displacement at a later time $t$.

Ignoring for the moment the boundary conditions at $x = 0$ and $x = l$, the general solution of the wave equation (10.36) involves two arbitrary *functions*, which may be determined by the initial values of $y$ and $\dot{y}$. It is easy to verify by direct substitution that, for any function $f$,

$$y(x,t) = f(x - ct)$$

is a solution. It represents a wave travelling along the string with velocity $c$ to the right; for, the shape of the displacement $y$ is the same at time $t$ as at time 0, but shifted to the right by a distance $ct$ (see Fig. 10.3). Similarly,

$$y(x,t) = f(x + ct)$$

is a solution, representing a wave travelling to the left along the string. The general solution is

$$y = f(x + ct) + g(x - ct),$$

where $f$ and $g$ are arbitrary functions.

To satisfy the boundary conditions at $x = 0$ and $x = l$ both terms in the solution must be present. For a string of finite length, a wave cannot travel indefinitely in one direction. When it gets to the end of the string it must be reflected back.

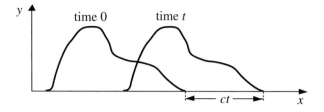

Fig. 10.3

The condition $y = 0$ at $x = 0$ requires that

$$f(ct) + g(-ct) = 0$$

for all $t$. Thus the functions $f$ and $g$ differ only in sign, and in the sign of the argument, and we can write the solution as

$$y = f(x + ct) - f(ct - x). \qquad (10.37)$$

The condition $y = 0$ at $x = l$ requires that

$$f(ct + l) - f(ct - l) = 0$$

for all $t$. In other words, $f$ must be a *periodic* function, of period $2l$, satisfying

$$f(x + 2l) = f(x). \qquad (10.38)$$

To establish the connection between the periodic function $f$ and the initial conditions, let us write $f$ as the sum of an even function and an odd function:

$$f(x) = \tfrac{1}{2}[h(x) + k(x)], \qquad (10.39)$$

where

$$h(x) = f(x) + f(-x),$$
$$k(x) = f(x) - f(-x).$$

To fix $f$, we have to specify its values in the interval from $-l$ to $l$, or, equivalently, to specify both functions $h$ and $k$ in the interval from $0$ to $l$. Now from (10.37), the initial value of $y$ is

$$y(x, 0) = f(x) - f(-x) = k(x). \qquad (10.40)$$

This therefore determines the odd part of $f$. Moreover, the initial value of $\dot{y}$ is

$$\dot{y}(x,0) = cf'(x) - cf'(-x) = ch'(x). \tag{10.41}$$

(Remember that $[f(-x)]' = -f'(-x)$.) Thus the initial value of $\dot{y}$ determines the even part of $f$ (up to an irrelevant additive constant, which cancels in (10.37)).

## 10.7  Summary

The position of every part of a system may be fixed by specifying the values of a set of generalized co-ordinates. If these co-ordinates can all vary independently, the system is holonomic. This is the case in all the examples we have considered. The system is natural if the functions specifying the positions of particles in terms of the generalized co-ordinates do not involve the time explicitly. In that case, the kinetic energy is a homogeneous quadratic function of the $\dot{q}_\alpha$. For a forced system, on the other hand, $T$ may contain linear and constant terms. In either case, the equations of motion are given by Lagrange's equations. If the forces are conservative (and sometimes in other cases too), all we need is the Lagrangian function $L = T - V$. In general, for dissipative forces, the generalized forces $F_\alpha$ corresponding to the generalized co-ordinates $q_\alpha$ must be found by evaluating the work done in a small displacement.

---

## Problems

1. Masses $m$ and $2m$ are joined by a light inextensible string which runs without slipping over a uniform circular pulley of mass $2m$ and radius $a$. Using the angular position of the pulley as generalized co-ordinate, write down the Lagrangian function, and Lagrange's equation. Find the acceleration of the masses.

2. A uniform cylindrical drum of mass $M$ and radius $a$ is free to rotate about its axis, which is horizontal. A cable of negligible mass and length $l$ is wound on the drum, and carries on its free end a mass $m$. Write down the Lagrangian function in terms of an appropriate generalized co-ordinate, assuming no slipping or stretching of the cable. If the cable

is initially fully wound up, and the system is released from rest, find the angular velocity of the drum when it is fully unwound.

3. Treat the system of Problem 2 as one with two generalized co-ordinates, the angular position of the drum and the free length of cable, with an appropriate constraint. Hence find the tension in the cable. (Show that it is equal to the Lagrange multiplier.)

4. *Find the Lagrangian function for the system of Problem 2 if the cable is elastic, with elastic potential energy $\frac{1}{2}kx^2$, where $x$ is the extension of the cable. Show that the motion of the mass $m$ is a uniform acceleration at the same rate as before, with a superimposed oscillation of angular frequency given by $\omega^2 = k(M + 2m)/Mm$. Find the amplitude of this oscillation if the system is released from rest with the cable unextended.

5. Write down the kinetic energy of a particle in cylindrical polar co-ordinates in a frame rotating with angular velocity $\omega$ about the $z$-axis. Show that the terms proportional to $\omega$ and $\omega^2$ reproduce the Coriolis force and centrifugal force respectively.

6. A light inextensible string passes over a light smooth pulley, and carries a mass $4m$ on one end. The other end supports a second pulley with a string over it carrying masses $3m$ and $m$ on the two ends. Using a suitable pair of generalized co-ordinates, write down the Lagrangian function for the system, and Lagrange's equations. Find the downward accelerations of the three masses.

7. Find the tensions in the strings in Problem 6. Explain why the first pulley turns, although the total mass on each side is the same.

8. Evaluate accurately the two possible precessional angular velocities of the top described in Chapter 9, Problem 13, if the axis makes an angle of $30°$ with the vertical. Compare the slower value with the approximate result found earlier. Find also the minimum angular velocity $\omega_3$ for which steady precession at this angle is possible.

9. A simple pendulum of mass $m$ and length $l$ hangs from a trolley of mass $M$ running on smooth horizontal rails. The pendulum swings in a plane parallel to the rails. Using the position $x$ of the trolley and the angle of inclination $\theta$ of the pendulum as generalized co-ordinates, write down the Lagrangian function, and Lagrange's equations. Obtain an equation of motion for $\theta$ alone. If the system is released from rest with the pendulum inclined at $30°$ to the vertical, use energy conservation to find its angular velocity when it reaches the vertical, given that $M = 2\,\text{kg}$, $m = 1\,\text{kg}$, and $l = 2\,\text{m}$.

10. *Show that the kinetic energy of the gyroscope described in Chapter 9, Problem 21, is

$$T = \tfrac{1}{2}I_1(\Omega \sin \lambda \cos \varphi)^2 + \tfrac{1}{2}I_1(\dot\varphi + \Omega \cos \lambda)^2 + \tfrac{1}{2}I_3(\dot\psi + \Omega \sin \lambda \sin \varphi)^2.$$

From Lagrange's equations, show that the angular velocity $w_3$ about the axis is constant, and obtain the equation for $\varphi$ without neglecting $\Omega^2$. Show that motion with the axis pointing north becomes unstable for very small values of $w_3$, and find the smallest value for which it is stable. What are the stable positions when $w_3 = 0$? Interpret this result in terms of a non-rotating frame.

11. *Find the Lagrangian function for a symmetric top whose pivot is free to slide on a smooth horizontal table, in terms of the generalized co-ordinates $X, Y, \varphi, \theta, \psi$, and the principal moments $I_1^*, I_1^*, I_3^*$ about the centre of mass. (Note that $Z$ is related to $\theta$.) Show that the horizontal motion of the centre of mass may be completely separated from the rotational motion. What difference is there in the equation (10.15) for steady precession? Are the precessional angular velocities greater or less than in the case of a fixed pivot? Show that steady precession at a given value of $\theta$ can occur for a smaller value of $w_3$ than in the case of a fixed pivot.

12. *A uniform plank of length $2a$ is placed with one end on a smooth horizontal floor and the other against a smooth vertical wall. Write down the Lagrangian function, using two generalized co-ordinates, the distance $x$ of the foot of the plank from the wall, and its angle $\theta$ of inclination to the horizontal, with a suitable constraint between the two. Given that the plank is initially at rest at an inclination of $60°$, find the angle at which it loses contact with the wall. (*Hint:* First write the co-ordinates of the centre of mass in terms of $x$ and $\theta$. Note that the reaction at the wall is related to the Lagrange multiplier.)

13. Use Hamilton's principle to show that if $F$ is any function of the generalized co-ordinates, then the Lagrangian functions $L$ and $L + dF/dt$ must yield the same equations of motion. Hence show that the equations of motion of a charged particle in an electromagnetic field are unaffected by the 'gauge transformation' (A.42). (*Hint:* Take $F = -q\Lambda$.)

14. The stretched string of §10.6 is released from rest with its mid-point displaced a distance $a$, and each half of the string straight. Find the function $f(x)$. Describe the shape of the string after (a) a short time, (b) a time $l/2c$, and (c) a time $l/c$.

15. *Two bodies of masses $M_1$ and $M_2$ are moving in circular orbits of radii $a_1$ and $a_2$ about their centre of mass. The *restricted three-body problem* concerns the motion of a third small body of mass $m$ ($\ll M_1$ or $M_2$) in their gravitational field (*e.g.*, a spacecraft in the vicinity of the Earth–Moon system). Assuming that the third body is moving in the plane of the first two, write down the Lagrangian function of the system, using a rotating frame in which $M_1$ and $M_2$ are fixed. Find the equations of motion. (*Hint*: The identities $GM_1 = \omega^2 a^2 a_2$ and $GM_2 = \omega^2 a^2 a_1$ may be useful, with $a = a_1 + a_2$ and $\omega^2 = GM/a^3$.)

16. *For the system of Problem 15, find the equations that must be satisfied for 'equilibrium' in the rotating frame (*i.e.*, circular motion with the same angular velocity as $M_1$ and $M_2$). Consider 'equilibrium' positions on the line of centres of $M_1$ and $M_2$. By roughly sketching the effective potential energy curve, show that there are three such positions, but that all three are unstable. (*Note*: The positions are actually the solutions of a quintic equation.) Show also that there are two 'equilibrium' positions *off* the line of centres, in each of which the three bodies form an equilateral triangle. (The stability of these so-called *Lagrangian points* is the subject of Problem 12, Chapter 12. There is further consideration of this important problem in §14.4.)

# Chapter 11

# Small Oscillations and Normal Modes

In this chapter, we shall discuss a generalization of the harmonic oscillator problem treated in §2.2 — the oscillations of a system of several degrees of freedom near a position of equilibrium. We consider only conservative, holonomic systems, described by $n$ generalized co-ordinates $q_1, q_2, \ldots, q_n$. Without loss of generality, we may choose the position of equilibrium to be $q_1 = q_2 = \cdots = q_n = 0$. We shall begin by investigating the form of the kinetic and potential energy functions near this point.

## 11.1 Orthogonal Co-ordinates

We shall restrict our attention to natural systems, for which the kinetic energy is a homogeneous quadratic function of $\dot{q}_1, \dot{q}_2, \ldots, \dot{q}_n$. (More generally, we could include also those forced systems — like that of §10.4 — for which $L$ can be written as a sum of a quadratic term $T'$ and a term $-V'$ independent of the time derivatives. The only essential restriction is that there should be no linear terms.)

For example, for $n = 2$, we have

$$T = \tfrac{1}{2}a_{11}\dot{q}_1^2 + a_{12}\dot{q}_1\dot{q}_2 + \tfrac{1}{2}a_{22}\dot{q}_2^2. \tag{11.1}$$

In general, the coefficients $a_{11}, a_{12}, a_{22}$ will be functions of $q_1$ and $q_2$. However, if we are interested only in small values of $q_1$ and $q_2$, we may neglect this dependence, and treat them as constants (equal to their values at $q_1 = q_2 = 0$).

For a particle described by curvilinear co-ordinates, the co-ordinates are called *orthogonal* if the co-ordinate curves always intersect at right angles (see §3.5). In that case, the kinetic energy contains terms in $\dot{q}_1^2, \dot{q}_2^2, \dot{q}_3^2$, but no cross products like $\dot{q}_1\dot{q}_2$. By an extension of this terminology, the generalized co-ordinates $q_1, q_2, \ldots, q_n$ are called *orthogonal* if $T$ is a sum of squares, with no cross products — for example, if $a_{12}$ above is zero.

It is a considerable simplification to choose the co-ordinates to be orthogonal, and this can always be done. For instance, we may set

$$q_1' = q_1 + \frac{a_{12}}{a_{11}} q_2,$$

so that, in terms of $q_1'$ and $q_2$, (11.1) becomes

$$T = \tfrac{1}{2} a_{11} \dot{q}_1'^2 + \tfrac{1}{2} a_{22}' \dot{q}_2^2, \qquad \text{with} \qquad a_{22}' = a_{22} - \frac{a_{12}^2}{a_{11}}. \qquad (11.2)$$

We can even go further. Since $T$ is necessarily positive for all possible $\dot{q}_1'$, $\dot{q}_2$, the coefficients in (11.2) must be positive numbers. Hence we can define new co-ordinates

$$q_1'' = \sqrt{a_{11}} q_1', \qquad q_2'' = \sqrt{a_{22}'} q_2,$$

so that $T$ is reduced to a simple sum of squares:

$$T = \tfrac{1}{2} \dot{q}_1''^2 + \tfrac{1}{2} \dot{q}_2''^2. \qquad (11.3)$$

A similar procedure can be used in the general case. (It is called the *Gram–Schmidt* orthogonalization procedure.) We may first eliminate the cross products involving $\dot{q}_1$ by means of the transformation to

$$q_1' = q_1 + \frac{a_{12}}{a_{11}} q_2 + \cdots + \frac{a_{1n}}{a_{11}} q_n,$$

then those involving $\dot{q}_2$, and so on. Thus we can always reduce $T$ to the standard form

$$T = \sum_{\alpha=1}^{n} \tfrac{1}{2} \dot{q}_\alpha^2, \qquad (11.4)$$

where we now drop any primes.

As an illustration of these ideas, let us consider the double pendulum illustrated in Fig. 11.1.

*Example:* **Kinetic energy of the double pendulum**

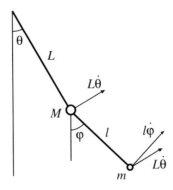

Fig. 11.1

A double pendulum consists of a simple pendulum of mass $M$ and length $L$, with a second simple pendulum of mass $m$ and length $l$ suspended from it. Consider only motion in a vertical plane, so that the system has two degrees of freedom. Find a standard set of co-ordinates in terms of which the kinetic energy takes the form (11.4).

As generalized co-ordinates, we may initially choose the inclinations $\theta$ and $\varphi$ to the downward vertical. The velocity of the upper pendulum bob is $L\dot{\theta}$. That of the lower bob has two components — the velocity $L\dot{\theta}$ of its point of support, and the velocity $l\dot{\varphi}$ relative to that point. The angle between these is $\varphi - \theta$. Hence the kinetic energy is

$$T = \tfrac{1}{2}ML^2\dot{\theta}^2 + \tfrac{1}{2}m[L^2\dot{\theta}^2 + l^2\dot{\varphi}^2 + 2Ll\dot{\theta}\dot{\varphi}\cos(\varphi - \theta)].$$

For small values of $\theta$ and $\varphi$, we may approximate $\cos(\varphi - \theta)$ by 1. Since there is a term in $\dot{\theta}\dot{\varphi}$, these co-ordinates are not orthogonal, but we can make them so by adding an appropriate multiple of $\varphi$ to $\theta$, or, equally well, of $\theta$ to $\varphi$. In fact, it is easy to see physically that a pair of orthogonal co-ordinates is provided by the displacements of the two bobs, which for small angles are

$$x = L\theta, \qquad y = L\theta + l\varphi. \tag{11.5}$$

In terms of these, $T$ becomes

$$T = \tfrac{1}{2}M\dot{x}^2 + \tfrac{1}{2}m\dot{y}^2. \tag{11.6}$$

To complete the reduction to the standard form (11.4), we may define the new co-ordinates

$$q_1 = \sqrt{M}x, \qquad q_2 = \sqrt{m}y. \tag{11.7}$$

In practice, this is not always essential, though it makes the discussion of the general case much easier.

## 11.2  Equations of Motion for Small Oscillations

Now let us consider the potential energy function $V$. With $T$ given by (11.4), the equations of motion are simply

$$\ddot{q}_\alpha = -\frac{\partial V}{\partial q_\alpha}, \qquad \text{for} \qquad \alpha = 1, 2, \ldots, n. \tag{11.8}$$

Thus the condition for equilibrium is that all $n$ partial derivatives of $V$ should vanish at the equilibrium position.

For small values of the co-ordinates, we can expand $V$ as a series, just as we did for a single co-ordinate in §2.2. For example, for $n = 2$,

$$V = V_0 + (b_1 q_1 + b_2 q_2) + (\tfrac{1}{2}k_{11}q_1^2 + k_{12}q_1 q_2 + \tfrac{1}{2}k_{22}q_2^2) + \cdots .$$

The equilibrium conditions require that the linear terms should be zero, $b_1 = b_2 = 0$, just as in §2.2. Moreover, the constant term $V_0$ is arbitrary, and may be set equal to zero without changing the equations of motion. Thus the leading terms are the quadratic ones, and for small values of $q_1$ and $q_2$, we may approximate $V$ by

$$V = \tfrac{1}{2}k_{11}q_1^2 + k_{12}q_1 q_2 + \tfrac{1}{2}k_{22}q_2^2. \tag{11.9}$$

Then the equations of motion (11.8) become

$$\begin{aligned}
\ddot{q}_1 &= -k_{11}q_1 - k_{12}q_2, \\
\ddot{q}_2 &= -k_{21}q_1 - k_{22}q_2,
\end{aligned} \tag{11.10}$$

where for the sake of symmetry we have written $k_{21} = k_{12}$ in the second equation.

In the general case, $V$ may be taken to be a homogeneous quadratic function of the co-ordinates, which can be written

$$V = \sum_{\alpha=1}^{n} \sum_{\beta=1}^{n} \tfrac{1}{2} k_{\alpha\beta} q_\alpha q_\beta, \tag{11.11}$$

with $k_{\beta\alpha} = k_{\alpha\beta}$. (Notice that each term with $\alpha \neq \beta$ appears twice, for example $\tfrac{1}{2} k_{12} q_1 q_2$ and $\tfrac{1}{2} k_{21} q_2 q_1$, which are of course equal.) Then the equations of motion are

$$\ddot{q}_\alpha = -\sum_{\beta=1}^{n} k_{\alpha\beta} q_\beta, \qquad \text{for} \qquad \alpha = 1, 2, \ldots, n, \tag{11.12}$$

or, in matrix notation,

$$\begin{bmatrix} \ddot{q}_1 \\ \ddot{q}_2 \\ \vdots \\ \ddot{q}_n \end{bmatrix} = - \begin{bmatrix} k_{11} & k_{12} & \ldots & k_{1n} \\ k_{21} & k_{22} & \ldots & k_{2n} \\ \vdots & \vdots & & \vdots \\ k_{n1} & k_{n2} & \ldots & k_{nn} \end{bmatrix} \begin{bmatrix} q_1 \\ q_2 \\ \vdots \\ q_n \end{bmatrix}. \tag{11.13}$$

*Example:* **Equations of motion of the double pendulum**

Find the equations of motion of the double pendulum, in terms of the orthogonal co-ordinates $x$ and $y$.

The potential energy of the double pendulum is easily seen to be

$$V = (M + m)gL(1 - \cos\theta) + mgl(1 - \cos\varphi).$$

For small angles, we may approximate $(1 - \cos\theta)$ by $\tfrac{1}{2}\theta^2$. Hence in terms of $x$ and $y$, we have

$$V = \tfrac{1}{2}(M + m)gL\theta^2 + \tfrac{1}{2}mgl\varphi^2 = \frac{(M + m)g}{2L}x^2 + \frac{mg}{2l}(y - x)^2.$$

Thus the equations of motion are

$$\begin{bmatrix} M\ddot{x} \\ m\ddot{y} \end{bmatrix} = \begin{bmatrix} -\dfrac{(M + m)g}{L} - \dfrac{mg}{l} & \dfrac{mg}{l} \\ \dfrac{mg}{l} & -\dfrac{mg}{l} \end{bmatrix} \begin{bmatrix} x \\ y \end{bmatrix}. \tag{11.14}$$

Note the appearance of the masses on the left hand side, because we have not gone over to the normalized variables of (11.7).

## 11.3   Normal Modes

The general solution of a pair of second-order differential equations like
(11.14) must involve four arbitrary constants, which may be fixed by the
initial values of $q_1, q_2, \dot{q}_1, \dot{q}_2$. Similarly, the general solution of (11.13) must
involve $2n$ arbitrary constants. To find the general solution, we adopt a
generalization of the method used for the damped harmonic oscillator in
§2.5: we look first for solutions in which all the co-ordinates are oscillating
with the same frequency $\omega$, of the form

$$q_\alpha = A_\alpha e^{i\omega t}, \tag{11.15}$$

where the $A_\alpha$ are complex constants. (As in §2.3 and §2.5, the physical
solution may be taken to be the real part of (11.15).) Such solutions are
called *normal modes* of the system.

Substituting (11.15) into (11.13), we obtain a set of $n$ simultaneous
equations for the $n$ amplitudes $A_n$,

$$-\omega^2 A_\alpha = -\sum_{\beta=1}^{n} k_{\alpha\beta} A_\beta. \tag{11.16}$$

Let us consider first the case $n = 2$. Then these equations are

$$\begin{bmatrix} k_{11} & k_{12} \\ k_{21} & k_{22} \end{bmatrix} \begin{bmatrix} A_1 \\ A_2 \end{bmatrix} = \omega^2 \begin{bmatrix} A_1 \\ A_2 \end{bmatrix}. \tag{11.17}$$

This is what is known as an *eigenvalue equation*. The values of $\omega^2$ for which
non-zero solutions exist are called the *eigenvalues* of the $2 \times 2$ matrix with
elements $k_{\alpha\beta}$. The column vector formed by the $A_\alpha$ is an *eigenvector* of
the matrix.

The equations (11.17) can alternatively be written as

$$\begin{bmatrix} k_{11} - \omega^2 & k_{12} \\ k_{21} & k_{22} - \omega^2 \end{bmatrix} \begin{bmatrix} A_1 \\ A_2 \end{bmatrix} = \begin{bmatrix} 0 \\ 0 \end{bmatrix}.$$

These equations have a non-zero solution if and only if the *determinant* of
the coefficient matrix vanishes, *i.e.*, if and only if

$$\begin{vmatrix} k_{11} - \omega^2 & k_{12} \\ k_{21} & k_{22} - \omega^2 \end{vmatrix} = (k_{11} - \omega^2)(k_{22} - \omega^2) - k_{12}^2 = 0. \tag{11.18}$$

This is called the *characteristic equation* for the system. It determines the frequencies $\omega$ of the normal modes, which are the square roots of the eigenvalues $\omega^2$.

Equation (11.18) is a quadratic equation for $\omega^2$. Its discriminant may be written $(k_{11} - k_{22})^2 + 4k_{12}^2$, which is clearly positive. Hence it always has two real roots. The condition for stability is that both roots should be positive. A negative root, say $-\gamma^2$, would yield a solution of the form

$$q_\alpha = A_\alpha e^{\gamma t} + B_\alpha e^{-\gamma t},$$

where both the $A_\alpha$ and the $B_\alpha$ coefficients constitute eigenvectors of the $2 \times 2$ matrix, corresponding to the eigenvalue $-\gamma^2$. Except in the degenerate case of two equal eigenvalues, this means that the $A_\alpha$ and $B_\alpha$ coefficients must be proportional, since the eigenvector is unique up to an overall factor. In general, therefore, the solution yields an exponential increase in the displacements with time.

This stability condition, that all the eigenvalues be positive, is a natural generalization of the requirement, in the one-dimensional case, that the second derivative of the potential energy function be positive (see §2.2).

If $\omega^2$ is chosen equal to one of the two roots of (11.18), then either of the two equations in (11.17) fixes the ratio $A_1/A_2$. Since the coefficients are real numbers, the ratio is obviously real. (This is a special case of a general theorem about symmetric matrices — those satisfying $k_{\beta\alpha} = k_{\alpha\beta}$. See §A.10.) This means that $A_1$ and $A_2$ have the same phase (or phases differing by $\pi$), so that $q_1$ and $q_2$ not only oscillate with the same frequency, but actually in (or directly out of) phase. The ratio of $q_1$ to $q_2$ remains fixed throughout the motion.

There remains in $A_1$ and $A_2$ a common arbitrary complex factor, which serves to fix the overall amplitude and phase of the normal mode solution. Thus each normal mode solution contains two arbitrary real constants.

Since the equations of motion (11.10) are linear, any superposition of solutions is again a solution. Hence the general solution is simply a super-position of the two normal mode solutions. If $\omega^2$ and $\omega'^2$ are the roots of (11.18), it may be written as the real part of

$$\begin{aligned} q_1 &= A_1 e^{i\omega t} + A_1' e^{i\omega' t}, \\ q_2 &= A_2 e^{i\omega t} + A_2' e^{i\omega' t}, \end{aligned} \tag{11.19}$$

in which the ratios $A_1/A_2$ and $A_1'/A_2'$ are fixed by (11.17).

*Example:* **Normal modes of the double pendulum**

Find the normal mode frequencies of the double pendulum.

Here the equations (11.17) may be written

$$
\begin{bmatrix}
\dfrac{(M+m)g}{ML} + \dfrac{mg}{Ml} & -\dfrac{mg}{Ml} \\[2ex]
-\dfrac{g}{l} & \dfrac{g}{l}
\end{bmatrix}
\begin{bmatrix} A_x \\[1ex] A_y \end{bmatrix}
= \omega^2 \begin{bmatrix} A_x \\[1ex] A_y \end{bmatrix}.
\tag{11.20}
$$

The characteristic equation (11.18) simplifies to

$$
\omega^4 - \frac{M+m}{M}\left(\frac{g}{L}+\frac{g}{l}\right)\omega^2 + \frac{M+m}{M}\frac{g^2}{Ll} = 0.
\tag{11.21}
$$

The roots of this equation determine the frequencies of the two normal modes.

It is interesting to examine special limiting cases. First, let us suppose that the upper pendulum is very heavy ($M \gg m$). Then, provided that $l$ is not too close to $L$, the two roots, with the corresponding ratios determined by (11.20), are, approximately

$$
\omega^2 \approx \frac{g}{l}, \qquad \frac{A_x}{A_y} \approx \frac{m}{M}\frac{L}{l-L},
$$

and

$$
\omega^2 \approx \frac{g}{L}, \qquad \frac{A_x}{A_y} \approx \frac{L-l}{L}.
$$

In the first mode, the upper pendulum is practically stationary, while the lower is swinging with its natural frequency. In the second mode, whose frequency is that of the upper pendulum, the amplitudes are comparable.

At the other extreme, if $M \ll m$, the two normal modes are

$$
\omega^2 \approx \frac{g}{L+l}, \qquad \frac{A_x}{A_y} \approx \frac{L}{L+l},
$$

and

$$
\omega^2 \approx \frac{m}{M}\left(\frac{g}{L}+\frac{g}{l}\right), \qquad \frac{A_x}{A_y} \approx -\frac{m}{M}\frac{L+l}{L}.
$$

In the first mode, the pendulums swing almost like a single rigid pendulum of length $L + l$. In the second, the lower bob remains almost stationary, while the upper one executes a very rapid oscillation.

The normal modes of a system with $n$ degrees of freedom may be found by a very similar method. The condition for consistency of the simultaneous equations (11.16) is again that the determinant of the coefficients should vanish. For example, for $n = 3$, we require

$$\begin{vmatrix} k_{11} - \omega^2 & k_{12} & k_{13} \\ k_{21} & k_{22} - \omega^2 & k_{23} \\ k_{31} & k_{32} & k_{33} - \omega^2 \end{vmatrix} = 0. \qquad (11.22)$$

This is a cubic equation for $\omega^2$. It can be proved that its three roots are all real (see §A.10). As before, the condition for stability is that all three roots should be positive. The roots then determine the frequencies of the three normal modes.

For each normal mode, the ratios of the amplitudes are fixed by the equations (11.16). As in the case $n = 2$, these ratios are all real, so that in a normal mode all the co-ordinates oscillate in phase (or 180° out of phase). Each normal mode solution involves just two arbitrary real constants, and the general solution is a superposition of all the normal modes.

## 11.4 Coupled Oscillators

One often encounters examples of physical systems that may be described as two (or more) harmonic oscillators, which are approximately independent, but with some kind of relatively weak coupling between the two. (As a specific example, we shall consider below the system shown in Fig. 11.2, which consists of a pair of identical pendulums coupled by a spring.)

If the co-ordinate $q$ of a harmonic oscillator is normalized so that $T = \frac{1}{2}\dot{q}^2$, then $V = \frac{1}{2}\omega^2 q^2$, where $\omega$ is the angular frequency. Hence for a pair of uncoupled oscillators, the coefficients in (11.10) are

$$k_{11} = \omega_1^2, \qquad k_{12} = 0, \qquad k_{22} = \omega_2^2.$$

When the oscillators are weakly coupled, these equalities will still be approximately true, so that in particular $k_{12}$ is small in comparison to $k_{11}$ and $k_{22}$. Thus from (11.18) it is clear that the characteristic frequencies of the system are given by $\omega^2 \approx k_{11}$ and $\omega^2 \approx k_{22}$; as one might expect, they are close to the frequencies of the uncoupled oscillators. Then from (11.17) we see that, in the first normal mode, the ratio $A_2/A_1$ is approximately $k_{12}/(k_{11} - k_{22})$. Thus, *unless* the frequencies of the two normal modes are nearly equal, the normal modes differ very little from those of the uncoupled system, and the coupling is not of great importance. The interesting case, in which even a weak coupling can be important, is that in which the two frequencies are equal, or nearly so.

As a specific example of this case, we consider a pair of pendulums, each of mass $m$ and length $l$, coupled by a weak spring (see Fig. 11.2). We shall

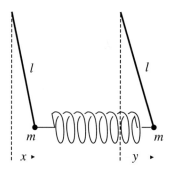

Fig. 11.2

use the displacements $x$ and $y$ as generalized co-ordinates. Then, in the absence of coupling, the potential energy is approximately $\frac{1}{2}m\omega_0^2(x^2+y^2)$, where $\omega_0^2 = g/l$ gives the free oscillation frequency. For small values of $x$ and $y$, the potential energy of the spring has the form $\frac{1}{2}k(x-y)^2$. It will be convenient to introduce another frequency, $\omega_s$ defined by $\omega_s^2 = k/m$. In fact, $\omega_s$ is the angular frequency of the spring if one end is held fixed and the other attached to a mass $m$. Thus we take

$$T = \tfrac{1}{2}m(\dot{x}^2 + \dot{y}^2), \qquad V = \tfrac{1}{2}m(\omega_0^2 + \omega_s^2)(x^2 + y^2) - m\omega_s^2 xy. \qquad (11.23)$$

The normal mode equations now read

$$\begin{bmatrix} \omega_0^2 + \omega_s^2 & -\omega_s^2 \\ -\omega_s^2 & \omega_0^2 + \omega_s^2 \end{bmatrix} \begin{bmatrix} A_x \\ A_y \end{bmatrix} = \omega^2 \begin{bmatrix} A_x \\ A_y \end{bmatrix}. \qquad (11.24)$$

The two solutions of the characteristic equation are easily seen to be

$$\begin{aligned} \omega^2 &= \omega_0^2 & \text{with} \quad & A_x/A_y = 1, \\ \omega^2 &= \omega_0^2 + 2\omega_s^2 & \text{with} \quad & A_x/A_y = -1. \end{aligned}$$

In the first normal mode, the two pendulums oscillate together in phase, with equal amplitude (see Fig. 11.3(a)). Since the spring is neither expanded nor compressed in this motion, it is not surprising that the frequency is just that of the uncoupled pendulums. In the second normal mode, which has a somewhat higher frequency, the pendulums swing in opposite directions, alternately expanding and compressing the spring (see Fig. 11.3(b)).

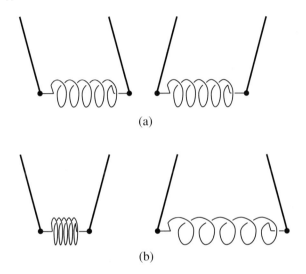

Fig. 11.3

The general solution is a superposition of these two normal modes, and may be written as the real part of

$$x = Ae^{i\omega_0 t} + A'e^{i\omega' t},$$
$$y = Ae^{i\omega_0 t} - A'e^{i\omega' t}, \qquad (11.25)$$

where $\omega'^2 = \omega_0^2 + 2\omega_s^2$. The constants $A$ and $A'$ may be determined by the initial conditions.

### *Example:* Motion of coupled pendulums

Find the solution for the positions of the bobs if the system is released from rest with one bob displaced a distance $a$ from its equilibrium position. Describe the motion.

Here, at $t = 0$, we have $x = a, y = 0, \dot{x} = 0, \dot{y} = 0$, whence we find $A = A' = a/2$, so the solution is

$$x = \tfrac{1}{2}a\cos\omega_0 t + \tfrac{1}{2}a\cos\omega' t = a\cos\omega_- t\,\cos\omega_+ t,$$
$$y = \tfrac{1}{2}a\cos\omega_0 t - \tfrac{1}{2}a\cos\omega' t = a\sin\omega_- t\,\sin\omega_+ t,$$

where $\omega_\pm = \tfrac{1}{2}(\omega' \pm \omega_0)$, and we have used standard trigonometric identities.

Since the spring is weak, $\omega'$ is only slightly greater than $\omega_0$, and therefore $\omega_- \ll \omega_+$. We have beats between two nearly equal frequencies. Thus we may describe the motion as follows. Initially, the first spring swings with angular frequency $\omega_+$ and gradually decreasing amplitude, $a\cos\omega_- t$. Meanwhile, the second pendulum starts to swing with the same angular frequency, but 90° out of phase, and with gradually increasing amplitude $a\sin\omega_- t$. After a time $\pi/2\omega_-$, the first pendulum has come momentarily to rest, and the second is oscillating with amplitude $a$. Then its amplitude starts to decrease, while the first increases again. The whole process is then repeated indefinitely (though in practice there will of course be some damping).

This behaviour should be contrasted with that of a pair of coupled oscillators of very different frequencies. In such a case, if one is started oscillating, one of the two normal modes will have a much larger amplitude than the other. Thus only a very small oscillation will be set up in the second oscillator, and the amplitude of the first will be practically constant.

### Normal co-ordinates

The two normal modes of this system (or the $n$ normal modes in the general case) are completely independent. We can make this fact explicit by introducing new 'normal' co-ordinates. In the case of the coupled pendulums, we introduce, in place of $x$ and $y$, the new co-ordinates

$$q_1 = \sqrt{\frac{m}{2}}(x + y), \qquad q_2 = \sqrt{\frac{m}{2}}(x - y). \tag{11.26}$$

In terms of these co-ordinates, the solution (11.25) is

$$\begin{aligned}
q_1 &= A_1 e^{i\omega_0 t}, & A_1 &= \sqrt{2m}A, \\
q_2 &= A_2 e^{i\omega' t}, & A_2 &= \sqrt{2m}A'.
\end{aligned} \tag{11.27}$$

Thus, in each normal mode, one co-ordinate only is oscillating. Co-ordinates with this property are called *normal co-ordinates.*

The independence of the two normal co-ordinates may also be seen by examining the Lagrangian function. From (11.23) and (11.26), we find

$$T = \tfrac{1}{2}\dot{q}_1^2 + \tfrac{1}{2}\dot{q}_2^2, \qquad V = \tfrac{1}{2}\omega_0^2(q_1^2 + q_2^2) + \omega_s^2 q_2^2,$$

whence the Lagrangian function is

$$L = \tfrac{1}{2}(\dot{q}_1^2 - \omega_0^2 q_1^2) + \tfrac{1}{2}(\dot{q}_2^2 - \omega'^2 q_2^2). \tag{11.28}$$

In effect, we have reduced the Lagrangian to that for a pair of uncoupled oscillators, with angular frequencies $\omega_0$ and $\omega'$.

The normal co-ordinates are very useful in studying the effect on the system of a prescribed external force.

### *Example:* **Forced oscillation of coupled pendulums**

Find the amplitudes of forced oscillations of the coupled pendulums if one of them is subjected to a periodic force $F(t) = F_1 \cos \omega_1 t$.

We write the force as the real part of $F_1 e^{i\omega_1 t}$. To find the equations of motion in the presence of this force, we must evaluate the work done in a small displacement (see §10.2). This is

$$F(t)\delta x = \frac{F(t)}{\sqrt{2m}}(\delta q_1 + \delta q_2).$$

Hence the equations of motion are

$$\begin{aligned}
\ddot{q}_1 &= -\omega_0^2 q_1 + \frac{F_1 e^{i\omega_1 t}}{\sqrt{2m}}, \\
\ddot{q}_2 &= -\omega'^2 q_2 + \frac{F_1 e^{i\omega_1 t}}{\sqrt{2m}}.
\end{aligned} \tag{11.29}$$

These independent oscillator equations may be solved exactly as in §2.6. In particular, the amplitudes of the forced oscillations are given by

$$A_1 = \frac{F_1/\sqrt{2m}}{\omega_0^2 - \omega_1^2}, \qquad A_2 = \frac{F_1/\sqrt{2m}}{\omega'^2 - \omega_1^2}.$$

The corresponding forced oscillation amplitudes for $x$ and $y$ are found by solving (11.26), and are

$$A_x = \frac{A_1 + A_2}{\sqrt{2m}}, \qquad A_y = \frac{A_1 - A_2}{\sqrt{2m}}.$$

Note that if the forcing frequency is very close to $\omega_0$, the first normal mode will predominate, and the pendulums will swing in the same direction; while if it is close to $\omega'$ the second will be more important. (Of course, we should really include the effects of damping, so that the amplitudes do not become infinite at resonance, and so that transient effects disappear in time.)

## 11.5   Oscillations of Particles on a String

Consider a light string of length $(n + 1)l$, stretched to a tension $F$, with $n$ equal masses $m$ placed along it at regular intervals $l$. We shall consider transverse oscillations of the particles, and use as our generalized co-ordinates the displacements $y_1, y_2, \ldots, y_n$. (See Fig. 11.4, drawn for the case $n = 3$.) Since the kinetic energy is

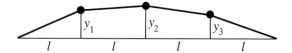

Fig. 11.4

$$T = \tfrac{1}{2}m(\dot{y}_1^2 + \dot{y}_2^2 + \cdots + \dot{y}_n^2), \tag{11.30}$$

these co-ordinates are orthogonal.

Next, we must calculate the potential energy. Let us consider the length of string between the $j$th and $(j + 1)$th particles. In equilibrium, its length is $l$, but when the particles are displaced it is

$$l + \delta l = \sqrt{l^2 + (y_{j+1} - y_j)^2} \approx l \left[ 1 + \frac{(y_{j+1} - y_j)^2}{2l^2} \right],$$

assuming the displacements are small. This calculation also applies to the segments of string at each end, provided we set $y_0 = y_{n+1} = 0$. The work done against the tension in increasing the length of the string to this extent is $F\delta l$. Hence, adding the contributions from each segment of string, we find that the potential energy is

$$V = \frac{F}{2l}\left[y_1^2 + (y_2 - y_1)^2 + \cdots + (y_n - y_{n-1})^2 + y_n^2\right]. \tag{11.31}$$

It is worth noting that the potential energy of a continuous string may be obtained as a limiting case, as $n \to \infty$ and $l \to 0$. For small $l$, $(y_{j+1}-y_j)^2/l^2$ is approximately $y'^2$, so we recover the expression (10.32).

From (11.30) and (11.31), we find that Lagrange's equations are

$$\ddot{y}_1 = \frac{F}{ml}(-2y_1 + y_2),$$

$$\ddot{y}_2 = \frac{F}{ml}(y_1 - 2y_2 + y_3),$$

$$\vdots$$

$$\ddot{y}_n = \frac{F}{ml}(y_{n-1} - 2y_n). \tag{11.32}$$

It will be convenient to write $\omega_0^2 = F/ml$. Then, substituting the normal mode solution $y_j = A_j e^{i\omega t}$, we obtain the equations

$$\begin{bmatrix} 2\omega_0^2 & -\omega_0^2 & 0 & \cdots & 0 \\ -\omega_0^2 & 2\omega_0^2 & -\omega_0^2 & \cdots & 0 \\ 0 & -\omega_0^2 & 2\omega_0^2 & \cdots & 0 \\ \vdots & \vdots & \vdots & & \vdots \\ 0 & 0 & 0 & \cdots & 2\omega_0^2 \end{bmatrix} \begin{bmatrix} A_1 \\ A_2 \\ A_3 \\ \vdots \\ A_n \end{bmatrix} = \omega^2 \begin{bmatrix} A_1 \\ A_2 \\ A_3 \\ \vdots \\ A_n \end{bmatrix}. \tag{11.33}$$

Let us look at the first few values of $n$. For $n = 1$, there is of course just one normal mode, with $\omega^2 = 2\omega_0^2$. For $n = 2$, the characteristic equation is

$$(2\omega_0^2 - \omega^2)^2 - \omega_0^4 = 0,$$

and we obtain two normal modes:

$$\omega^2 = \omega_0^2, \qquad A_1/A_2 = 1,$$
$$\omega^2 = 3\omega_0^2, \qquad A_1/A_2 = -1.$$

Now consider $n = 3$. The characteristic equation here is

$$\begin{vmatrix} 2\omega_0^2 - \omega^2 & -\omega_0^2 & 0 \\ -\omega_0^2 & 2\omega_0^2 - \omega^2 & -\omega_0^2 \\ 0 & -\omega_0^2 & 2\omega_0^2 - \omega^2 \end{vmatrix} = 0. \tag{11.34}$$

Expanding this determinant by the usual rules, we obtain a cubic equation for $\omega^2$:

$$(2\omega_0^2 - \omega^2)^3 - 2\omega_0^4(2\omega_0^2 - \omega^2) = 0.$$

The roots of this equation are $2\omega_0^2$ and $(2 \pm \sqrt{2})\omega_0^2$. Hence we obtain the three normal modes

$$\omega^2 = (2 - \sqrt{2})\omega_0^2, \qquad A_1 : A_2 : A_3 = 1 : \sqrt{2} : 1;$$
$$\omega^2 = 2\omega_0^2, \qquad A_1 : A_2 : A_3 = 1 : 0 : -1;$$
$$\omega^2 = (2 + \sqrt{2})\omega_0^2, \qquad A_1 : A_2 : A_3 = 1 : -\sqrt{2} : 1.$$

These normal modes are illustrated in Fig. 11.5.

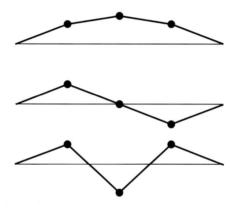

Fig. 11.5

Higher values of $n$ may be treated similarly. For $n = 4$, the characteristic equation requires the vanishing of a $4 \times 4$ determinant, which may be expanded by similar rules to yield

$$(2\omega_0^2 - \omega^2)^4 - 3\omega_0^4(2\omega_0^2 - \omega^2)^2 + \omega_0^8 = 0.$$

The roots of this equation are given by

$$(2\omega_0^2 - \omega^2)^2 = \frac{3 \pm \sqrt{5}}{2}\omega_0^4 = \left(\frac{\sqrt{5} \pm 1}{2}\omega_0^2\right)^2.$$

Thus we obtain the four normal modes:

$$\omega^2 = 0.38\,\omega_0^2, \qquad A_1 : A_2 : A_3 : A_4 = 1 : 1.62 : 1.62 : 1;$$
$$\omega^2 = 1.38\,\omega_0^2, \qquad A_1 : A_2 : A_3 : A_4 = 1.62 : 1 : -1 : -1.62;$$
$$\omega^2 = 2.62\,\omega_0^2, \qquad A_1 : A_2 : A_3 : A_4 = 1.62 : -1 : -1 : 1.62;$$
$$\omega^2 = 3.62\,\omega_0^2, \qquad A_1 : A_2 : A_3 : A_4 = 1 : -1.62 : 1.62 : -1.$$

(See Fig. 11.6.)

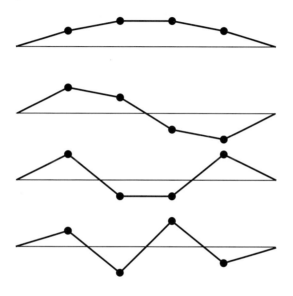

Fig. 11.6

For every value of $n$, the slowest mode is the one in which all the masses are oscillating in the same direction, while the fastest is one in which alternate masses oscillate in opposite directions. For large values of $n$, the normal modes approach those of a continuous stretched string, which we discuss in the following section.

## 11.6 Normal Modes of a Stretched String

We now wish to discuss the problem treated in §10.6 from the point of view of normal modes. We start from the equations of motion, (10.36),

$$\ddot{y} = c^2 y'', \qquad c^2 = F/\mu, \qquad (11.35)$$

and look for normal mode solutions of the form

$$y(x,t) = A(x)e^{i\omega t}. \tag{11.36}$$

Substituting in (11.35), we obtain

$$A''(x) + k^2 A(x) = 0, \qquad k = \omega/c.$$

Thus, in place of a set of simultaneous equations for the amplitudes $A_j$, we obtain a differential equation for the amplitude function $A(x)$.

The general solution of this equation is

$$A(x) = a \cos kx + b \sin kx.$$

However, because the ends of the string are fixed, we must impose the boundary conditions $A(0) = A(l) = 0$. (Note that $l$ here is the full length of the string, denoted by $(n + 1)l$ in the preceding section.) Thus $a = 0$ and moreover $\sin kl = 0$. The possible values of $k$ are

$$k = \frac{n\pi}{l}, \qquad n = 1, 2, 3, \ldots. \tag{11.37}$$

Each of these values corresponds to a normal mode of the string. The corresponding angular frequencies are

$$\omega = \frac{n\pi c}{l}, \qquad n = 1, 2, 3, \ldots. \tag{11.38}$$

Note that they are all multiples of the *fundamental frequency*

$$\omega_1 = \frac{\pi c}{l} = \pi \sqrt{\frac{F}{Ml}},$$

where $M$ is the total mass of the string.

The solution for the $n$th normal mode can be written as

$$y(x,t) = \mathrm{Re}\left(A_n e^{in\pi ct/l}\right) \sin \frac{n\pi x}{l}, \tag{11.39}$$

where $A_n$ is an arbitrary complex constant. It represents a 'standing wave' of wavelength $2l/n$, with $n - 1$ *nodes*, or points where $y = 0$. The first few normal modes are illustrated in Fig. 11.7.

The general solution for the stretched string is a superposition of all the normal modes (11.39). It is easy to establish the connection with the general solution (10.37) obtained in the previous chapter. According to

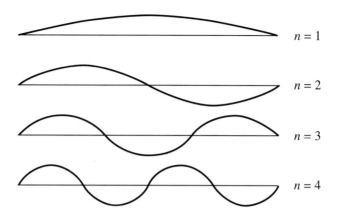

Fig. 11.7

(10.38), $f(x)$ is a periodic function of $x$, with period $2l$. Hence it may be expanded in a Fourier series (see Eq. (2.45)),

$$f(x) = \sum_{n=-\infty}^{+\infty} f_n e^{in\pi x/l}.$$

Thus the solution (10.37) is

$$y(x,t) = \sum_{n=-\infty}^{+\infty} f_n \left( e^{in\pi(ct+x)/l} - e^{in\pi(ct-x)/l} \right)$$

$$= \sum_{n=-\infty}^{+\infty} 2i f_n e^{in\pi ct/l} \sin\frac{n\pi x}{l}.$$

Now recall that $f_n$ and $f_{-n}$ must be complex conjugates (so that the two terms add to give a real contribution to $y$). Thus we may restrict the sum to positive values of $n$. (Note that there is no contribution from $n = 0$ because the sine function then vanishes.) If we define $A_n = 2if_n$, we can write the solution as

$$y(x,t) = 2\text{Re} \sum_{n=1}^{+\infty} A_n e^{in\pi ct/l} \sin\frac{n\pi x}{l}.$$

## 11.7   Summary

Near a position of equilibrium of any natural, conservative system, the kinetic energy may be taken to be a homogeneous quadratic function of the $\dot{q}_\alpha$, with constant coefficients, and the potential energy to be a homogeneous quadratic function of the $q_\alpha$. We can always find a set of orthogonal co-ordinates, in terms of which $T$ is reduced to a sum of squares. Lagrange's equations then take on a simple form. To find the normal modes of oscillation, we substitute solutions of the form $q_\alpha = A_\alpha e^{i\omega t}$, and obtain a set of simultaneous linear equations for the coefficients. The condition for consistency of these equations is the characteristic equation, which determines the frequencies of the normal modes. The stability condition is that all the roots of this equation for $\omega^2$ should be positive.

The problem of finding the normal modes is equivalent to that of finding normal co-ordinates, which reduce not only $T$ but also $V$ to a sum of squares. In terms of normal co-ordinates, the system is reduced to a set of uncoupled harmonic oscillators, whose frequencies are the characteristic frequencies of the system. The general solution to the equations of motion is a superposition of all the normal modes. In it, each normal co-ordinate is oscillating at its own frequency, and with amplitude and phase determined by the initial conditions.

The linearized analysis of small amplitude oscillations near to a position of stable equilibrium in the form of normal modes is a technique which is applicable generally. For some systems, which are special but important, the idea of a normal mode may be generalized. Such systems may then be analyzed as a combination of 'nonlinear' normal modes, where no small amplitude approximation needs to be made — see §14.1.

---

### Problems

1. A double pendulum, consisting of a pair, each of mass $m$ and length $l$, is released from rest with the pendulums displaced but in a straight line. Find the displacements of the pendulums as functions of time.

2. Find the normal modes of a pair of coupled pendulums (like those of Fig. 11.2) if the two are of different masses $M$ and $m$, but still the same length $l$. Given that the pendulum of mass $M$ is started oscillating with amplitude $a$, find the maximum amplitude of the other pendulum in

the subsequent motion. Does the amplitude of the first pendulum ever fall to zero?

3. A spring of negligible mass, and spring constant (force/extension) $k$, supports a mass $m$, and beneath it a second, identical spring, carrying a second, identical mass. Using the vertical displacements $x$ and $y$ of the masses from their positions with the springs unextended as generalized co-ordinates, write down the Lagrangian function. Find the position of equilibrium, and the normal modes and frequencies of vertical oscillations.

4. Three identical pendulums are coupled, as in Fig. 11.2, with springs between the first and second and between the second and third. Find the frequencies of the normal modes, and the ratios of the amplitudes.

5. The first of the three pendulums of Problem 4 is initially displaced a distance $a$, while the other two are vertical. The system is released from rest. Find the maximum amplitudes of the second and third pendulums in the subsequent motion.

6. Three identical springs, of negligible mass, spring constant $k$, and natural length $a$ are attached end-to-end, and a pair of particles, each of mass $m$, are fixed to the points where they meet. The system is stretched between fixed points a distance $3l$ apart ($l > a$). Find the frequencies of normal modes of (a) longitudinal, and (b) transverse oscillations.

7. *A bead of mass $m$ slides on a smooth circular hoop of mass $M$ and radius $a$, which is pivoted at a point on its rim so that it can swing freely in its plane. Write down the Lagrangian in terms of the angle of inclination $\theta$ of the diameter through the pivot and the angular position $\varphi$ of the bead relative to a fixed point on the hoop. Find the frequencies of the normal modes, and sketch the configuration of hoop and bead at the extreme point of each.

8. *The system of Problem 7 is released from rest with the centre of the hoop vertically below the pivot and the bead displaced by a small angle $\varphi_0$. Given that $M = 8m$ and that $2a$ is the length of a simple pendulum of period 1 s, find the angular displacement $\theta$ of the hoop as a function of time. Determine the maximum value of $\theta$ in the subsequent motion, and the time at which it first occurs.

9. A simple pendulum of mass $m$, whose period when suspended from a rigid support is 1 s, hangs from a supporting block of mass $2m$ which can move along a horizontal line (in the plane of the pendulum), and is restricted by a harmonic-oscillator restoring force. The period of the

oscillator (with the pendulum removed) is 0.1 s. Find the periods of the
two normal modes. When the pendulum bob is swinging in the slower
mode with amplitude 100 mm, what is the amplitude of the motion of
the supporting block?

10. *The system of Problem 9 is initially at rest, and the pendulum bob is
given an impulsive blow which starts it moving with velocity $0.5\,\mathrm{m\,s^{-1}}$.
Find the position of the support as a function of time in the subsequent
motion.

11. *A particle of charge $q$ and mass $m$ is free to slide on a smooth horizontal table. Two fixed charges $q$ are placed at $\pm aj$, and two fixed
charges $12q$ at $\pm 2ai$. Find the electrostatic potential near the origin
(see §6.2). Show that this is a position of stable equilibrium, and find
the frequencies of the normal modes of oscillation near it.

12. *A rigid rod of length $2a$ is suspended by two light, inextensible strings
of length $l$ joining its ends to supports also a distance $2a$ apart and
level with each other. Using the longitudinal displacement $x$ of the
centre of the rod, and the transverse displacements $y_1, y_2$ of its ends,
as generalized co-ordinates, find the Lagrangian function (for small
$x, y_1, y_2$). Determine the normal modes and frequencies. (*Hint*: First
find the height by which each end is raised, the co-ordinates of the
centre of mass and the angle through which the rod is turned.)

13. *Each of the pendulums in Fig 11.2 is subjected to a damping force, of
magnitude $\alpha\dot{x}$ and $\alpha\dot{y}$ respectively, while there is a damping force $\beta(\dot{x} -
\dot{y})$ in the spring. Show that the equations for the normal co-ordinates $q_1$
and $q_2$ are still uncoupled. Find the amplitudes of the forced oscillations
obtained by applying a periodic force to one pendulum. Given that the
forcing frequency is that of the uncoupled pendulums, and that $\beta$ is
negligible, find the range of values of $\alpha$ for which the amplitude of the
second pendulum is less than half that of the first.

14. *Show that a stretched string is equivalent mathematically to an infinite
number of uncoupled oscillators, described by the co-ordinates

$$q_n(t) = \sqrt{\frac{2}{l}} \int_0^l y(x,t) \sin \frac{n\pi x}{l}\, dx.$$

Determine the amplitudes of the various normal modes in the motion
described in Chapter 10, Problem 14. Why, physically, are the modes
for even values of $n$ not excited?

15. Show that a typical equation of the set (11.33) may be satisfied by
setting $A_\alpha = \sin \alpha k$ ($\alpha = 1, 2, \ldots, n$), provided that $\omega = 2\omega_0 \sin \frac{1}{2}k$.

Hence show by considering the required condition when $\alpha = n + 1$ that the frequencies of the normal modes are $\omega_r = 2\omega_0 \sin[r\pi/2(n+1)]$, with $r = 1, 2, \ldots, n$. Why may we ignore values of $r$ greater than $n+1$? Show that, in the limit of large $n$, the frequency of the $r$th normal mode tends to the corresponding frequency of the continuous string with the same total length and mass.

16. A particle moves under a conservative force with potential energy $V(\boldsymbol{r})$. The point $\boldsymbol{r} = \mathbf{0}$ is a position of equilibrium, and the axes are so chosen that $x, y, z$ are normal co-ordinates. Show that, if $V$ satisfies Laplace's equation, $\boldsymbol{\nabla}^2 V = 0$ (see §6.7), then the equilibrium is necessarily unstable, and hence that stable equilibrium under purely gravitational and electrostatic forces is impossible. (Of course, *dynamic* equilibrium — stable periodic motion — can occur. Note also that the two-dimensional stable equilibrium of Problem 11 does not contradict this result because there is another force imposed, confining the charge to the horizontal plane.)

# Chapter 12

# Hamiltonian Mechanics

We have already seen, in several examples, the value of the Lagrangian method, which allows us to find equations of motion for any system in terms of an arbitrary set of generalized co-ordinates. In this chapter, we shall discuss an extension of the method, due to Hamilton. Its principal feature is the use of the generalized momenta $p_1, p_2, \ldots, p_n$ in place of the generalized velocities $\dot{q}_1, \dot{q}_2, \ldots, \dot{q}_n$. It is particularly valuable when, as often happens, some of the generalized momenta are constants of the motion. More generally, it is well suited to finding conserved quantities, and making use of them.

## 12.1 Hamilton's Equations

The Lagrangian function $L$ for a *natural* system is a function of $q_1, q_2, \ldots, q_n$ and $\dot{q}_1, \dot{q}_2, \ldots, \dot{q}_n$. For brevity, we shall indicate this dependence by writing $L(q, \dot{q})$, where $q$ stands for all the generalized coordinates, and $\dot{q}$ for all their time derivatives.

Lagrange's equations may be written in the form

$$\dot{p}_\alpha = \frac{\partial L}{\partial q_\alpha},\tag{12.1}$$

where the generalized momenta are defined by

$$p_\alpha = \frac{\partial L}{\partial \dot{q}_\alpha}.\tag{12.2}$$

Here and in the following equations, $\alpha$ runs over $1, 2, \ldots, n$.

The instantaneous position and velocity of every part of our system may be specified by the values of the $2n$ variables $q$ and $\dot{q}$. However, we can alternatively solve the equations (12.2) for the $\dot{q}$ in terms of $q$ and $p$, obtaining, say,

$$\dot{q}_\alpha = \dot{q}_\alpha(q, p), \tag{12.3}$$

so the $2n$ variables $q$ and $p$ also serve equally well to specify the instantaneous position and velocity of every particle.

For example, for a particle moving in a plane, and described by polar coordinates, the generalized momenta are given by $p_r = m\dot{r}$ and $p_\theta = mr^2\dot{\theta}$. In this case, Eqs. (12.3) read

$$\dot{r} = \frac{p_r}{m}, \qquad \dot{\theta} = \frac{p_\theta}{mr^2}. \tag{12.4}$$

The instantaneous position and velocity of the particle may be fixed by the values of $r, \theta, p_r$ and $p_\theta$.

We now define a function of $q$ and $p$, the *Hamiltonian function*, by

$$H(q, p) = \sum_{\beta=1}^{n} p_\beta \dot{q}_\beta(q, p) - L(q, \dot{q}(q, p)). \tag{12.5}$$

Next, we compute the derivatives of $H$ with respect to its $2n$ independent variables. We differentiate first with respect to $p_\alpha$. One term in this derivative is the coefficient of $p_\alpha$ in the sum $\sum p\dot{q}$, namely $\dot{q}_\alpha$. Other terms arise from the dependence of each $\dot{q}_\beta$, either in the sum or as an argument of $L$, on $p_\alpha$. Altogether, we obtain

$$\frac{\partial H}{\partial p_\alpha} = \dot{q}_\alpha + \sum_{\beta=1}^{n} p_\beta \frac{\partial \dot{q}_\beta}{\partial p_\alpha} - \sum_{\beta=1}^{n} \frac{\partial L}{\partial \dot{q}_\beta} \frac{\partial \dot{q}_\beta}{\partial p_\alpha}.$$

Now, by (12.2), the second and third terms cancel. Hence we are left with

$$\frac{\partial H}{\partial p_\alpha} = \dot{q}_\alpha. \tag{12.6}$$

We now examine the derivative with respect to $q_\alpha$. Again, there are two kinds of terms, the term coming from the explicit dependence of $L$ on $q_\alpha$, and those from the dependence of each $\dot{q}_\beta$ on $q_\alpha$. We find

$$\frac{\partial H}{\partial q_\alpha} = -\frac{\partial L}{\partial q_\alpha} + \sum_{\beta=1}^{n} p_\beta \frac{\partial \dot{q}_\beta}{\partial q_\alpha} - \sum_{\beta=1}^{n} \frac{\partial L}{\partial \dot{q}_\beta} \frac{\partial \dot{q}_\beta}{\partial q_\alpha}.$$

As before, the second and third terms cancel. Thus, using Lagrange's equations (12.1), we obtain

$$\frac{\partial H}{\partial q_\alpha} = -\dot{p}_\alpha. \tag{12.7}$$

The equations (12.6) and (12.7) together constitute *Hamilton's equations*. Note that whereas Lagrange's equations are a set of $n$ second-order differential equations, Hamilton's constitute a set of $2n$ first-order equations.

Let us consider, for example, a particle moving in a plane under a central, conservative force.

### *Example:* Central conservative force

Obtain Hamilton's equations for a particle moving in a plane with potential energy function $V(r)$.

Here the Lagrangian is

$$L = \tfrac{1}{2}m\dot{r}^2 + \tfrac{1}{2}mr^2\dot{\theta}^2 - V(r).$$

Thus the Hamiltonian function is

$$H = (p_r\dot{r} + p_\theta\dot{\theta}) - [\tfrac{1}{2}m\dot{r}^2 + \tfrac{1}{2}mr^2\dot{\theta}^2 - V(r)],$$

or, using (12.4) to eliminate the velocities,

$$H = \frac{p_r^2}{2m} + \frac{p_\theta^2}{2mr^2} + V(r). \tag{12.8}$$

Note that this is the expression for the total energy, $T + V$. This is no accident, but a general property of natural systems, as we shall see below.

The first pair of Hamilton's equation, (12.6), are

$$\dot{r} = \frac{\partial H}{\partial p_r} = \frac{p_r}{m}, \qquad \dot{\theta} = \frac{\partial H}{\partial p_\theta} = \frac{p_\theta}{mr^2}. \tag{12.9}$$

They simply reproduce the relations (12.4) between velocities and momenta. The second pair, (12.7), are

$$-\dot{p}_r = \frac{\partial H}{\partial r} = -\frac{p_\theta^2}{mr^3} + \frac{\mathrm{d}V}{\mathrm{d}r}, \qquad -\dot{p}_\theta = \frac{\partial H}{\partial \theta} = 0. \tag{12.10}$$

The second of these two equations yields the law of conservation of angular momentum,

$$p_\theta = J = \text{constant.} \qquad (12.11)$$

The first gives the radial equation of motion,

$$\dot{p}_r = m\ddot{r} = \frac{J^2}{mr^3} - \frac{dV}{dr}.$$

It may be integrated to give what we termed the 'radial energy equation' (4.12) in Chapter 4.

## 12.2   Conservation of Energy

We saw in §10.1 that a natural system is characterized by the fact that the kinetic energy contains no explicit dependence on the time, and is a homogeneous quadratic function of the time derivatives $\dot{q}$. This latter condition may be expressed algebraically by the equation

$$\sum_{\alpha=1}^{n} \frac{\partial T}{\partial \dot{q}_\alpha} \dot{q}_\alpha = 2T.$$

For example, for $n = 2$, $T$ has the form (11.1). Thus

$$\frac{\partial T}{\partial \dot{q}_1} \dot{q}_1 + \frac{\partial T}{\partial \dot{q}_2} \dot{q}_2 = (a_{11}\dot{q}_1 + a_{12}\dot{q}_2)\dot{q}_1 + (a_{21}\dot{q}_1 + a_{22}\dot{q}_2)\dot{q}_2 = 2T.$$

Since $p_\alpha = \partial T/\partial \dot{q}_\alpha$, we therefore have

$$H = \sum_{\beta=1}^{n} p_\beta \dot{q}_\beta - L = \sum_{\beta=1}^{n} \frac{\partial T}{\partial \dot{q}_\beta} \dot{q}_\beta - (T - V)$$

$$= 2T - (T - V) = T + V.$$

Thus, for a natural system, the value of the Hamiltonian function is equal to the total energy of the system.

For a forced system, the Lagrangian can sometimes be written, as we saw in §10.4, in the form $L = T' - V'$, where $T'$ is a homogeneous quadratic in the variables $\dot{q}$, and $V'$ is independent of them. In such a case, the Hamiltonian function is equal to $T' + V'$, which is in general *not* the total energy.

Now let us examine the time derivative of $H$. We shall now allow for the possibility that $H$ may contain an explicit time dependence (as it does for some forced systems), and write $H = H(q, p, t)$. Then the value of $H$ varies with time for two reasons: firstly, because of its explicit time dependence, and, secondly, because the variables $q$ and $p$ are themselves functions of time. Thus the total time derivative is

$$\frac{\mathrm{d}H}{\mathrm{d}t} = \frac{\partial H}{\partial t} + \sum_{\alpha=1}^{n} \frac{\partial H}{\partial q_\alpha} \dot{q}_\alpha + \sum_{\alpha=1}^{n} \frac{\partial H}{\partial p_\alpha} \dot{p}_\alpha.$$

Now, if we express $\dot{q}$ and $\dot{p}$ in terms of derivatives of $H$, using Hamilton's equations (12.6) and (12.7), we obtain

$$\frac{\mathrm{d}H}{\mathrm{d}t} = \frac{\partial H}{\partial t} + \sum_{\alpha=1}^{n} \left( \frac{\partial H}{\partial q_\alpha} \frac{\partial H}{\partial p_\alpha} - \frac{\partial H}{\partial p_\alpha} \frac{\partial H}{\partial q_\alpha} \right).$$

Obviously, the terms in parentheses cancel, whence

$$\frac{\mathrm{d}H}{\mathrm{d}t} = \frac{\partial H}{\partial t}. \tag{12.12}$$

This equation asserts that the value of $H$ changes in time *only* because of its explicit time dependence. The net change induced by the fact that $q$ and $p$ vary with time is zero.

In particular, for a natural, conservative system, neither $T$ nor $V$ contains any explicit dependence on the time. Thus $\partial H / \partial t = 0$, and it follows that

$$\frac{\mathrm{d}H}{\mathrm{d}t} = 0. \tag{12.13}$$

Thus there is a law of *conservation of energy*,

$$H = T + V = E = \text{constant}. \tag{12.14}$$

In a forced system, if $H = T' + V'$, and is time-independent, we again have a conservation law — like (10.20) — though not for $T + V$.

Even when $H$ is not of the form $T' + V'$, an energy conservation law may exist. The prime example of this is that of a charged particle moving in static electric and magnetic fields $\boldsymbol{E}$ and $\boldsymbol{B}$. Since the magnetic force is perpendicular to $\dot{\boldsymbol{r}}$, it does no work, and the sum of the kinetic and electrostatic potential energies is a constant. It is interesting to see how this emerges from the Hamiltonian formalism.

*Example:* **Particle in electric and magnetic fields**

Find the Hamiltonian for a charged particle in electric and magnetic fields. Show that, if the fields are time-independent, there is an energy conservation law.

We begin with the Lagrangian (10.27):

$$L = \tfrac{1}{2}m\dot{\boldsymbol{r}}^2 + q\dot{\boldsymbol{r}} \cdot \boldsymbol{A}(\boldsymbol{r},t) - q\phi(\boldsymbol{r},t),$$

from which, as in (10.28), it follows that the generalized momentum is $\boldsymbol{p} = m\dot{\boldsymbol{r}} + q\boldsymbol{A}$. Thus we find

$$H = \boldsymbol{p} \cdot \dot{\boldsymbol{r}} - L = \frac{(\boldsymbol{p} - q\boldsymbol{A})^2}{2m} + q\phi. \qquad (12.15)$$

Now, if $\boldsymbol{E}$ and $\boldsymbol{B}$ are time-independent, it is possible to choose the scalar and vector potentials $\phi$ and $\boldsymbol{A}$ also to be so (see §A.7). Thus $H$ has no explicit time-dependence, and therefore is conserved: $H = E =$ constant.

Note that in this case, $L$ does have a term linear in $\dot{\boldsymbol{r}}$: it has the form $L = L_2 + L_1 + L_0$, where $L_k$ is of degree $k$ in $\dot{\boldsymbol{r}}$. In such a case, it turns out that $H = L_2 - L_0$. The 'magnetic' term $L_1$ apparently does not appear. In fact, $\boldsymbol{B}$ appears in $H$ only via the relation between $\dot{\boldsymbol{r}}$ and $\boldsymbol{p}$.

The Hamiltonian formalism is particularly well suited to finding conservation laws, or constants of the motion. The conservation law for energy is the first of a large class of conservation laws which we shall discuss in the following sections.

## 12.3  Ignorable Co-ordinates

It sometimes happens that one of the generalized co-ordinates, say $q_\alpha$, does not appear in the Hamiltonian function (though the corresponding momentum, $p_\alpha$, does). In that case, the co-ordinate $q_\alpha$ is said to be *ignorable* — for a reason we shall explain in a moment.

For an ignorable co-ordinate, Hamilton's equation (12.7) yields

$$-\dot{p}_\alpha = \frac{\partial H}{\partial q_\alpha} = 0. \qquad (12.16)$$

It leads immediately to a conservation law for the corresponding generalized momentum,

$$p_\alpha = \text{constant.} \qquad (12.17)$$

For example, for a particle moving in a plane under a central, conservative force, $H$ is independent of the angular co-ordinate $\theta$ and we therefore have the law of conservation of angular momentum, (12.11).

The term 'ignorable co-ordinate' means just what it says: that for many purposes we can ignore the co-ordinate $q_\alpha$, and treat the corresponding $p_\alpha$ simply as a constant appearing in the Hamiltonian function. This is, in effect, what we did for the central force problem in Chapter 4. Because of the conservation law for angular momentum, we were able to deal with an effectively one-dimensional problem involving only the radial co-ordinate $r$. The generalized momentum $p_\theta = J$ was simply a constant appearing in the equation of motion or the energy conservation equation.

Let us re-examine this problem from the Hamiltonian point of view. Since the Hamiltonian (12.8) is independent of $\theta$, so $\theta$ is ignorable. Thus, we may regard (12.8) as the Hamiltonian for a system with one degree of freedom, described by the co-ordinate $r$ and its corresponding momentum $p_r$, in which a constant $p_\theta$ appears. It is identical with the Hamiltonian for a particle moving in one dimension under a conservative force with potential energy function

$$U(r) = \frac{p_\theta^2}{2mr^2} + V(r). \qquad (12.18)$$

This is precisely the 'effective potential energy function' of (4.13).

Hamilton's equations for $r$ and $p_r$ are

$$\dot{r} = \frac{\partial H}{\partial p_r} = \frac{p_r}{m}, \qquad -\dot{p}_r = \frac{\partial H}{\partial r} = \frac{dU}{dr}.$$

To solve the central force problem, we solve first this one-dimensional problem (for example, using its energy conservation equation, the 'radial energy equation'). Our solution gives us complete information about the radial motion — it gives $\dot{r}$ as a function of $r$, and therefore $r$ as a function of $t$, by integrating.

Any required information about the angular part of the motion can then be found from the remaining pair of Hamilton's equations, one of which is the angular momentum conservation equation, $\dot{p}_\theta = 0$, while the other gives $\dot{\theta}$ in terms of $p_\theta$, in the form $\dot{\theta} = p_\theta/mr^2$. Clearly, though we did not then introduce the Hamiltonian, this is essentially just the method we used in Chapter 4.

We shall use the same method in the following section to discuss the general motion of a symmetric top.

*Example:* **The symmetric top**

Show that for the symmetric top two of the three Euler angles are ignorable co-ordinates, and find the 'effective potential energy function' for the remaining co-ordinate.

We start from the Lagrangian function (10.11):

$$L = \tfrac{1}{2} I_1 \dot{\varphi}^2 \sin^2 \theta + \tfrac{1}{2} I_1 \dot{\theta}^2 + \tfrac{1}{2} I_3 (\dot{\psi} + \dot{\varphi} \cos \theta)^2 - MgR \cos \theta.$$

The corresponding generalized momenta are

$$p_\varphi = I_1 \dot{\varphi} \sin^2 \theta + I_3 (\dot{\psi} + \dot{\varphi} \cos \theta) \cos \theta,$$

$$p_\theta = I_1 \dot{\theta},$$

$$p_\psi = I_3 (\dot{\psi} + \dot{\varphi} \cos \theta).$$

Solving these equations for $\dot{\varphi}, \dot{\theta}, \dot{\psi}$, we obtain

$$\dot{\varphi} = \frac{p_\varphi - p_\psi \cos \theta}{I_1 \sin^2 \theta},$$

$$\dot{\theta} = \frac{p_\theta}{I_1}, \qquad\qquad (12.19)$$

$$\dot{\psi} = \frac{p_\psi}{I_3} - \frac{p_\varphi - p_\psi \cos \theta}{I_1 \sin^2 \theta} \cos \theta.$$

The simplest way to construct the Hamiltonian function is to use the fact that $H = T + V$, and express $T$ in terms of the generalized momenta using (12.19). In this way, we find

$$H = \frac{(p_\varphi - p_\psi \cos \theta)^2}{2 I_1 \sin^2 \theta} + \frac{p_\theta^2}{2 I_1} + \frac{p_\psi^2}{2 I_3} + MgR \cos \theta. \qquad (12.20)$$

It is easy to verify that the first set of Hamilton's equations, (12.6), correctly reproduce (12.19).

It is clear that the two co-ordinates $\varphi$ and $\psi$ here are ignorable, and there are two corresponding conservation laws, $p_\varphi = $ constant and $p_\psi = $ constant. Thus the problem can be reduced to that of a system with one degree of freedom only, described by the co-ordinate $\theta$. The Hamiltonian function (12.20) may be written

$$H = \frac{p_\theta^2}{2I_1} + U(\theta),$$

where the effective potential energy function $U(\theta)$ is

$$U(\theta) = \frac{(p_\varphi - p_\psi \cos \theta)^2}{2I_1 \sin^2 \theta} + \frac{p_\psi^2}{2I_3} + MgR \cos \theta. \tag{12.21}$$

## 12.4 General Motion of the Symmetric Top

Hamilton's equations for $\theta$ and $p_\theta$ give

$$-I_1 \ddot{\theta} = -\dot{p}_\theta = \frac{\partial H}{\partial \theta} = \frac{dU}{d\theta}. \tag{12.22}$$

This is obviously a rather complicated equation to solve. However, the qualitative features of the motion can be found from the energy conservation equation,

$$\frac{p_\theta^2}{2I_1} + U(\theta) = E = \text{constant}. \tag{12.23}$$

In particular, the angles $\theta$ at which $\dot{\theta} = 0$ are given by the equation $U(\theta) = E$, and the motion is confined to the region where $U(\theta) \le E$.

Now let us examine the function $U(\theta)$. We exclude for the moment the special case where $p_\varphi = \pm p_\psi$. (We return to this important special case in the next section.) Then it is clear from (12.21) that as $\theta$ approaches either 0 or $\pi$, $U(\theta) \to +\infty$. Hence it has roughly the form shown in Fig. 12.1, with a minimum at some value of $\theta$, say $\theta_0$, between 0 and $\pi$. It can be shown that there is only one minimum (see Problem 7). When $E$ is equal to this minimum value, we have an 'equilibrium' situation, and $\theta$ remains fixed at $\theta_0$. This corresponds to steady precession. For any larger value of $E$, the angle $\theta$ oscillates between a minimum $\theta_1$ and a maximum $\theta_2$.

It is not hard to describe the motion of the top. We note that, according to (12.19), the angular velocity of the axis about the vertical, $\dot{\varphi}$, is zero when $\cos \theta = p_\varphi / p_\psi$. If this angle lies outside the range between $\theta_1$ and $\theta_2$ (or if $|p_\varphi / p_\psi| > 1$), then $\dot{\varphi}$ never vanishes and the axis precesses round the vertical in a fixed direction, and wobbles up and down between $\theta_1$ and $\theta_2$.

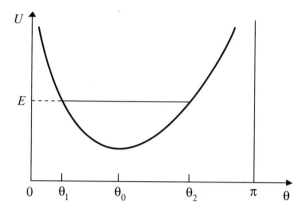

Fig. 12.1

This motion is illustrated in Fig. 12.2, which shows the track of the end of the axis on a sphere.

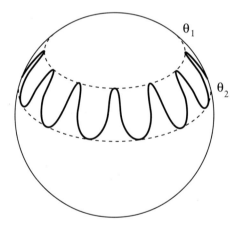

Fig. 12.2

On the other hand, if $\theta_1 < \arccos(p_\varphi/p_\psi) < \theta_2$, the axis moves in loops, as shown in Fig. 12.3. The angular velocity has one sign near the top of the loop, and the opposite near the bottom.

The limiting case between the two kinds of motion occurs when $\arccos(p_\varphi/p_\psi) = \theta_1$. Then the loops shrink to cusps, as shown in Fig. 12.4. The axis of the top comes instantaneously to rest at the top of each loop.

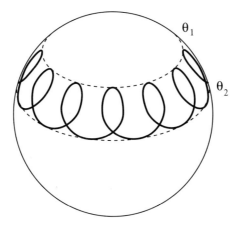

Fig. 12.3

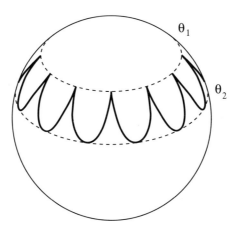

Fig. 12.4

This kind of motion will occur if the top is set spinning with its axis initially at rest. (It is impossible to have cusped motion with the cusps at the bottom, for they correspond to points of minimum kinetic energy, and the motion must always be below such points. A top set spinning with its axis stationary cannot rise without increasing its energy.)

It is easy to observe these kinds of motion using a small gyroscope. In practice, because of frictional effects, the type of motion will change slowly with time.

*Example:* **Stability of a vertical top**

A top is set spinning with angular velocity $\omega_3$ with its axis vertical. How will it move?

If the axis of the top passes through the vertical, $\theta = 0$, then it is clear that $U(0)$ must be finite. This is possible only if $p_\varphi = p_\psi$, and both generalized momenta must be equal to $I_3\omega_3$. (We could also consider types of motion for which the axis passes through the downward vertical, $\theta = \pi$. In that case, we require $p_\varphi = -p_\psi = -I_3\omega_3$. The treatment is entirely similar — but less interesting because a downward-pointing top is always stable.)

If we set $p_\varphi = p_\psi = I_3\omega_3$, the effective potential energy function (12.21) becomes

$$U(\theta) = \frac{I_3^2\omega_3^2}{2I_1}\tan^2\tfrac{1}{2}\theta + \tfrac{1}{2}I_3\omega_3^2 + MgR\cos\theta, \qquad (12.24)$$

where we have used the identity $(1 - \cos\theta)/\sin\theta = \tan\tfrac{1}{2}\theta$.

For small values of $\theta$, we may expand $U(\theta)$, and retain only the terms up to order $\theta^2$, obtaining

$$U(\theta) \approx (\tfrac{1}{2}I_3\omega_3^2 + MgR) + \frac{1}{2}\left(\frac{I_3^2\omega_3^2}{4I_1} - MgR\right)\theta^2. \qquad (12.25)$$

Since there is no linear term, $\theta = 0$ is always a position of equilibrium. It is a position of *stable* equilibrium if $U(\theta)$ has a minimum at $\theta = 0$, that is, if the coefficient of $\theta^2$ is positive. Thus there is a minimum value of $\omega_3$ for which the vertical top is stable, given by

$$\omega_3^2 = \frac{4I_1 MgR}{I_3^2} = \omega_0^2, \quad \text{say.} \qquad (12.26)$$

If the top is set spinning with angular velocity greater than this critical value, it will remain vertical. When the angular velocity falls below the critical value (as it eventually will, because of friction), the top will begin to wobble. The energy of the vertical top is

$$E = \tfrac{1}{2}I_3\omega_3^2 + MgR.$$

Thus the angles at which $\dot{\theta} = 0$ are given by $U(\theta) = E$ (see Fig. 12.5), or, from (12.24), by

$$\frac{I_3^2 \omega_3^2}{2I_1} \tan^2 \tfrac{1}{2}\theta = MgR(1 - \cos\theta) = 2MgR\sin^2 \tfrac{1}{2}\theta.$$

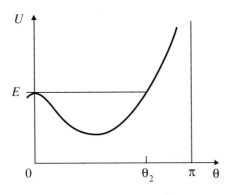

Fig. 12.5

They are easily seen to be $\theta_1 = 0$ and $\theta_2 = 2\arccos(\omega_3/\omega_0)$. Thus if the top is set spinning with its axis vertical and almost stationary, with angular velocity less than the critical value $\omega_0$, it will oscillate in the subsequent motion between the vertical and the angle $\theta_2$. Note that $\theta_2$ increases as $\omega_3$ is decreased, and tends to $\pi$ as $\omega_3$ approaches zero. When $\omega_3 = 0$, the top behaves like a compound pendulum, and swings in a circle through both the upward and downward verticals.

## 12.5   Liouville's Theorem

This theorem really belongs to statistical mechanics, but it is interesting to consider it here because it is a very direct consequence of Hamilton's equations.

The instantaneous position and velocity of every particle in our system is specified by the $2n$ variables $(q_1, \ldots, q_n, p_1, \ldots, p_n)$. It is convenient to think of these as co-ordinates in a $2n$-dimensional space, called the *phase space* of the system. Symbolically, we may write

$$\boldsymbol{r} = (q_1, \ldots, q_n, p_1, \ldots, p_n).$$

As time progresses, the changing state of the system can be described by a curve $r(t)$ in the phase space. Hamilton's equations, (12.6) and (12.7), prescribe the rates of change $(\dot{q}_1, \ldots, \dot{q}_n, \dot{p}_1, \ldots, \dot{p}_n)$. These may be regarded as the components of a $2n$-dimensional velocity vector,

$$v = \dot{r} = (\dot{q}_1, \ldots, \dot{q}_n, \dot{p}_1, \ldots, \dot{p}_n).$$

Now suppose that we have a large number of copies of our system, starting out with slightly different initial values of the co-ordinates and momenta. For example, we may repeat many times an experiment on the same system, but with small random variations in the initial conditions. Each copy of the system is represented by a point in the phase space, moving according to Hamilton's equations. We thus have a swarm of points, occupying some volume in phase space, rather like the particles in a fluid.

Liouville's theorem concerns how this swarm moves. What it says is a very simple, but very remarkable, result, namely that the representative points in phase space move as though they formed an *incompressible* fluid. The $2n$-dimensional volume occupied by the swarm does not change with time, though of course its shape may, and usually does, change in very complicated ways.

To prove this, we need to apply a generalization of the divergence operation. In three dimensions, it is shown in Appendix A (see (A.24)) that the fluid velocity in an incompressible fluid satisfies the condition $\nabla \cdot v = 0$. It is easy to see that the argument generalizes to any number of dimensions. The condition that the phase-space volume does not change with time in the flow described by the velocity field $v$ is simply that the $2n$-dimensional divergence of $v$ is zero, *i.e.*,

$$\nabla \cdot v = \frac{\partial \dot{q}_1}{\partial q_1} + \cdots + \frac{\partial \dot{q}_n}{\partial q_n} + \frac{\partial \dot{p}_1}{\partial p_1} \cdots + \frac{\partial \dot{p}_n}{\partial p_n} = 0. \tag{12.27}$$

But, by (12.6) and (12.7), the 1st and (n+1)st terms are

$$\frac{\partial \dot{q}_1}{\partial q_1} + \frac{\partial \dot{p}_1}{\partial p_1} = \frac{\partial}{\partial q_1}\left(\frac{\partial H}{\partial p_1}\right) + \frac{\partial}{\partial p_1}\left(-\frac{\partial H}{\partial q_1}\right),$$

which is indeed zero. Similarly, all the other terms in (12.27) cancel in pairs. Thus the theorem is proved.

In general, except for special cases, the motion in phase space is complicated. The phase-space volume containing the swarm of representative points maintains its volume, but becomes extremely distorted, rather like a drop of immiscible coloured liquid in a glass of water which is then stirred.

The points cannot just go anywhere in phase space, because of the energy conservation equation. They must remain on the same constant-energy surface, $H(q, p) = E$, and of course there might be other conservation laws that also restrict the accessible region of phase space. But one might expect that in time the phase-space volume would become thinly distributed throughout almost all the accessible parts of phase space. Such behaviour is called *ergodic*, and is commonly assumed in statistical mechanics. Averaged properties of the system over a long time can then be estimated by averaging over the accessible phase space.

Remarkably, however, many quite complicated systems do *not* behave in this way, but show surprising almost-periodic behaviour, to be discussed in Chapter 14. The study of which systems do and do not behave ergodically, and particularly the transition between one type and another as the parameters of the system are varied, is now one of the most active fields of mathematical physics. It has revealed an astonishing range of possibilities.

## 12.6  Symmetries and Conservation Laws

In §§12.2 and 12.3, we found some examples of conserved quantities, but so far we have not discussed the physical reasons for their existence. In fact, they are expressions of symmetry properties possessed by the system.

For example, the conservation law of angular momentum for the central force problem, (12.11), arises from the fact that the Hamiltonian is independent of $\theta$. This is an expression of the rotational symmetry of the system — in other words, of the fact that there is no preferred orientation in the plane. Explicitly, the equation $\partial H / \partial \theta = 0$ means that the energy of the system is unchanged if we rotate it to a new position, replacing $\theta$ by $\theta + \delta \theta$, without changing $r, p_r$ or $p_\theta$. Thus angular momentum is conserved ($p_\theta = $ constant) for systems possessing this rotational symmetry. Of course, if the force is non-central, it does determine a preferred orientation is space, and angular momentum is not conserved.

In a three-dimensional problem, every component of the angular momentum $\boldsymbol{J}$ is conserved if the force is purely central. If the force is non-central, but still possesses axial symmetry — so that $H$ depends on $\theta$ but not on $\varphi$ — then only the component of $\boldsymbol{J}$ along the axis of symmetry, namely $p_\varphi$, is conserved.

Similarly, for the symmetric top, the equation $\partial H / \partial \varphi = 0$ is an expression of the rotational symmetry of the system about the vertical. The corresponding conserved quantity $p_\varphi$ is the vertical component of $\boldsymbol{J}$; for, by (9.43) and (9.45),

$$J_z = \boldsymbol{k} \cdot \boldsymbol{J} = I_1 \dot\varphi \sin^2 \theta + I_3 (\dot\psi + \dot\varphi \cos \theta) \cos \theta = p_\varphi.$$

The equation $\partial H/\partial\psi = 0$ expresses the rotational symmetry of the top itself about its own axis. The energy is clearly unchanged by rotating the top about its axis. In this case, we see from (9.45) that the conserved quantity $p_\psi$ is the component of $\boldsymbol{J}$ along the axis of the top, $p_\psi = J_3 = \boldsymbol{e}_3 \cdot \boldsymbol{J}$.

Now of course not all symmetries are expressible simply by saying that $H$ is independent of some particular co-ordinate. For example, we might consider the central force problem in terms of the Cartesian co-ordinates $x$ and $y$. Then the Hamiltonian is

$$H = \frac{p_x^2 + p_y^2}{2m} + V \left( \sqrt{x^2 + y^2} \right).$$

Since it depends on both $x$ and $y$, neither co-ordinate is ignorable. It does, however, possess a symmetry under rotations. If we make a small rotation through an angle $\delta\theta$, the changes in the co-ordinates and momenta are (see Fig. A.4)

$$\begin{aligned} \delta x &= -y\,\delta\theta, & \delta y &= x\,\delta\theta, \\ \delta p_x &= -p_y\,\delta\theta, & \delta p_y &= p_x\,\delta\theta. \end{aligned} \tag{12.28}$$

Under this transformation, $\delta(x^2 + y^2) = 0$ and $\delta(p_x^2 + p_y^2) = 0$, so clearly $\delta H = 0$.

Now we know from our earlier discussion in terms of polar co-ordinates that this symmetry is related to the conservation of angular momentum,

$$J = x p_y - y p_x = \text{constant}.$$

The problem is to understand the relationship between the transformation (12.28) and the conserved quantity $J$.

Let us consider a general function of the co-ordinates, momenta, and time, $G(q, p, t)$. We define the transformation *generated by* $G$ to be

$$\delta q_\alpha = \frac{\partial G}{\partial p_\alpha}\,\delta\lambda, \qquad \delta p_\alpha = -\frac{\partial G}{\partial q_\alpha}\,\delta\lambda, \tag{12.29}$$

where $\delta\lambda$ is an infinitesimal parameter. For example, the function $G = p_1$ generates the transformation in which $\delta q_1 = \delta\lambda$, while all the remaining co-ordinates and momenta are unchanged. Using Hamilton's equations (12.6) and (12.7), we see that the transformation generated by the Hamiltonian is

$$\delta q_\alpha = \dot{q}_\alpha \, \delta\lambda, \qquad \delta p_\alpha = \dot{p}_\alpha \, \delta\lambda. \tag{12.30}$$

If $\delta\lambda$ is interpreted as a small time interval, this represents the time development of the system.

We can now return to the function $J$. The transformation it generates is given by

$$\delta x = \frac{\partial J}{\partial p_x} \, \delta\lambda \; = -y \, \delta\lambda, \qquad \delta y = \frac{\partial J}{\partial p_y} \, \delta\lambda \; = x \, \delta\lambda,$$

$$\delta p_x = -\frac{\partial J}{\partial x} \, \delta\lambda = -p_y \, \delta\lambda, \qquad \delta p_y = -\frac{\partial J}{\partial y} \, \delta\lambda = p_x \, \delta\lambda.$$

This is clearly identical to the infinitesimal rotation (12.28). Thus we have established a connection between $J$ and this transformation (12.28).

The next problem is to understand why the fact that this transformation represents a symmetry property of the system should lead to a conservation law. To this end, we return to a general function $G$, and consider the effect of the transformation (12.29) on some other function $F(q, p, t)$. The change in $F$ is

$$\delta F = \sum_{\alpha=1}^{n} \left( \frac{\partial F}{\partial q_\alpha} \delta q_\alpha + \frac{\partial F}{\partial p_\alpha} \delta p_\alpha \right) = \sum_{\alpha=1}^{n} \left( \frac{\partial F}{\partial q_\alpha} \frac{\partial G}{\partial p_\alpha} - \frac{\partial F}{\partial p_\alpha} \frac{\partial G}{\partial q_\alpha} \right) \delta\lambda.$$

This kind of sum, involving the derivatives of two functions, appears quite frequently, and it is convenient to introduce an abbreviated notation. We define the *Poisson bracket* of $F$ and $G$ to be

$$[F, G] = \sum_{\alpha=1}^{n} \left( \frac{\partial F}{\partial q_\alpha} \frac{\partial G}{\partial p_\alpha} - \frac{\partial F}{\partial p_\alpha} \frac{\partial G}{\partial q_\alpha} \right). \tag{12.31}$$

Then we can write the change in $F$ under the transformation generated by $G$ in the form

$$\delta F = [F, G] \, \delta\lambda. \tag{12.32}$$

A particular example is provided by the transformation (12.30) generated by $H$. The rate of change of $F$ is

$$\frac{\mathrm{d}F}{\mathrm{d}t} = \frac{\partial F}{\partial t} + \sum_{\alpha=1}^{n} \left( \frac{\partial F}{\partial q_\alpha} \dot{q}_\alpha + \frac{\partial F}{\partial p_\alpha} \dot{p}_\alpha \right) = \frac{\partial F}{\partial t} + [F, H]. \tag{12.33}$$

The extra term here arises from the fact that we have now allowed $F$ to have an explicit dependence on the parameter $t$, in addition to the dependence via $q$ and $p$.

Now an obvious property of the Poisson bracket is its *antisymmetry*. If we interchange $F$ and $G$, we merely change the sign:

$$[G, F] = -[F, G]. \tag{12.34}$$

This has the important consequence that, if $F$ is unchanged by the transformation generated by $G$, then reciprocally $G$ is unchanged by the transformation generated by $F$.

We are now finally in a position to apply this discussion to the case of a symmetry property of the system. Let us suppose that there exists a transformation of the co-ordinates and momenta which leaves the Hamiltonian unaffected, and which is generated by a function $G$. From (12.29) we see that the generator $G$ is unique, apart from an arbitrary additive function of $t$, independent of $q$ and $p$. In particular, if the transformation does not involve the time explicitly, then $G$ may be chosen to contain no explicit $t$ dependence. The condition that $H$ should be unchanged is

$$\delta H = [H, G]\, \delta\lambda = 0. \tag{12.35}$$

It then follows from the reciprocity relation (12.34) that $[G, H] = 0$ also. Hence if $\partial G/\partial t = 0$, we find from (12.33) that

$$\frac{\mathrm{d}G}{\mathrm{d}t} = [G, H] = 0. \tag{12.36}$$

Thus we have shown that if $H$ is unaltered by a $t$-independent transformation of this type, then the corresponding generator is conserved.

The number of independent symmetries possessed by a system, and hence the number of conserved quantities, has a profound effect on the way that the system may behave. By exploiting the conservation laws, the complexity of problems may be reduced progressively. If the number of independent conserved quantities for a system, *i.e.*, constants or integrals of the motion, is at least equal to the number of degrees of freedom (the number of independent co-ordinates), then the reduction may be complete, and the system is termed *integrable* (in the sense of Liouville — see §14.1). The motion is then ordered and 'regular'. If there are insufficient conserved quantities to bring this about, then the motion may exhibit disorder or 'chaos'.

The central, conservative force problems considered in Chapter 4, and the symmetric top considered in §10.3 and §§12.3, 12.4, are examples of integrable systems, since they possess respectively two and three conserved quantities, equal in each case to the number of degrees of freedom. The restricted three-body problem, considered in Problems 15 and 16 of Chapter 10, and in Problems 12 and 13 at the end of this chapter, does not possess such a complete set of conserved quantities, and so is not integrable.

The formal description of the connection between symmetry properties and invariance is contained in a famous theorem due to Emmy Noether (1918).

## 12.7 Galilean Transformations

To illustrate the ideas of the preceding section, we shall consider a general, isolated system of $N$ particles, and investigate the symmetry properties implied by the relativity principle of §1.1.

The system has $3N$ degrees of freedom, and may be described by the particle positions $r_i$ and momenta $p_i$ $(i = 1, 2, \ldots, N)$. We shall consider four distinct symmetry properties, associated with the requirements that there should be no preferred zero of the time scale, origin in space, orientation of axes, or standard of rest. The corresponding symmetry transformations are translations in time, spatial translations, rotations, and transformations between frames moving with uniform relative velocity (sometimes called *boosts*). A combination of these four types of transformation is the most general transformation which takes one inertial frame into another. They are known collectively as *Galilean* transformations. We consider them in turn.

### Time translations

The changes in $r$ and $p$ in an infinitesimal time $\delta t$ are generated by the Hamiltonian function $H$. The condition for invariance of $H$ under this transformation is $[H, H] = 0$, which is certainly true because of (12.34). Thus, as we showed in §12.2, if $H$ contains no explicit time dependence, then it is in fact conserved,

$$\frac{\mathrm{d}H}{\mathrm{d}t} = 0. \tag{12.37}$$

### Spatial translations

An infinitesimal translation of the system through a distance $\delta x$ in the $x$ direction is represented by the transformation

$$\begin{aligned} \delta x_i = \delta x, && \delta y_i = 0, && \delta z_i = 0, \\ \delta p_{xi} = 0, && \delta p_{yi} = 0, && \delta p_{zi} = 0. \end{aligned} \qquad (12.38)$$

The corresponding generator is easily seen to be the total $x$-component of momentum,

$$P_x = \sum_{i=1}^{N} p_{xi}.$$

The condition for $H$ to be invariant under this transformation is

$$0 = [H, P_x]\,\delta x = \sum_{i=1}^{N} \frac{\partial H}{\partial x_i}\,\delta x.$$

It is satisfied if $H$ depends only on the co-ordinate differences $x_i - x_j$; for then, changing each $x_i$ by the same amount cannot affect $H$. When this condition holds, we obtain a conservation law for the $x$-component of momentum, $dP_x/dt = 0$.

More generally, a translation in the direction of the unit vector $\boldsymbol{n}$ is generated by the component of the total momentum $\boldsymbol{P}$ in this direction, $\boldsymbol{n} \cdot \boldsymbol{P}$. When the system possesses translational invariance in all directions, then all components of $\boldsymbol{P}$ are conserved:

$$\frac{d\boldsymbol{P}}{dt} = \boldsymbol{0}. \qquad (12.39)$$

This is physically very reasonable. We know from our earlier work that momentum is conserved for an isolated system, but not for a system subjected to external forces which determine a preferred origin (for example, a centre of force).

### Rotations

An infinitesimal rotation through an angle $\delta\varphi$ about the $z$-axis yields

$$\begin{aligned} \delta x_i = -y_i\,\delta\varphi, && \delta y_i = x_i\,\delta\varphi, && \delta z_i = 0, \\ \delta p_{xi} = -p_{yi}\,\delta\varphi, && \delta p_{yi} = p_{xi}\,\delta\varphi, && \delta p_{zi} = 0. \end{aligned} \qquad (12.40)$$

The corresponding generator is the $z$-component of the total angular momentum,

$$J_z = \sum_{i=1}^{N} (x_i p_{yi} - y_i p_{xi}).$$

The condition for $H$ to be rotationally symmetric, $[H, J_z] = 0$, is satisfied provided that $H$ involves the $x$ and $y$ co-ordinates and momenta only through invariant combinations like $x_i x_j + y_i y_j$.

In general, a rotation through an angle $\delta\varphi$ about an axis in the direction of the unit vector $\boldsymbol{n}$ may be written in the form

$$\delta\boldsymbol{r}_i = \boldsymbol{n} \wedge \boldsymbol{r}_i \, \delta\varphi, \qquad \delta\boldsymbol{p}_i = \boldsymbol{n} \wedge \boldsymbol{p}_i \, \delta\varphi. \tag{12.41}$$

It is generated by the appropriate component of the angular momentum, $\boldsymbol{n} \cdot \boldsymbol{J}$.

The Hamiltonian function is invariant under this transformation provided that it is a *scalar* function of $\boldsymbol{r}_i$ and $\boldsymbol{p}_i$, involving only the squares and scalar products of vectors. It is easy to verify that scalar quantities are indeed unchanged by (12.41). For example,

$$\delta(\boldsymbol{r}_i \cdot \boldsymbol{r}_j) = (\delta\boldsymbol{r}_i) \cdot \boldsymbol{r}_j + \boldsymbol{r}_i \cdot (\delta\boldsymbol{r}_j) = [(\boldsymbol{n} \wedge \boldsymbol{r}_i) \cdot \boldsymbol{r}_j + \boldsymbol{r}_i \cdot (\boldsymbol{n} \wedge \boldsymbol{r}_j)]\delta\varphi = 0,$$

by the symmetry of the scalar triple product (see (A.13)).

When the Hamiltonian possesses complete rotational symmetry, then every component of $\boldsymbol{J}$ is conserved, and so

$$\frac{\mathrm{d}\boldsymbol{J}}{\mathrm{d}t} = \boldsymbol{0}. \tag{12.42}$$

So far, we have shown that the conservations laws of energy, momentum, and angular momentum are expressions of symmetry properties required by the relativity principle. They have, therefore, a much more general validity than the specific assumptions used in their original derivation. For example, we have not assumed that the forces in our system are all two-body forces, or that they are central or conservative. All we have assumed is the existence of the Hamiltonian function, and the relativity principle.

There remains one type of Galilean transformation, which is in some ways rather different from the others.

### Transformations to moving frames

Let us consider the effect of giving our system a small overall velocity $\delta v$ in the $x$ direction — a small 'boost'. The corresponding transformations are

$$\delta x_i = t\,\delta v, \qquad \delta y_i = 0, \qquad \delta z_i = 0,$$
$$\delta p_{xi} = m_i\,\delta v, \qquad \delta p_{yi} = 0, \qquad \delta p_{zi} = 0. \tag{12.43}$$

This transformation differs from the others we have considered in that $t$ appears explicitly in (12.43). The generator of this transformation is

$$G_x = \sum_{i=1}^{N} (p_{xi}t - m_i x_i) = P_x t - MX,$$

where $M$ is the total mass, and $X$ is the $x$ co-ordinate of the centre of mass. This generator is also explicitly time-dependent. It is clearly the $x$ component of the vector

$$\boldsymbol{G} = \boldsymbol{P}t - M\boldsymbol{R}. \tag{12.44}$$

Transformations to frames moving in other directions are generated by appropriate components of this vector.

We must now be careful. For, because of the explicit time-dependence, it is no longer true that $H$ must be invariant under (12.43). Indeed, we know that the energy of a system *does* depend on the choice of reference frame, though not on the choice of origin or axes. What the relativity principle actually requires is that the *equations of motion* should be unchanged by the transformation. It can be shown (see Problem 17) that the condition for this is still

$$\frac{dG}{dt} = 0. \tag{12.45}$$

However, this no longer implies the invariance of $H$. In fact, using (12.33) and (12.34), we find

$$\delta H = [H, G]\,\delta\lambda = -[G, H]\,\delta\lambda = \frac{\partial G}{\partial t}\,\delta\lambda. \tag{12.46}$$

Thus the change in $H$ is related to the explicit time-dependence of $G$.

The fourth of the basic conservation laws, for the quantity (12.44), is

$$\frac{d\boldsymbol{G}}{dt} = \frac{d}{dt}(\boldsymbol{P}t - M\boldsymbol{R}) = \boldsymbol{0}. \tag{12.47}$$

Though this is an unfamiliar form, the equation is actually quite familiar. For, since $d\boldsymbol{P}/dt = \boldsymbol{0}$, it may be written

$$\boldsymbol{P} - M\frac{d\boldsymbol{R}}{dt} = \boldsymbol{0},$$

which is simply the relation (8.7) between the total momentum and the centre-of-mass velocity. This relation too is therefore a consequence of the relativity principle.

Let us now turn to the question of what this symmetry implies for the form of the Hamiltonian function. For a transformation in the $x$ direction, the change in $H$, given by (12.46), is

$$\delta H = \frac{\partial G_x}{\partial t}\,\delta v = P_x\,\delta v. \tag{12.48}$$

Now, if we write

$$H = T + V, \qquad \text{where} \qquad T = \sum_{i=1}^{N}\frac{p_i^2}{2m_i},$$

then we find directly from (12.43) that

$$\delta T = \sum_{i=1}^{N}p_{xi}\,\delta v = P_x\,\delta v.$$

Thus the change in kinetic energy is exactly what is demanded by (12.48), and this relation reduces to

$$\delta V = 0. \tag{12.49}$$

It is interesting to examine the conditions imposed on $V$ by this requirement. We have already seen that $V$ must be a scalar function, and must involve the particle positions only through the differences $\boldsymbol{r}_{ij} = \boldsymbol{r}_i - \boldsymbol{r}_j$. Since these are unaffected by (12.43), the requirement (12.49) imposes no further restrictions on the $\boldsymbol{r}$ dependence. Moreover, $V$ must contain no explicit dependence on time. However, none of the conditions imposed so far requires it to be independent of the momenta. To satisfy (12.49), it must contain them only through the combinations

$$\boldsymbol{v}_{ij} = \frac{\boldsymbol{p}_i}{m_i} - \frac{\boldsymbol{p}_j}{m_j},$$

which are easily seen to be invariant under (12.43), but this is the only new requirement. Thus the most general form of interaction in an $N$-particle system which is invariant under all Galilean transformations is one described by a 'potential energy function' which is an arbitrary scalar function of the relative position vectors *and* the relative velocity vectors.

### Reflections; parity

There is one other type of co-ordinate transformation which might be mentioned. All those we have considered so far have the property that if we start with a right-handed set of axes, then the transformation takes us to another right-handed set. However, we could also consider transformations like reflections (say $x \to -x, y \to y, z \to z$) or inversions ($\boldsymbol{r} \to -\boldsymbol{r}$) which lead from a right-handed set of axes to a left-handed set. These are called *improper* co-ordinate transformations. They differ from the proper transformations, such as rotations, in being discrete rather than continuous — no continuous change can ever take a right-handed set of axes into a left-handed set.

The condition for the Hamiltonian to be unchanged also under improper co-ordinate transformations is that it should be a true scalar function, like $\boldsymbol{r}_i \cdot \boldsymbol{r}_j$, rather than a *pseudoscalar*, like $(\boldsymbol{r}_i \wedge \boldsymbol{r}_j) \cdot \boldsymbol{r}_k$, which changes sign under inversion. If this condition is fulfilled, the equations of motion will have the same form in right-handed and left-handed frames of reference.

Because of the discontinuous nature of these transformations, this symmetry does not lead to a conservation law for some continuous variable. In fact, in classical mechanics, it does not lead to a conservation law at all. However, in quantum mechanics, it yields a conservation law for a quantity known as the *parity*, which has only two possible values $\pm 1$. Until 1957, it was believed that all physical laws were unchanged by reflections, but it was then discovered by Wu *et al*, following a theoretical prediction by Lee and Yang, that parity is in fact not conserved in the process of radioactive decay of atomic nuclei. The laws describing such processes do not have the same form in right-handed and left-handed frames of reference.

## 12.8   Summary

The Hamiltonian method is an extremely powerful tool in dealing with complex problems. In particular, when the Hamiltonian function is independent of some particular co-ordinate $q_\alpha$, then the corresponding generalized momentum $p_\alpha$ is conserved. In such a case, the number of degrees of freedom is effectively reduced by one.

More generally, we have seen that any symmetry property of the system leads to a corresponding conservation law. This can be of great importance in practice, since the amount of labour involved in solving a complicated problem can be greatly reduced by making full use of all the available symmetries. If there is a sufficient number of symmetries, then the system is 'integrable' (in the sense of Liouville) and the conservation laws may then be exploited to produce (in principle) the complete solution to the problem.

The Hamiltonian function is also of great importance in quantum mechanics, and many of the features of our discussion carry over to that case. We have seen that the variables appear in pairs. To each co-ordinate $q_\alpha$ there corresponds a momentum $p_\alpha$. Such pairs are called *canonically conjugate*. This relationship between pairs of variables is of central importance in quantum mechanics, where there is an 'uncertainty principle' according to which it is impossible to measure both members of such a pair simultaneously with arbitrary accuracy.

The relationship between symmetries and conservation laws also applies to quantum mechanics. In relativity, the transformations we consider are slightly different (Lorentz transformations rather than Galilean), but the same principles apply, and lead to very similar conservation laws.

The relationship between the relativity principle and the familiar conservation laws (including the 'conservation law' $\boldsymbol{P} = M\dot{\boldsymbol{R}}$) is of the greatest importance for the whole of physics. It is the basic reason for the universal character of these laws, which were originally derived as rather special consequences of Newton's laws, but can now be seen as having a far more fundamental role.

---

## Problems

1. A particle of mass $m$ slides on the inside of a smooth cone of semi-vertical angle $\alpha$, whose axis points vertically upwards. Obtain the Hamiltonian function, using the distance $r$ from the vertex, and the azimuth angle $\varphi$ as generalized co-ordinates. Show that stable circular motion is possible for any value of $r$, and determine the corresponding angular velocity, $\omega$. Find the angle $\alpha$ if the frequency of small oscillations about this circular motion is also $\omega$.

2. Find the Hamiltonian function for the forced pendulum considered in §10.4, and verify that it is equal to $T' + V'$. Determine the frequency of small oscillations about the stable 'equilibrium' position when $\omega^2 > g/l$.

3. A light, inextensible string passes over a small pulley and carries a mass $2m$ on one end. On the other end is a mass $m$, and beneath it, supported by a spring with spring constant $k$, a second mass $m$. Find the Hamiltonian function, using the distance $x$ of the first mass beneath the pulley, and the extension $y$ in the spring, as generalized co-ordinates. Show that $x$ is ignorable. To what symmetry property does this correspond? (In other words, what operation can be performed on the system without changing its energy?) If the system is released from rest with the spring unextended, find the positions of the particles at any later time.

4. A particle of mass $m$ moves in three dimensions under the action of a central, conservative force with potential energy $V(r)$. Find the Hamiltonian function in terms of spherical polar co-ordinates, and show that $\varphi$, but not $\theta$, is ignorable. Express the quantity $\boldsymbol{J}^2 = m^2 r^4(\dot\theta^2 + \sin^2\theta\,\dot\varphi^2)$ in terms of the generalized momenta, and show that it is a second constant of the motion.

5. *Find the Hamiltonian for the pendulum hanging from a trolley described in Chapter 10, Problem 9. Show that $x$ is ignorable. To what symmetry does this correspond?

6. *Obtain the Hamiltonian function for the top with freely sliding pivot described in Chapter 10, Problem 11. Find whether the minimum angular velocity required for stable vertical rotation is greater or less than in the case of a fixed pivot. Can you explain this result physically?

7. *To prove that the effective potential energy function $U(\theta)$ of the symmetric top (see §12.4) has only a single minimum, show that the equation $U(\theta) = E$ can be written as a cubic equation in the variable $z = \cos\theta$, with three roots in general. Show, however, that $f(z)$ has the same sign at both $z = \pm 1$, and hence that there are either two roots or none between these points: for every $E$ there are at most two values of $\theta$ for which $U(\theta) = E$.

8. Find the Hamiltonian for a charged particle in electric and magnetic fields in cylindrical polars, starting from the Lagrangian function (10.29). Show that in the case of an axially symmetric, static magnetic field, described by the single component $A_\varphi(\rho, z)$ of the vector

potential, it takes the form

$$H = \frac{1}{2m}\left(p_z^2 + p_\rho^2 + \frac{(p_\varphi - q\rho A_\varphi)^2}{\rho^2}\right).$$

(*Note*: Remember that the subscripts $\varphi$ on the generalized momentum $p_\varphi$ and on the component $A_\varphi$ mean different things.)

9. A particle of mass $m$ and charge $q$ is moving around a fixed point charge $-q'$ ($qq' > 0$), and in a uniform magnetic field $\boldsymbol{B}$. The motion is confined to the plane perpendicular to $\boldsymbol{B}$. Write down the Lagrangian function in polar co-ordinates rotating with the Larmor angular velocity $\omega_{\mathrm{L}} = -qB/2m$ (see §5.5). Hence find the Hamiltonian function. Show that $\varphi$ is ignorable, and interpret the conservation law. (Note that $J_z$ is *not* a constant of the motion.)

10. Consider a system like that of Problem 9, but with a charge $+q'$ at the origin. By examining the effective radial potential energy function, find the radius of a stable circular orbit with angular velocity $\omega_{\mathrm{L}}$, and determine the angular frequency of small oscillations about it.

11. *A particle of mass $m$ and charge $q$ is moving in the equatorial plane $z = 0$ of a magnetic dipole of moment $\mu$, described (see Appendix A, Problem 12) by a vector potential with the single non-zero component $A_\varphi = \mu_0 \mu \sin\theta/4\pi r^2$. Show that it will continue to move in this plane. Initially, it is approaching from a great distance with velocity $v$ and impact parameter $b$, whose sign is defined to be that of $p_\varphi$. Show that $v$ and $p_\varphi$ are constants of the motion, and that the distance of closest approach to the dipole is $\frac{1}{2}(\sqrt{b^2 \mp a^2} \pm b)$, according as $b > a$ or $b < a$, where $a^2 = \mu_0 q\mu/\pi m v$. (Here $q\mu$ is assumed positive.) Find also the range of values of $b$ for which the velocity can become purely radial, and the distances at which it does so. Describe qualitatively the appearance of the orbits for different values of $b$. (*Hint*: It may be useful to sketch the effective radial potential energy function.)

12. *Find the Hamiltonian for the restricted three-body problem described in Chapter 10, Problems 15 and 16. Investigate the stability of one of the Lagrangian 'equilibrium' positions off the line of centres by assuming a solution where $x - x_0, y - y_0, p_x + m\omega y_0$ and $p_y - m\omega x_0$ are all small quantities proportional to $e^{pt}$, with $p$ constant. Show that the possible values for $p$ are given by

$$p^4 + \omega^2 p^2 + \frac{27 M_1 M_2 \omega^4}{4(M_1 + M_2)^2} = 0,$$

and hence that the points are stable provided that the masses $M_1$ and $M_2$ are sufficiently different. Specifically, given that $M_1 > M_2$ show that the minimum possible ratio for stability is slightly less than 25.

13. The stability condition of Problem 12 is well satisfied for the case of the Sun and Jupiter, for which $M_1/M_2 = 1047$. Indeed, in that case these positions are occupied by the so-called *Trojan* asteroids, whose orbital periods are the same as Jupiter's, 11.86 years. Find for this case the periods of small oscillations about the 'equilibrium' points (in the plane of the orbit).

14. *The magnetic field in a particle accelerator is axially symmetric (as in Problem 8), and in the plane $z = 0$ has only a $z$ component. Defining $J = p_\varphi - qpA_\varphi$, show, using (A.40) and (A.55), that $\partial J/\partial \rho = -q\rho B_z$, and $\partial J/\partial z = q\rho B_\rho$. What is the relation between $\dot\varphi$ and $J$? Treat the third term of the Hamiltonian in Problem 8 as an effective potential energy function $U(\rho, z) = J^2/2m\rho^2$, compute its derivatives, and write down the 'equilibrium' conditions $\partial U/\partial \rho = \partial U/\partial z = 0$. Hence show that a particle of mass $m$ and charge $q$ can move in a circle of any given radius $a$ in the plane $z = 0$ with angular velocity equal to the cyclotron frequency for the field at that radius (see §5.2).

15. *To investigate the stability of the motion described in the preceding question, evaluate the second derivatives of $U$ at $\rho = a, z = 0$, and show that they may be written

$$\frac{\partial^2 U}{\partial \rho^2} = \frac{q^2}{m}\left[B_z\left(B_z + \rho\frac{\partial B_z}{\partial \rho}\right)\right]_{\rho=a,z=0},$$

$$\frac{\partial^2 U}{\partial \rho\, \partial z} = 0, \qquad \frac{\partial^2 U}{\partial z^2} = -\frac{q^2}{m}\left[B_z\rho\frac{\partial B_z}{\partial \rho}\right]_{\rho=a,z=0}.$$

(*Hint:* You will need to use the $\varphi$ component of the equation $\boldsymbol{\nabla}\wedge\boldsymbol{B} = \boldsymbol{0}$, and the fact that, since $B_\rho = 0$ for all $\rho$, $\partial B_\rho/\partial \rho = 0$ also.) Given that the dependence of $B_z$ on $\rho$ near the equilibrium orbit is described by $B_z \propto (a/\rho)^n$, show that the orbit is stable if $0 < n < 1$.

16. Show that the Poisson brackets of the components of angular momentum are

$$[J_x, J_y] = J_z$$

(together with two other relations obtained by cyclic permutation of $x, y, z$). Interpret this result in terms of the transformation of one component generated by another.

17. *Show that the condition that Hamilton's equations remain unchanged under the transformation generated by $G$ is $dG/dt = 0$ even in the case when $G$ has an explicit time-dependence, in addition to its dependence via $q(t)$ and $p(t)$. Proceed as follows. The first set of Hamilton's equations, (12.6), will be unchanged provided that

$$\frac{d}{dt}(\delta q_\alpha) = \delta\left(\frac{\partial H}{\partial p_\alpha}\right).$$

Write both sides of this equation in terms of $G$ and use (12.33) applied both to $\partial G/\partial p_\alpha$ and to $G$ itself to show that it is equivalent to the condition

$$\frac{\partial}{\partial p_\alpha}\left(\frac{dG}{dt}\right) = 0.$$

Thus $dG/dt$ is independent of each $p_\alpha$. Similarly, by using the other set of Hamilton's equations, (12.7), show that it is independent of each $q_\alpha$. Thus $dG/dt$ must be a function of $t$ alone. But since we can always add to $G$ any function of $t$ alone without affecting the transformation it generates, this means we can choose it so that $dG/dt = 0$.

## Chapter 13

# Dynamical Systems and Their Geometry

In this chapter we will begin to look at *continuous* dynamical problems in a new, powerful and largely geometrical way, which had its origins in the work of Poincaré.

This approach enables us to move on from the solution of the differential equations for a system having one set of parameters and initial conditions to the consideration of the geometry of *all* the possible solutions. Not only does this new perspective bring out important features of systems which are already familiar, but it also enables us to gain qualitative (and some quantitative) information about systems in which there is complex interaction, which may be highly nonlinear and accessible by other methods only with difficulty, if at all.

The ideas are of very wide applicability in many fields, not only in mechanics, other branches of physics, engineering and applied mathematics, but also in, for example, biology, chemistry and economics.

[For some systems it is appropriate and useful to observe at *discrete* time intervals (not necessarily equal). Differential equations are then replaced by *maps* — see §14.2 and appendix D.]

## 13.1  Phase Space and Phase Portraits

Many systems have a finite number $n$ of distinct dynamic elements, which may be represented by functions $x_i(t)$ of time $t$ where $i = 1, 2, \ldots, n$. It should be noted that $n$ is not necessarily the same as the number of degrees of freedom if, for example, the system is mechanical. Here we are dealing with the number of elements which describe the dynamic state and these might therefore include, for example, positions and velocities or momenta (as in §12.1).

We may denote the state of the system by the vector $\boldsymbol{x}(t)$ and the $n$ equations of motion can be written in vector form

$$\dot{\boldsymbol{x}} \equiv \frac{\mathrm{d}\boldsymbol{x}}{\mathrm{d}t} = \boldsymbol{F}(\boldsymbol{x}, t), \qquad (13.1)$$

describing the evolution of the dynamic elements with 'velocity' $\boldsymbol{F}$.

In practice $\boldsymbol{F}$ usually contains a set $\boldsymbol{c}$ of *control parameters* which are characteristic of the physical problem, like masses, spring constants and so on.

The idea is that we fix $\boldsymbol{c}$ and consider the solutions of (13.1) for various different prescribed values of the $x_i$ at a particular initial time, so determining the evolution of these different initial states.

It is extremely useful to view (13.1) as the prescription for the way in which the point $\boldsymbol{x}$ evolves with time along *trajectories* (or orbits) in an $n$-dimensional space — the *phase space* of the system. The function $\boldsymbol{F}$ in (13.1) is then called the *phase velocity*. The set of all trajectories in this space for various initial prescribed values of $\boldsymbol{x}$, together with the velocity directions upon them, constitutes the *phase portrait* of the system. The evolutions of a continuous dynamical system pictured in this manner are often described as a *flow*, by analogy with fluid motion (see Fig. 13.1).

The phase space is usually taken to be Euclidean in its geometry. While this is not absolutely necessary, it is useful to do so, although there are obvious difficulties in graphical representation when $n > 3$.

Fig. 13.1

It is important to note that the first-order system (13.1) is not as special as it appears, since any system having higher-order derivatives can be put into this standard form by a suitable change of variables. If we have

$$\frac{\mathrm{d}^n x}{\mathrm{d}t^n} = G\left(t, x, \frac{\mathrm{d}x}{\mathrm{d}t}, \ldots, \frac{\mathrm{d}^{n-1}x}{\mathrm{d}t^{n-1}}\right), \tag{13.2}$$

then we may write $x_i = \mathrm{d}^{i-1}x/\mathrm{d}t^{i-1}$ for $i = 1, 2, \ldots, n$, and recover the form (13.1). Simple examples of this technique (for $n = 2$) will be given below.

In (13.1) the functions $F_i$ contained within the phase velocity $\boldsymbol{F}$ may not depend explicitly on time, and the system is then said to be *autonomous*. Of course, it is immediately apparent that any non-autonomous system may be made autonomous through the substitution $x_{n+1} = t$, but at the expense of raising by one the dimension of the phase space. We will not be concerned further with this matter here, but will concentrate almost exclusively on autonomous systems [cf. also 'natural systems' of §10.1].

Evidently the phase portrait for a particular dynamical system may look very different for different values of the control parameters $\boldsymbol{c}$ in (13.1) and, for many such systems, there is considerable interest in the dynamical transitions which occur in their behaviour at particular parameter values. The importance of this lies in what is called the *structural stability* of systems; abrupt changes at particular $\boldsymbol{c}$ are *bifurcations* or *catastrophes*.

## 13.2  First-order Systems — the Phase Line ($n = 1$)

Let us consider the *logistic* differential equation for the dynamic state variable $x(t)$:

$$\dot{x} \equiv \frac{\mathrm{d}x}{\mathrm{d}t} = kx - \sigma x^2, \qquad \text{with} \qquad x(0) = x_0. \tag{13.3}$$

Here $k, \sigma$ are positive control parameters.

The equation, which is due to Verhulst, is a model of population growth (among many other applications) and the general motivation is that the growth rate $\dot{x}/x$ decreases as the population $x(t)$ increases on account of overcrowding or lack of resources. Naturally the interest for this biological application is in $x_0 \geq 0$.

Now this equation (13.3) is first-order separable and it may be solved explicitly to give

$$x(t) = \frac{kx_0}{[\sigma x_0 + (k - \sigma x_0)\exp(-kt)]}, \quad \text{for } x_0 > 0,$$

$$x(t) \equiv 0, \qquad\qquad\qquad\qquad \text{for } x_0 = 0.$$

(13.4)

This solution may be plotted as a function of $t$ as in Fig. 13.2.

It is evident from (13.4) and from Fig. 13.2 that, for all $x_0 > 0$, the evolution of $x$ with $t$ is such that $x \to k/\sigma$. For $0 < x_0 < k/\sigma$ the 'sigmoid' curves obtained represent the way in which the exponential growth (which would be present if $\sigma = 0$) is inhibited by overcrowding. For $x_0 > k/\sigma$ there is steady decay of the population towards the value $k/\sigma$. Since $x \to k/\sigma$ as $t \to \infty$ in each case, then the equilibrium population $x(t) \equiv k/\sigma$ is said to be *asymptotically stable* and it is an *attractor*. The equilibrium population $x(t) \equiv 0$ is *unstable*, in that it is only realizable if $x_0 = 0$. Any positive change in $x_0$, however small, results in explosive growth until the overcrowding takes effect and $x$ is attracted towards $k/\sigma$, as above. So $x(t) \equiv 0$ is a *repeller*.

Now here we have been fortunate to be able to arrive at the complete solution of (13.3) in the form (13.4). The comments detailed above result from interpretation of (13.4) and of Fig. 13.2 derived from it. However, the important qualitative behaviour of the solutions is obtainable more directly from (13.3) by considering the phase space and the phase portrait.

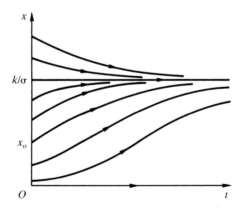

Fig. 13.2

### *Example:* **Phase portrait of the logistic system**

Find the qualitative features of the logistic system by considering the phase portrait.

The phase space is here one-dimensional (a phase line). (See Fig. 13.3.) In constructing the phase portrait for this example the *equilibria* $x = 0$, $x = k/\sigma$ are found by putting $\dot{x} \equiv 0$ in (13.3). It is important to note that these equilibria, or *critical points*, are the only values of $x$ at which $\dot{x}$ can change sign. The arrow directions are those of the phase velocity and, here, the complete portrait can be found by testing just a single $x$ value away from the equilibria. From this we see immediately that $x = k/\sigma$ is an attractor and $x = 0$ is a repeller.

$$O \qquad\qquad k/\sigma \qquad\qquad x$$

Fig. 13.3

The above is sufficient argument, but with an eye to developments for higher-order systems we can consider a small-amplitude approximation near to each critical point, very similar to the analysis carried out in, for example, §2.2. If we write $x(t) = k/\sigma + \xi(t)$, where $\xi(t)$ is small enough for quadratic terms involving it to be neglected, then we find that

$$\dot{\xi} = -k\xi \qquad\qquad (13.5)$$

to first order in $\xi$ and $\xi(t) \to 0$ exponentially as $t \to \infty$, since $k$ is positive, confirming of course that $x = k/\sigma$ is an attractor. Note that this solution of (13.5) is consistent with our assumption that quadratic terms in $\xi$ could be neglected. The other critical point $x = 0$ may be treated similarly and confirmed as a repeller.

An important point to note here is that for this system and for other such first-order systems the qualitative behaviour is completely determined by the position of the critical points, by the local behaviour on the phase line near to them and through the strong geometrical requirement that $\dot{x}$ cannot change sign at any other value of $x$.

We might consider the equation

$$\dot{x} = kx - \sigma x^2 - f, \qquad\qquad (13.6)$$

where $f$ represents a positive constant extraction rate. For example, $x(t)$ could be the population of fish in a lake and $f$ the rate of fishing this population. The exact solution is now rather more awkward to obtain and the use of the one-dimensional phase space provides clear advantage. In particular, the extra control parameter $f$ which has been introduced can be varied and predictions made about the evolution of $x$ from resulting changes in the phase portrait. (See Problem 1.)

The analysis given above will be generalized in following sections to deal with higher-order systems.

Naturally for many biological systems a discrete, rather than continuous, model is more appropriate [see Appendix D].

## 13.3   Second-order Systems — the Phase Plane ($n = 2$)

Again let us proceed by considering a particular and familiar example.

*Example:* **Simple harmonic oscillator**

Determine the phase-space trajectories of the simple harmonic oscillator.

The oscillator equation (2.13) may be written

$$\ddot{x} = -\omega^2 x, \qquad \omega = \sqrt{k/m}, \tag{13.7}$$

which is in the form (13.2) and which may be converted to the form (13.1) as

$$\dot{x} = y,$$
$$\dot{y} = -\omega^2 x. \tag{13.8}$$

The trajectories of this autonomous system in the $(x, y)$ *phase plane* are evidently given by the solutions of

$$\frac{\mathrm{d}y}{\mathrm{d}x} = \frac{\dot{y}}{\dot{x}} = -\frac{\omega^2 x}{y}, \tag{13.9}$$

which may be solved easily to give

$$\omega^2 x^2 + y^2 = A^2, \qquad \text{with} \qquad A^2 \text{ constant.} \tag{13.10}$$

The phase portrait is a family of ellipses centred at the origin. (See Fig. 13.4.) Here, along each trajectory the energy $E$ of the oscillator is

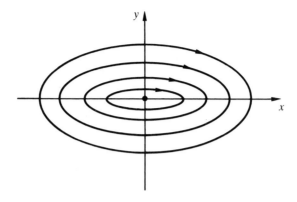

Fig. 13.4

constant and given by $E = \frac{1}{2}mA^2$. Evidently $(x, y) = (0, 0)$ is a critical point of the system (13.8) and any small perturbation leads to the system traversing an elliptical trajectory around this equilibrium point. Each trajectory is closed and is therefore a periodic orbit.

In this case, we were able to find the trajectories by direct integration of (13.9). In the case of a general system this may well not be easy, or possible, except by numerical computation.

The general second-order autonomous system with which we are presented is

$$\dot{x} = F(x, y),$$
$$\dot{y} = G(x, y),$$

(13.11)

where $F, G$ are suitable functions. It is apparent that the equation

$$\frac{dy}{dx} = \frac{G}{F}$$

(13.12)

certainly determines the slope of the unique trajectory in the phase plane at each point except at the *critical points* $(x_0, y_0)$, for which $F(x_0, y_0) = 0 = G(x_0, y_0)$. Each such $(x_0, y_0)$ is an *equilibrium solution* of the system and trajectories can only intersect at such points and nowhere else.

The interesting question to ask, from the physical point of view, concerns the fate of the system when it is displaced slightly from such a critical point. The system could, for general small displacements:

1. Move so as to tend towards $(x_0, y_0)$ as $t \to \infty$, in which case we say the system is *asymptotically stable*.
2. Move increasingly away from $(x_0, y_0)$, in which case we say it is *unstable*.
3. Move in an arbitrarily small neighbourhood of $(x_0, y_0)$ without necessarily tending towards $(x_0, y_0)$ as $t \to \infty$, in which case we say it is *stable*.

The simple harmonic oscillator is therefore stable at $x = 0$, $y \equiv \dot{x} = 0$ — hardly a surprising conclusion!

The behaviour of the trajectories near to each critical point $(x_0, y_0)$ in the phase plane can be found by an analysis similar to that introduced in §13.2 using $x = x_0 + \xi$, $y = y_0 + \eta$, in which $\xi, \eta$ are small and we expand functions $F(x, y), G(x, y)$ locally. We obtain

$$\dot{\xi} = \xi \left( \frac{\partial F}{\partial x} \right)_0 + \eta \left( \frac{\partial F}{\partial y} \right)_0 + F_1(\xi, \eta),$$

$$\dot{\eta} = \xi \left( \frac{\partial G}{\partial x} \right)_0 + \eta \left( \frac{\partial G}{\partial y} \right)_0 + G_1(\xi, \eta).$$

(13.13)

Here the partial derivatives are evaluated *at* the critical point $(x_0, y_0)$ and $F_1, G_1$ are generally quadratic in $\xi, \eta$. In matrix notation, these equations are

$$\begin{bmatrix} \dot{\xi} \\ \dot{\eta} \end{bmatrix} \equiv \frac{\mathrm{d}}{\mathrm{d}t} \begin{bmatrix} \xi \\ \eta \end{bmatrix} = M \begin{bmatrix} \xi \\ \eta \end{bmatrix} + \text{higher-order terms},$$

(13.14)

and the $2 \times 2$ Jacobian matrix $M$ has constant entries which are special to each critical point. In nearly all cases the correct behaviour of the *almost linear system* near $\xi = 0 = \eta$ is given by the behaviour of the *linear system* which is obtained by neglecting the higher-order terms $F_1, G_1$ in the above. The behaviour near $(x_0, y_0)$ of the linear system may be classified using the *eigenvalues* $\lambda$ of $M$. We seek a solution of the linear system which is a linear combination of 'modes', in each of which $\xi, \eta$ are constant multiples of $\mathrm{e}^{\lambda t}$. The analysis appears similar to that involved in the calculation of the frequencies of the normal modes in §11.3, but we are dealing with matrices $M$ which are not necessarily symmetric, so that the roots of the quadratic *characteristic equation* for $\lambda$ need not be real. Evidently, the signs of the real parts of these roots are crucial in determining whether trajectories near to $(x_0, y_0)$ go inwards towards this critical point or go outwards from it.

Details of the classification of critical points in the various cases are given in Appendix C.

The curves which go directly in and directly out of a critical point are called the *stable* and *unstable manifolds* respectively and, in the linear and almost linear systems, they correspond directly and are tangent to one another. We note that the critical point $(x_0, y_0) = (0, 0)$ for the simple harmonic oscillator is a *centre*, *i.e.* nearby trajectories form closed curves around it. (See Appendix C, Fig. C.6.)

The importance of the local behaviours near critical points of system (13.11) is that we can usually obtain a qualitative picture of the full phase portrait of the system by this means. The trajectories are constrained in that they cannot intersect elsewhere and they are locally parallel to the $x, y$ axes respectively when crossing the curves $G(x, y) = 0$, $F(x, y) = 0$.

The fate of particular initial states becomes apparent from the phase portrait. Detailed quantitative information can, if required, be obtained from exact computational plots.

Apart from systems for which (13.12) may be integrated in straightforward fashion to give the curve family $f(x, y) = $ constant, there are some special cases of particular interest in mechanics.

### Conservative systems with one degree of freedom

For the systems considered in §2.1 we have

$$\dot{x} = y, \qquad \dot{y} = \frac{F}{m}, \qquad \text{with} \quad F = -\frac{dV(x)}{dx}, \tag{13.15}$$

and we find the trajectories are given by

$$\tfrac{1}{2}my^2 + V(x) = E, \qquad y = \pm\sqrt{\frac{2}{m}[E - V(x)]}. \tag{13.16}$$

When the potential $V(x)$ is known the phase portrait may be sketched immediately. (See Fig. 13.5.)

Since

$$M = \begin{bmatrix} 0 & 1 \\ -\dfrac{V''(x_0)}{m} & 0 \end{bmatrix}$$

at equilibria $(x_0, 0)$, the eigenvalues of $M$ when $V(x)$ is a local minimum are pure imaginary, since $V''(x_0)$ is positive, and $(x_0, 0)$ is a *centre*, which is stable. Similarly, local maxima of $V(x)$ lead to *saddles* in the phase

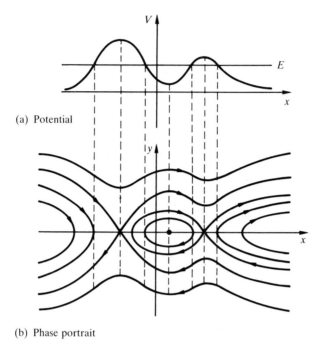

(a) Potential

(b) Phase portrait

Fig. 13.5

portrait and these are unstable, *i.e.* most trajectories approach the critical point, but then turn away. (See Appendix C, Fig. C.2.)

An important physical example is provided by the simple pendulum of §2.1.

### *Example:* Simple pendulum

Determine the possible types of motion of a simple pendulum by considering the phase portrait.

From the $2\pi$-periodic potential function $V(\theta)$ sketched in Fig. 2.1 we can sketch the phase portrait. (See Fig. 13.6.)

The phase portrait is also $2\pi$-periodic. The simple oscillatory motions corresponding to closed orbits around $(0,0)$ are called *librations* and, since $V(\theta) = mgl(1 - \cos\theta)$, it is necessary that $E < 2mgl$ for these

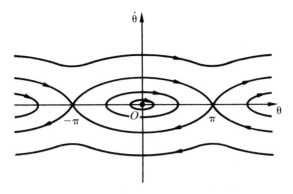

Fig. 13.6

to occur ($E_{(1)}$ in Fig. 2.1). If $E > 2mgl$ ($E_{(2)}$ in Fig. 2.1) the pendulum makes complete revolutions or *rotations*, either in the positive or negative sense. In the special case $E = 2mgl$ the trajectory is a *separatrix* connecting the saddles $(-\pi, 0)$, $(\pi, 0)$, and corresponds to motions of the pendulum in which it is vertically above the point of support at $t \to \pm\infty$. The period of libration motion of the pendulum increases with increasing amplitude and becomes infinite as the phase-plane trajectory corresponds more and more closely to the separatrices from the inside.

It should be noted that (13.15) in general, and hence the simple harmonic oscillator and simple pendulum in particular, are Hamiltonian systems with one degree of freedom. For such systems the critical points in the phase plane are either centres or saddles — spirals and nodes do not occur!

### Gradient systems

These systems have equations of the form

$$\dot{x} = -\frac{\partial U(x, y)}{\partial x}, \qquad \dot{y} = -\frac{\partial U(x, y)}{\partial y}, \qquad (13.17)$$

so that in (13.1) $\boldsymbol{F}$ takes the form $-\boldsymbol{\nabla} U$. The trajectories in the phase plane are everywhere perpendicular to the curves $U(x, y) = $ constant, *i.e.* they are orthogonal trajectories of the level curves (contours) of $U$. Here almost all trajectories tend towards local minima of the function $U(x, y)$ and away from local maxima. For these systems the critical points in the phase plane are nodes and saddles — spirals and centres do not occur.

A physical example of a gradient system is the 'runoff' problem of water flowing downhill through surface soil. (See Fig. 13.7 and Problem 5.)

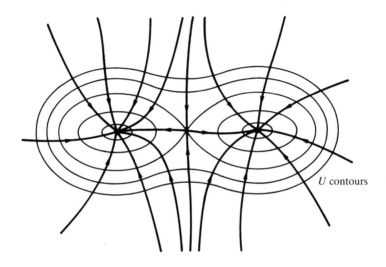

Fig. 13.7

## 13.4    Prey–Predator, Competing-species Systems and War

The general method of analysis in the phase plane introduced in §13.3 finds very wide application.

An interesting example, which is deceptively simple, leads to the equations

$$\dot{x} = ax - bxy,$$
$$\dot{y} = -cy + dxy,$$

$$(13.18)$$

where $a, b, c, d$ are positive constants. These equations have been used by Volterra (1926) to model biological populations and by Lotka (1920) to model chemical reactions. They are usually known as the *Lotka–Volterra* system.

The biological application is to a simple ecological model of population dynamics, where $x, y$ are prey, predators respectively (*e.g.* rabbits, foxes) and the terms proportional to $xy$ model the effect of interaction between these species through encounters — advantageous to the predators, disadvantageous to the prey. Without the interactions the prey would proliferate exponentially in time, while the predators would die out. The parameters $a, b, c, d$ are therefore rates of growth, decay and competitive efficiency.

The ecology, even of such isolated sets of species, is, of course, not this simple, but the consequences of idealized models prove useful as a basis for more extensive investigations.

*Example:* **Lotka–Volterra system**

Find the critical points and their nature for the Lotka–Volterra system.

In the local linear analysis there are two critical points for the system (13.18):

- $(0,0)$ with $M = \begin{bmatrix} a & 0 \\ 0 & -c \end{bmatrix}$, so that the eigenvalues are $\lambda_1 = a, \lambda_2 = -c$. We have a *saddle* with eigenvectors parallel to the $x, y$ co-ordinate axes.

- $\left(\dfrac{c}{d}, \dfrac{a}{b}\right)$ with $M = \begin{bmatrix} 0 & -\dfrac{bc}{d} \\ \dfrac{ad}{b} & 0 \end{bmatrix}$, so that the eigenvalues are imaginary: $\lambda_{1,2} = \pm i\sqrt{ac}$. We have a *centre*.

The question arises as to whether the centre is a true centre for the full system in this case (see Appendix C and Problem 6). That it is such is demonstrated by the realization that (13.18) leads to

$$\frac{dy}{dx} = \frac{y(-c + dx)}{x(a - by)}, \tag{13.19}$$

which is separable, so that it may be integrated to give

$$f(x, y) \equiv -c \ln |x| + dx - a \ln |y| + by = \text{constant}. \tag{13.20}$$

The phase portrait may be sketched as the set of contours of $f(x, y)$. (See Fig. 13.8.) Naturally, the biological interest is confined to the first

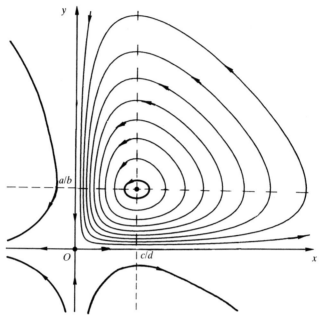

Fig. 13.8

quadrant $(x \geq 0, y \geq 0)$ and the trajectories are closed curves around the
critical point at $(c/d, a/b)$, so confirming the local linear analysis that
this is indeed a centre.

Small oscillations around the point of stable equilibrium have frequency
$\sqrt{ac}$ and period $T = 2\pi/\sqrt{ac}$, which increases with the size of the orbit,
although an explicit analytic demonstration of this is not easy and will not
be given here. However, since

$$\int_0^T \left(\frac{\dot{x}}{x}\right) \mathrm{d}t = \int_0^T (a - by)\,\mathrm{d}t = \big[\ln x(t)\big]_0^T = 0, \qquad (13.21)$$

we find that

$$\bar{y} = \frac{1}{T}\int_0^T y\,\mathrm{d}t = \frac{a}{b},$$

so that the mean value of $y$ around any periodic orbit is given by the
equilibrium value $a/b$. Similarly $\bar{x} = c/d$.

There are ecological consequences of this model in that there are cyclic variations in $x, y$, which are not in phase, and changes in parameter values can be investigated. The effect here of a proportional extraction rate is to reduce $a$ and increase $c$, so that the mean populations of prey/predators respectively increase/decrease — the paradoxical advantage to the prey of this extraction is due to the removal of some natural predators, and is known as *Volterra's principle*.

The model has one further interest for us in that it is, in effect, a *Hamiltonian system*, as detailed in Chapter 12. For the first quadrant we may substitute $x = e^p, y = e^q$ in order to obtain the Hamilton canonical form

$$\dot{q} = -c + d\,e^p = \frac{\partial H}{\partial p},$$
$$\dot{p} = a - b\,e^q = -\frac{\partial H}{\partial q}, \tag{13.22}$$

where $H(q, p) = -cp + d\,e^p - aq + b\,e^q$. For this autonomous system, $H$ is a conserved quantity. (See §12.2.) In the original variables $x, y$ the function $H(q, p)$ becomes equivalent to $f(x, y)$ in (13.20) and it is an 'energy', but not mechanical energy.

More complicated systems involving interactions are very common and the outcomes are even less intuitively transparent.

For example

$$\dot{x} = k_1 x - \sigma_1 x^2 - \alpha_1 xy,$$
$$\dot{y} = k_2 y - \sigma_2 y^2 - \alpha_2 xy, \tag{13.23}$$

where $k_1, k_2, \sigma_1, \sigma_2, \alpha_1, \alpha_2$ are positive control parameters, models logistic behaviour (see (13.3)) for each of $x, y$ with similarly destructive interactions. This system is certainly not usually Hamiltonian, but the phase portrait is accessible by the local linear analysis near the critical points, of which there may be four. The influence of the parameters is crucial. An example of such a portrait is given in Fig. 13.9, for which the parameters have been chosen to be $k_1 = \sigma_1 = \alpha_1 = 1, k_2 = \frac{1}{2}, \sigma_2 = \frac{1}{4}, \alpha_2 = \frac{3}{4}$. (See Problem 7 for the details.)

Here there are two asymptotically stable nodes at $(1, 0), (0, 2)$, an unstable node at $(0, 0)$ and a saddle (unstable) at $(\frac{1}{2}, \frac{1}{2})$. Nearly all initial conditions lead to extinction for one or other species and this is an example of the 'Principle of Competitive Exclusion' [Charles Darwin, *The Origin of Species* (1859)].

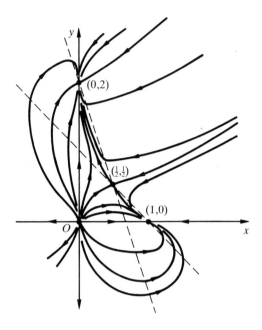

Fig. 13.9

In the general case (13.23) and where there is a critical point in the first quadrant, an extinction is inevitable when $k_1\sigma_2 < k_2\alpha_1$, $k_2\sigma_1 < k_1\alpha_2$ (so that $\sigma_1\sigma_2 < \alpha_1\alpha_2$). On the other hand, peaceful coexistence via an asymptotically stable node in the first quadrant is guaranteed when these inequalities are all reversed (again, see Problem 7).

In human conflict there are mathematical theories of war. The simplest such model involves forces $x(t)$, $y(t)$ which are effectively isolated apart from the confrontation between them.

### *Example:* Combat model

The simple system

$$\dot{x} = -ay,$$
$$\dot{y} = -bx,$$

(13.24)

where $a, b$ are positive constants, models the attrition rates of each force on the other. Determine how the final outcome of a conflict depends on the initial conditions in this model.

Evidently $(0,0)$ is the only critical point and it is a saddle. Indeed the trajectories are given by

$$ay^2 - bx^2 = ay_0^2 - bx_0^2 \equiv K, \qquad (13.25)$$

where $x_0, y_0$ are the initial strengths at $t = 0$. These are hyperbolic arcs in general (see Fig. 13.10).

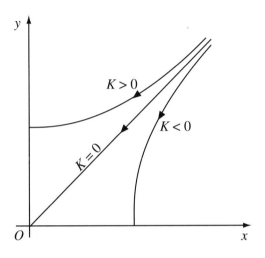

Fig. 13.10

Of course $x$ 'wins' if $K < 0$ and $y$ 'wins' if $K > 0$. One interesting feature is that the trajectories diverge — the discrepancy between the two sides becomes more marked as time progresses, so the likely outcome becomes clearer and the weaker party may be induced to surrender. The most damaging conflicts are those that start near the line $K = 0$ which leads to a 'draw' in which both sides are completely annihilated, since in this case the forces were initially perfectly balanced.

The measure of fighting effectiveness of *e.g.* force $y$ as $ay_0^2$ at $t = 0$, *i.e.* quadratic in numerical strength, is Lanchester's Law of Conventional Combat (1916). Through the ages, the general success of the empirical rule 'Divide and conquer!' is confirmation of the effect of this simple qualitative law.

One may also model not the conflict itself but the arms race leading up to it. Richardson (1939) introduced such a model in the form

$$\dot{x} = a_2 y - c_1 x + g_1,$$
$$\dot{y} = a_1 x - c_2 y + g_2.$$

(13.26)

Here $x(t), y(t)$ are war potentials/armaments of two nations/coalitions, $(a_1, a_2)$ are response parameters to the armament of the opponent (note that in contrast to the example above, these coefficients enter with positive sign), $(c_1, c_2)$ are cost deterrents and $g_1, g_2$ basic grievances. (See Problem 8.)

## 13.5 Limit Cycles

So far we have looked at phase portraits in which the attractors and repellers have been equilibria which are isolated critical points. For some systems there are other possibilities. Rayleigh's equation (1883)

$$\ddot{x} - \epsilon \dot{x}(1 - \dot{x}^2) + x = 0,$$

(13.27)

where $\epsilon$ is a positive control parameter, has been used to model various self-excited oscillations, for example the bowing of a violin string. By reference to the damped oscillator of §2.5 it is evident that (13.27) has positive or negative damping respectively according as $|\dot{x}| > 1$ or $|\dot{x}| < 1$. The sort of physical system modelled by this equation is illustrated schematically in Fig. 13.11, where a block is acted upon by a spring and by a dry frictional force at contact with a belt, which is itself moving uniformly. We may write

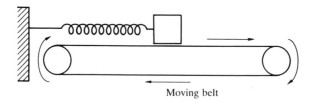

Moving belt

Fig. 13.11

(13.27) in the form

$$\dot{x} = y,$$
$$\dot{y} = -x + \epsilon y(1 - y^2).$$

(13.28)

The only critical point is at $(0,0)$ and is always unstable (spiral or node). However trajectories come inwards from large values of $x$ and $y$. The phase portraits for progressive $\epsilon$ values are sketched in Fig. 13.12.

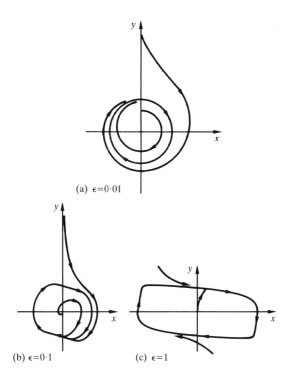

(a) $\epsilon=0\cdot01$

(b) $\epsilon=0\cdot1$

(c) $\epsilon=1$

Fig. 13.12

In each case there is a periodic orbit attractor, which is an asymptotically stable *limit cycle*. Other trajectories tend towards this cycle as $t \to \infty$ and the form of the cycle depends crucially on the parameter $\epsilon$. As $\epsilon$ increases, the variation of $x$ with time $t$ changes from essentially sinusoidal in Fig. 13.12(a) to a markedly non-symmetrical oscillation in Fig. 13.12(c). There is then a sharp rise and fall in $x$ as $t$ varies (see Fig. 13.13(a)). For this type of oscillation, which is called a *relaxation oscillation*, there is a mechanical analogy in the seesaw mechanism (see Fig. 13.13(b)). Here liquid is added at a constant rate to the reservoir on the right-hand side and periodically tips the balance, so that the reservoir is emptied suddenly and the cycle can be repeated.

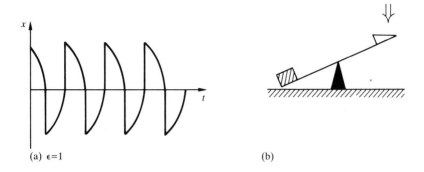

(a) $\epsilon=1$                                                    (b)

Fig. 13.13

There are many mechanical and electrical examples of Rayleigh's equation and of another which may be derived from it by the substitution $\dot{x} = v/\sqrt{3}$. The equation for $v$ is then (see Problem 10)

$$\ddot{v} - \epsilon\dot{v}(1 - v^2) + v = 0. \tag{13.29}$$

This equation was studied by Van der Pol (1926) in order to model various electrical circuits and in his study of the heartbeat.

A spectacular example of a limit cycle was in the behaviour of the Tacoma Narrows bridge near Seattle. The cycle of torsional oscillations, up to 1.5 m in the bridge deck, drew its energy from a constant wind flow across it. The bridge became known as 'Galloping Gertie', at least up to its spectacular collapse on 7 November 1940, fortunately and memorably recorded on film.

Relaxation oscillations have also been used to model natural phenomena, like geysers and earthquakes.

Analytic demonstration of the existence of limit cycles, both attractors and repellers, is beyond our scope here, but it is important to note that some systems exhibit bifurcation phenomena as a control parameter is varied.

### Hopf bifurcation

At the end of §13.1 we referred to the way in which critical points and phase portraits may have very different character when control parameters are varied. Consider now that the stability of a focus can change as a control parameter $c$ is varied, in such a way that the system dynamics 'gives birth' to a limit cycle. This is usually known as a Hopf bifurcation and it is

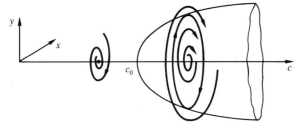

(a) Supercritical Hopf bifurcation

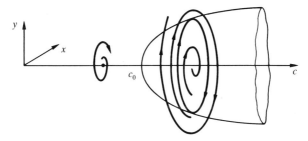

(b) Subcritical Hopf bifurcation

Fig. 13.14

termed supercritical or subcritical respectively according as the resulting limit cycle is an attractor or repeller. For the supercritical bifurcation an asymptotically stable focus becomes unstable at a critical value $c_0$ at which an attracting limit cycle appears. (See Fig. 13.14(a).) For the subcritical bifurcation an unstable focus becomes asymptotically stable at the critical value of $c$, at which a repelling limit cycle appears. (See Fig. 13.14(b).)

The effect of a supercritical Hopf bifurcation is relatively benign ('soft') in that the change results in the system remaining very much in the same neighbourhood of the phase plane when it is subject to a perturbation. However, near a subcritical Hopf bifurcation even a small perturbation can take the system outside the repelling limit cycle and then far away in the phase plane, despite the asymptotic stability of the critical point in the *linear* analysis — the effect is 'hard'.

The 'Brusselator' is a simple model of chemical oscillation proposed by Prigogine and Lefever (1968).

*Example:* **The 'Brusselator'**

The chemical concentrations $x(t), y(t)$ of two particular reactants are related by the equations:

$$\dot{x} = a - (1 + b)x + x^2 y,$$
$$\dot{y} = bx - x^2 y. \tag{13.30}$$

Here $a, b$ are concentrations of other reactants which, once chosen, are kept constant, with $a, b$ both positive. Find the nature of the attractors for this model.

There is only one critical point, at $(a, b/a)$, with Jacobian matrix

$$\begin{bmatrix} b - 1 & a^2 \\ -b & -a^2 \end{bmatrix}.$$

In consequence the critical point is asymptotically stable when $b < 1 + a^2$. When $b > 1 + a^2$ the critical point is unstable (a spiral or node according as $b < $ or $> (1 + a)^2$). At $b = 1 + a^2$ there is a supercritical Hopf bifurcation, so that the unstable critical point is accompanied by a limit-cycle attractor (see Fig. 13.15, where $a = 1, b = 3$).

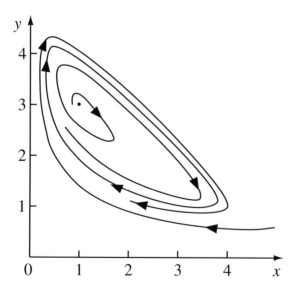

Fig. 13.15   Brusselator limit cycle attractor ($a = 1, b = 3$).

## 13.6 Systems of Third (and Higher) Order

We have seen that, for first-order systems, there are point attractors and repellers and, for second-order systems, there may also be limit cycles. Many higher-order systems are special, in that their dynamical behaviour is effectively lower-dimensional, through, for example, the existence of constants of the motion. Some important specific examples of this will be given in Chapter 14 for systems of fourth order and beyond, but in this section we will consider third-order systems which exhibit certain properties typical of systems of order higher than second.

### *Rigid-body rotation*

Consider the equations of rotational motion of a rigid body, which were introduced in §9.8. When the rotation is about the centre of mass and external forces have zero moment we may, for convenience, choose to express the equations (9.41) in terms of angular momentum components $(J_1, J_2, J_3) \equiv (I_1\omega_1, I_2\omega_2, I_3\omega_3)$ to obtain

$$\dot{J}_1 - \left(\frac{I_2 - I_3}{I_2 I_3}\right) J_2 J_3 = 0 \qquad (13.31)$$

and two similar equations obtained by cyclic interchange of $1, 2, 3$. In the $(J_1, J_2, J_3)$ phase space each trajectory is then the intersection of two surfaces (see Problem 12)

$$\text{sphere:} \quad J_1^2 + J_2^2 + J_3^2 = J^2,$$

$$\text{ellipsoid:} \quad \frac{J_1^2}{I_1} + \frac{J_2^2}{I_2} + \frac{J_3^2}{I_3} = 2T, \qquad (13.32)$$

with $J, T$ constant. The first of these relations expresses the constancy of the magnitude of the angular momentum vector $\boldsymbol{J}$ and the second expresses the constancy of the kinetic energy $T$. (See Fig. 13.16, with principal moments of inertia such that $I_1 < I_2 < I_3$.)

If we take $J$ to be fixed and look at the critical points on the corresponding phase sphere, it is evident that Fig 13.16 confirms the results obtained in §9.8, since the saddle-point geometry leading to instability is at the axis of the intermediate principal moment of inertia, while the centres are at the axes of the largest and the smallest moment. The result holds, of course, for all rigid bodies (from a match-box to a space station), but it is often known as the *tennis racquet theorem*.

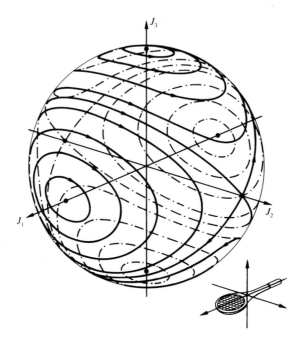

Fig. 13.16

When there is axial symmetry of the rigid body, as in §9.9, there is similar symmetry in the phase space. (See Fig. 13.17, with principal moments of inertia such that $I_1 = I_2 < I_3$.)

In this case $J_3$ is constant so that $J_3 = I_3\Omega$ (say), while the components $J_1, J_2$ of angular momentum are simple harmonic with frequency $[(I_3 - I_1)/I_1]\Omega$. Relative to axes *fixed in space* $\boldsymbol{J}$ is itself fixed and the angular velocity vector $\boldsymbol{\omega}$ precesses around $\boldsymbol{J}$ in the way described in §9.9. In the case $\omega_1, \omega_2 \ll \Omega$ then this precession rate is given approximately by $(I_3/I_1)\Omega$. Hence the precession rate for the Chandler wobble of the Earth is about $\Omega$ (*i.e.* once per day), while the precession rate is about $\frac{1}{2}\Omega$ for a spun pass with an American football and about $2\Omega$ for a spun coin, discus or dinner plate!

The problem of rotational motion, when the external force moment $\boldsymbol{G}$ is *not* special, can be very complicated, although it is of course an important one for many feedback control problems in, for example, the aerospace industry. Some of the complexity which may result is indicated by the following examples of other physical systems.

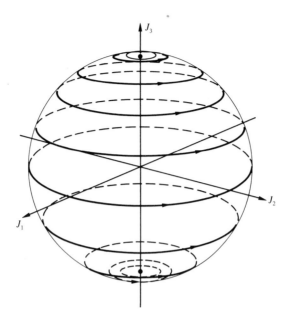

Fig. 13.17

## The Lorenz system

The system of equations

$$\dot{x} = \sigma(y - x),$$
$$\dot{y} = \rho x - y - xz, \qquad (13.33)$$
$$\dot{z} = -\beta z + xy,$$

where $\sigma, \rho, \beta$ are positive control parameters, is an important system which was originally put forward by Lorenz in 1963 to provide a very simple and idealized model of convection in a slab of fluid. The aim was to gain insight into the dynamics of weather systems. (See Fig. 13.18.)

For a general discussion of the origins of this system, see Lorenz, *The Essence of Chaos*, UCL Press (1993).

The parameters have the following physical meaning, after scaling:

- $\sigma$ quantifies the ratio of diffusion rates of momentum and heat in the fluid ($\sigma \equiv$ kinematic viscosity/thermal conductivity).
- $\rho$ quantifies the temperature difference applied across the slab of fluid.
- $\beta$ quantifies the slab geometry through an aspect ratio.

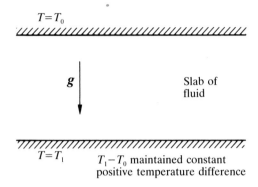

$T = T_0$

$g$

Slab of fluid

$T = T_1$    $T_1 - T_0$ maintained constant
positive temperature difference

Fig. 13.18

Despite its geophysical origin the system (13.33) and its wide range of behaviours have been the subject of extensive study, particularly because the system finds application in other physical contexts, for example in the study of lasers.

It is instructive (from both a practical and a theoretical point of view) to look at the $(x, y, z)$ phase space, say with $\sigma, \beta$ fixed and with a sequence of different values of $\rho$. We can find the critical points of the system and examine their local character through the methods of §13.3 and Appendix C, suitably extended. However the characteristic equation is now a cubic, so that there are three eigenvalues, although the general principles, with regard to stable and unstable manifolds according to the sign of real eigenvalues, still apply.

When $\sigma, \beta$ are indeed fixed the behaviour of the system (13.33) in the $(x, y, z)$ phase space can be summarized as follows (see Problem 14):

- $0 < \rho < 1$: The origin $P_1 \equiv (0, 0, 0)$ is the only critical point of the system, and it is a point attractor. This corresponds physically to the case of *no motion* and steady heat conduction across the slab (see Fig. 13.19(a)).
- $1 < \rho$: The origin $P_1$ is now unstable and two new critical points appear,

$$P_2, P_3 \equiv \left( \pm \sqrt{\beta(\rho - 1)}, \pm \sqrt{\beta(\rho - 1)}, (\rho - 1) \right).$$

It can be shown that when:

* $1 < \rho < \rho_{\text{crit}} = \dfrac{\sigma(\sigma + \beta + 3)}{(\sigma - \beta - 1)}$ in the practical case $\sigma > \beta + 1$,

all the eigenvalues in the analysis of $P_2, P_3$ are negative real or have negative real part, so that $P_2, P_3$ are point attractors. This corresponds physically to *steady convection rolls* in either of the two senses (see Fig. 13.19(b)).

* $\rho_{\text{crit}} < \rho$, the points $P_1, P_2, P_3$ are all unstable, so that a single point attractor does not exist. In fact the loss of asymptotic stability of $P_2, P_3$ can be shown to be *via* a subcritical Hopf bifurcation.

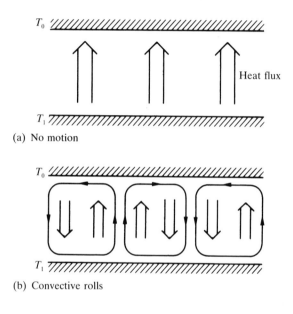

(a) No motion

(b) Convective rolls

Fig. 13.19

We can show that trajectories are attracted towards a bounded region containing the points $P_1, P_2, P_3$ and there should be an *attractor*, since

$$\nabla \cdot \dot{\boldsymbol{x}} = \frac{\partial \dot{x}}{\partial x} + \frac{\partial \dot{y}}{\partial y} + \frac{\partial \dot{z}}{\partial z} = -(\sigma + 1 + \beta) < 0 \qquad (13.34)$$

everywhere, which means that the phase-space volume elements are reduced progressively in volume by the flow $\dot{\boldsymbol{x}}$. (See §12.5 and relation (A.24).)

In fact, it turns out that there is a value $\bar{\rho}$, such that for $\bar{\rho} < \rho$ there is a highly disordered motion, which corresponds physically to *turbulent convection*. Trajectories in the phase space tend asymptotically towards a structure which is multi-leaved on a hierarchy of scales (a '*fractal*') and looped around the unstable critical points $P_2, P_3$. This is called a *chaotic* or *strange attractor*, on account of its peculiar properties. (See for example, Fig 13.20 in which $\sigma = 10, \beta = \frac{8}{3}$ and $\rho = 28$, so that $\rho_{\text{crit}} = 24.74$).

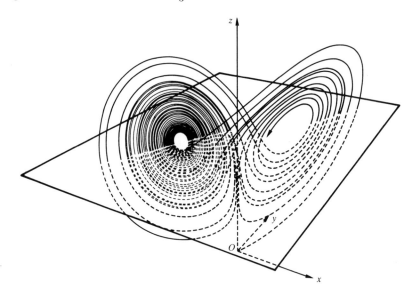

Fig. 13.20  Lorenz attractor [Lanford, (*Lecture Notes in Mathematics*, **615**, 114, Springer 1977)]

For the particular values of $\sigma, \beta$ chosen here a strange attractor of this general type is found to occur when $\rho > \bar{\rho} = 24.06$. Evidently in a (here) small range $\bar{\rho} < \rho < \rho_{\text{crit}}$ there are actually three attractors — two points (steady convection) and one strange (turbulent convection). In practice each of these will have its own *basin of attraction* of states attracted to it, so that the attractor which is observed to occur depends on the initial state of the system. The *strange attractor* has zero phase-space volume on account of (13.34) above, and a fractal geometric structure. However, states which are close initially are stretched out *along* the attractor exponentially with time. The latter property is characteristic of *chaotic* systems and there is more discussion of this property in the following section (§13.7).

### The Rikitake two-disc dynamo

In order to model irregular reversals of the Earth's magnetic field Rikitake put forward (in *Proceedings of the Cambridge Philosophical Society*, **54**, 89–

105, 1958) a system containing two identical coupled disc dynamos which does mimic this type of behaviour. (See Fig. 13.21.)

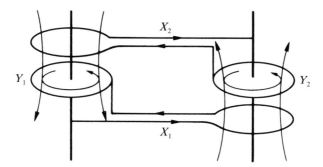

Identical discs driven by the
same constant torque

Fig. 13.21

If $X_1, X_2$ are scaled currents and $Y_1, Y_2$ are scaled rotation rates of the discs, which are driven by a constant external torque, the current in each disc feeds the coil of the other. Then the equations may be written

$$\dot{X}_1 = YX_2 - \mu X_1,$$
$$\dot{X}_2 = (Y - A)X_1 - \mu X_2, \qquad (13.35)$$
$$\dot{Y} = 1 - X_1 X_2,$$

where $Y \equiv Y_1 = Y_2 + A$ and $A, \mu$ are control parameters, with $\mu$ positive.

For this system there are two critical points in the $(X_1, X_2, Y)$ phase space, at $(\pm k, \pm 1/k, \mu k^2)$, where $A = \mu(k^2 - 1/k^2)$. (See Problem 15.) Although the three eigenvalues in the local linear approximation are $-2\mu, \pm i\sqrt{k^2 + 1/k^2}$, the apparent centre behaviour associated with the pure imaginary pair of eigenvalues is here not carried through for the full system. The critical points correspond to senses of North/South magnetic polarity and the irregular reversals which are associated with this model (see Fig. 13.22) mimic the field reversals of the Earth's magnetic field which are known, from the interpretation of geological records, to have occurred at irregular intervals during more than the last 160 million years.

There is, of course, no suggestion that the two-disc dynamo provides other than an interesting analogy for features of the magnetohydrodynamic

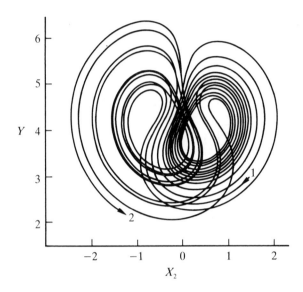

Fig. 13.22   Projection on the $(X_2, Y)$ plane of a trajectory in phase space for $\mu = 1, K = 2$ [Cook and Roberts (*Proceedings of the Cambridge Philosophical Society*, **68**, 547–569, 1970)]

motions in the Earth's core — it is a 'toy' system.

### The forced damped pendulum

If the simple pendulum of §2.1 and §13.3 is subjected to damping and forcing we may write the equation of motion in the form

$$ml\ddot{\theta} + \lambda\dot{\theta} + mg\sin\theta = F\cos\Omega t, \tag{13.36}$$

where $m, l, \lambda, g, F, \Omega$ are control parameters.

This equation may be scaled suitably, so that we study the equivalent third-order autonomous system

$$
\begin{aligned}
\dot{\theta} &= \omega, \\
\dot{\omega} &= -A\omega - \sin\theta + B\cos\phi, \\
\dot{\phi} &= \Omega,
\end{aligned}
\tag{13.37}
$$

where $A, B, \Omega$ are control parameters.

The nonlinearity of this system, through the $\sin\theta, \cos\phi$ terms, makes for a very rich variety of behaviour as $A, B, \Omega$ are changed.

The variety of behaviours instanced in this whole section is common for third- and higher-order continuous systems, but systems have to be 'special' in order to be effectively reducible to lower order *via*, for example, constants of the motion. Systems need to be of at least third order if they are to exhibit *chaotic* behaviour.

## 13.7  Sensitivity to Initial Conditions and Predictability

An important consequence of the *effective* order of a dynamical system is in the way that the resulting dimension of the phase space can restrict the evolution along trajectories of states which may be initially close together. In the line and plane, typically, states which are initially close, whose trajectories remain in a bounded region and do not intersect one another, move apart at a rate which is at most essentially uniform, on average.

If we examine a dynamical system with closed trajectories in phase space, we can examine the evolution of two neighbouring initial states

$$P \equiv (x_0, y_0), \qquad Q \equiv (x_0 + \Delta x_0, y_0 + \Delta y_0).$$

(See Fig. 13.23.)

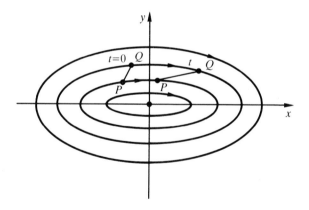

Fig. 13.23

Now, if the system is the simple harmonic oscillator (see §13.3) then, in this special case, the period along each closed orbit is the same and it can be shown that the ('Pythagorean') distance in phase space between $P$ and $Q$ remains bounded in time. This behaviour is not typical and, if Fig. 13.23

is taken to represent, for example, libration motions of a simple pendulum, it should be evident that different orbits have different periods. For this, the more typical situation, the distance between $P$ and $Q$ increases broadly linearly with time, as measured *along* the orbit. A useful analogy is to consider runners travelling at slightly different rates in neighbouring lanes of a running track (see Problems 16, 17).

In three- and higher-dimensional phase spaces the neighbouring trajectories can diverge *exponentially* from one another within a bounded region, without crossing one another. They can do this by wrapping over and under one another in a complex tangle, which may still have a layered structure. (See Fig. 13.20 and Fig. 13.22.) This geometrical behaviour is central to the 'strangeness' of a strange attractor to which we referred in §13.6. The states which are initially close move apart, in such a way that the distance $d$ in phase space between them increases with time, on average, like $e^{\lambda t}$, with $\lambda$ *positive*. (See Fig. 13.24.)

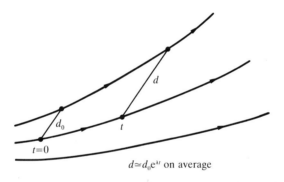

Fig. 13.24

The parameter $\lambda$ in the exponent is called a *Lyapunov exponent*. When a typical small *volume* of initial states in phase space is considered, it is evident that dynamical evolution will result, in general, in stretching and squeezing of this volume. There will be a number of principal deformations, which is equal to the dimension of the phase space, leading to a set of Lyapunov exponents. Each of the *positive* exponents leads to a corresponding strong divergence of initial states. We can evidently test flows in phase space for this divergence property. The strong divergence is called *sensitivity to initial conditions* and it is the prime property which identifies the irregular behaviour now called *chaos*.

The effect on the global behaviour of dynamical systems produced by sensitivity to initial conditions is often called the *butterfly effect*, which was first put forward by Lorenz in an address to the American Association for the Advancement of Science in December 1972 — 'Predictability: Does the flap of a butterfly's wings in Brazil set off a tornado in Texas?' — printed in his book *The Essence of Chaos* referred to above in the description of the Lorenz system (13.33). The point he was making is that chaotic regimes in weather systems can render predictions of them utterly useless, except on a very short time scale. The term 'butterfly effect' has become a much used metaphor for describing *chaos*, which has experienced an explosive growth in application.

The consequences for all dynamical systems and for the real situations they model are immense. A direct attack is made on the long-standing philosophical idea of *determinism* and on the concept of a 'clockwork universe'. The Laplacian point of view identifies the initial state of a system as the *cause*, from which (if it is known precisely) the *effects* may be calculated with *infinite precision* indefinitely far into the future, or into the past. There is, in this world view, no room for chance or free will.

The sensitivity to initial conditions, which is typical of many, if not most, systems and is made very apparent by the geometrical approach pioneered by Poincaré, leads to a *loss of predictability*. This comes about because the inevitable imprecision in the initial description of even a classical system, however small this may be, leads later to a vastly greater imprecision in its predicted fate. It is in this way that *deterministic systems*, where the mechanisms may be known completely and precisely, can have an essentially *stochastic* ('random') output, so that outcomes can be predicted only as a matter of probability.

A good illustration of this apparent paradox is provided by the Galton board (see Fig. 13.25) where small spheres of lead shot fall through a regular array of scattering pins.

The path of an individual piece of lead shot, which falls through such an array of pins, is impossible to predict, because of the sensitivity of the collisions at individual pins. However the typical outcome for the resulting distribution of shot along the channels at the base is, in this case, very like the normal probability distribution. In this way predictability of the fate of an individual piece of lead shot gives way to a collective predictability for many pieces in the form of *chance*. (See Problem 18.)

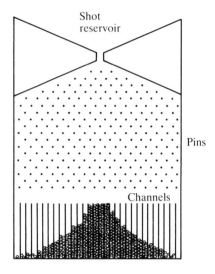

Fig. 13.25

## 13.8  Summary

The analysis of the geometry of dynamical systems in phase space is a
very powerful way of investigating their behaviour for all possible initial
conditions. The dynamics is dependent crucially on attractors and on their
stability.

For systems which are dynamically one-dimensional or two-dimensional
the only attractors are critical points (equilibria) in the phase line and the
phase plane and, for the latter, limit cycles. As control parameters are
varied, the character of critical points and limit cycles may change, so that
there is a corresponding change in the phase portraits and in the behaviour
of the system.

In systems of third (or higher) order additional attractors may arise
which can have a fractal and sensitive structure — strange (chaotic) attractors.

In the phase plane the system trajectories are constrained such that
the divergence of initial states takes place essentially at a uniform rate,
at most. For a strange attractor, in systems of third or higher order, the
divergence will be essentially exponential and this leads to a breakdown of
predictability. The consequence is that deterministic systems can produce
effectively stochastic output. This has profound implications for our view

of physical and other dynamical systems; it can also be expected to affect the relationship between classical physics and its quantum counterpart.

---

## Problems

1. The population $x(t)$ of fish in a pond is a function of time $t$ and when the fish are removed at a constant rate $f$ the function $x(t)$ satisfies equation (13.6).

   (a) Show that there is a critical fishing rate $\bar{f}$ such that for $f < \bar{f}$ there are two equilibrium fish populations, of which only one is asymptotically stable.

   (b) If the fish population is initially in stable equilibrium (with $f = 0$ and fishing commences at a rate $f > \bar{f}$, show that $x \to 0$ and in a finite time $t_0$ (which need not be found explicitly!).

   Why would it not be prudent to fish at a rate $f$ which is only just less than $\bar{f}$?

2. For the pendulum equation $\ddot{\theta} + (g/l)\sin\theta = 0$ find the equations of the trajectories in the phase plane and sketch the phase portrait of the system. Show that on the separatrices

$$\theta(t) = (2n+1)\pi \pm 4\arctan[\exp(\omega t + \alpha)],$$

   where $n$ is an integer, $\alpha = $ constant and $\omega = \sqrt{g/l}$. [*Hint*: Use the substitution $u = \tan\frac{1}{4}\{\theta - (2n+1)\pi\}$.]

3. The relativistic equivalent of the simple harmonic oscillator equation for a spring with constant $k$ and a rest mass $m_0$ attached is

$$\frac{d}{dt}\left(\frac{m_0 y}{\sqrt{1 - y^2/c^2}}\right) + kx = 0 \qquad \text{with} \qquad \dot{x} = y,$$

   where $c$ is the speed of light. Show that the phase trajectories are given by

$$m_0 c^2 / \sqrt{1 - y^2/c^2} + \tfrac{1}{2}kx^2 = \text{constant}$$

   and sketch the phase portrait for this system.

4. Draw the phase portrait of the damped linear oscillator, whose displacement $x(t)$ satisfies $\ddot{x} + \mu\dot{x} + \omega_0^2 x = 0$, in the phase plane $(x, y)$, where

$y = \dot{x}$. Distinguish the cases

(a)  under- (or light) damping      $0 < \mu < 2\omega_0$,
(b)  over-damping                   $\mu > 2\omega_0$,
(c)  critical damping               $\mu = 2\omega_0$.

5. Consider the gradient system (13.17) in the case $U(x, y) = x^2(x-1)^2+y^2$. Find the critical points and their character. Sketch the phase portrait for the system.

6. For the Lotka–Volterra system (13.18) show that the trajectories in the phase plane are given by $f(x, y) = $ constant as in (13.20). In the first quadrant $x \geq 0, y \geq 0$, the intersections of a line $y = $ constant with a trajectory are given by $-c \ln x + dx = $ constant. Hence show that there are 0, 1 or 2 such intersections, so that the equilibrium point $(c/d, a/b)$ fore this system is a true centre (*i.e.* it cannot be a spiral point). Using the substitution $x = e^p, y = e^q$, show that the system takes on the Hamiltonian canonical form (13.22).

7. Consider the 'competing species' system (13.23). For the case $k_1 = \sigma_1 = \alpha_1 = 1, k_2 = \frac{1}{2}, \sigma_2 = \frac{1}{4}, \alpha_2 = \frac{3}{4}$ find the critical points and their character in order to confirm all the features of Fig. 13.9. What happens in the case $k_1 = 1, \sigma_1 = 2, \alpha_1 = \frac{1}{2}, k_2 = 3, \sigma_2 = 2, \alpha_2 = 4$?

8. For the Arms-Race model system (13.26) with all parameters positive show that there is an asymptotically stable coexistence or a runaway escalation according as $c_1 c_2 > a_1 a_2$ or $c_1 c_2 < a_1 a_2$.

9. *A simple model for the dynamics of malaria due to Ross (1911) and Macdonald (1952) is

$$\dot{x} = \left(\tfrac{abM}{N}\right) y(1 - x) - rx,$$
$$\dot{y} = ax(1 - y) - \mu y,$$

where:
$x, y$ are the infected proportions of the human host, female mosquito populations,
$N, M$ are the numerical sizes of the human, female mosquito populations,
$a$ is the biting rate by a single mosquito,
$b$ is the proportion of infected bites that result in infection,
$r, \mu$ are *per capita* rates of recovery, mortality for humans, mosquitoes, respectively.
Show that the disease can maintain itself within these populations or

must die out according as

$$R = \frac{M}{N}\frac{a^2 b}{\mu r} > 1 \text{ or} < 1.$$

10. Show that Rayleigh's equation in the form (13.28) has a single critical point at $(0,0)$ and that this is always unstable. Making use of substitution $\dot{x} = v/\sqrt{3}$ show that $v$ satisfies the Van der Pol equation (13.29).

11. Show that the origin is the only critical point for the system

$$\dot{x} = -y + \alpha x(\beta - x^2 - y^2),$$
$$\dot{y} = x + \alpha y(\beta - x^2 - y^2),$$

where $\alpha, \beta$ are real parameters, with $\alpha$ fixed and positive and $\beta$ allowed to take different values. Show that the character of the critical point and the existence of a limit cycle depend on the parameter $\beta$, so that the system undergoes a supercritical Hopf bifurcation at $\beta = 0$. (*Hint:* Make the change from Cartesian co-ordinates to plane polars.)

12. *For the rotation of a rigid body about its centre of mass with zero torque the equations for the angular momentum components $J_1, J_2, J_3$ are given by (13.31).

   (a) Show that $J_1^2 + J_2^2 + J_3^2 = J^2$ (constant), so that the angular momentum $\boldsymbol{J}$ must lie on a sphere in $(J_1, J_2, J_3)$ phase space.
   (b) When $I_1 < I_2 < I_3$ show that there are six critical points on this phase sphere and show that, in local expansion, four of these are centres and two are saddles. (Hence the tennis racquet theorem of §13.6.)
   (c) Show that when $I_1 = I_2 \neq I_3$ then $J_3$ is constant ($\equiv I_3\Omega$) and that $J_1, J_2$ are simple harmonic (with frequency $[|I_3 - I_1|/I_1]\Omega$).
   (d) A space station with $I_1 < I_2 < I_3$ is executing a tumbling motion with $\omega_1, \omega_2, \omega_3$ nonzero. It is to be stabilized by reducing $\boldsymbol{\omega}$ to $\boldsymbol{0}$ with an applied torque $-|\mu|\boldsymbol{\omega}$, with $\mu$ constant, so that the right-hand side of (13.31) becomes $-|\mu|\omega_1$, etc. About which of its axes does the space station tend to be spinning as $\boldsymbol{\omega} \to \boldsymbol{0}$?

13. *The simple SIR model equations for the transmission of a disease are

$$\dot{S} = -aSI,$$
$$\dot{I} = aSI - bI,$$
$$\dot{R} = bI,$$

where $S(t), I(t), R(t)$ are respectively susceptibles, infectives, re-moved/recovered and $a, b$ are positive constants.

(a) Show that the overall population $N = S + I + R$ remains constant, so that we may consider $(S, I)$ in a projected phase plane. Hence show that a trajectory with initial values $(S_0, I_0)$ has equation $I(S) = I_0 + S_0 - S + (b/a) \ln(S/S_0)$.

(b) Using the function $I(S)$ show that an epidemic can occur only if the number of susceptibles $S_0$ in the population exceeds the threshold level $b/a$ and that the disease stops spreading through lack of infectives rather than through lack of susceptibles.

(c) For the trajectory which corresponds to $S_0 = (b/a) + \delta$, $I_0 = \epsilon$ with $\delta, \epsilon$ small and positive, show that, to a good approximation, there are $(b/a) - \delta$ susceptibles who escape infection [the Kermack–McKendrick theorem of epidemiology (1926/27)].

14. *Consider the Lorenz system (13.33).

(a) Show that the origin $P_1$ $(0, 0, 0)$ is a critical point and that its stability depends on eigenvalues $\lambda$ satisfying the cubic

$$(\lambda + \beta)[\lambda^2 + (\sigma + 1)\lambda + \sigma(1 - \rho)] = 0.$$

Hence show that $P_1$ is asymptotically stable only when $0 < \rho < 1$.

(b) Show that there are two further critical points

$$P_2, P_3 \equiv [\pm\sqrt{\beta(\rho - 1)}, \pm\sqrt{\beta(\rho - 1)}, (\rho - 1)],$$

when $\rho > 1$, and that their stability depends on eigenvalues $\lambda$ satisfying the cubic

$$\lambda^3 + \lambda^2(\sigma + \beta + 1) + \lambda\beta(\sigma + \rho) + 2\sigma\beta(\rho - 1) = 0.$$

(c) Show that when $\rho = 1$ the roots of the cubic in (b) are $0, -\beta, -(1+\sigma)$ and that in order for the roots to have the form $-\mu, \pm i\nu$ (with $\mu, \nu$ real) we must have

$$\rho = \rho_{\text{crit}} = \frac{\sigma(\sigma + \beta + 3)}{(\sigma - \beta - 1)} > 0.$$

(d) By considering how the roots of the cubic in (b) change continuously with $\rho$ (with $(\sigma, \beta)$ kept constant), show that $P_2, P_3$ are asymptotically stable for $1 < \rho < \rho_{\text{crit}}$ and unstable for $\rho > \rho_{\text{crit}}$.

(e) Show that if $\bar{z} = z - \rho - \sigma$, then

$$\frac{1}{2}\frac{d}{dt}(x^2 + y^2 + \bar{z}^2) = -\sigma x^2 - y^2 - \beta[\bar{z} + \tfrac{1}{2}(\rho + \sigma)]^2 + \tfrac{1}{4}\beta(\rho + \sigma)^2,$$

so that $(x^2 + y^2 + \bar{z}^2)^{1/2}$ decreases for all states outside any sphere which contains a particular ellipsoid (implying the existence of an attractor).

15. *For the Rikitake dynamo system (13.35):

   (a) Show that there are two real critical points at $(\pm k, \pm 1/k, \mu k^2)$ in the $(X_1, X_2, Y)$ phase space, where $k$ is given by $A = \mu(k^2 - 1/k^2)$.
   (b) Show that the stability of these critical points is determined by eigenvalues $\lambda$ satisfying the cubic

$$(\lambda + 2\mu)\left[\lambda^2 + \left(k^2 + \frac{1}{k^2}\right)\right] = 0,$$

   so that the points are not asymptotically stable in this approximation. (For the full system they are actually unstable.)
   (c) Show that the divergence of the phase-space flow velocity is negative, so that the flow causes volume to contract.
   (d) Given that $\bar{Y} = \sqrt{2}(Y - A/2)$, show that

$$\frac{1}{2}\frac{d}{dt}(X_1^2 + X_2^2 + \bar{Y}^2) = -\mu(X_1^2 + X_2^2) + \sqrt{2}\bar{Y}$$

   and use this result to determine in which region of the space the trajectories all have a positive inward component towards $X_1 = X_2 = \bar{Y} = 0$ on surfaces $X_1^2 + X_2^2 + \bar{Y}^2 = $ constant.

16. For the simple harmonic oscillator system (13.8) we may solve for the equation of the trajectory which passes through $x(x_0, y_0, t), y(x_0, y_0, t)$. If we now consider perturbations $\Delta x_0, \Delta y_0$ in the initial data $(x_0, y_0)$ at $t = 0$, find the resulting changes $\Delta x, \Delta y$ in $x, y$ and show that they remain bounded in time, so that the Pythagorean distance $d \equiv \sqrt{(\Delta x)^2 + (\Delta y)^2}$ then remains bounded.

17. *In contrast to Problem 16, consider the perfectly elastic bouncing of a ball vertically under gravity above the plane $x = 0$. We have $\dot{x} = y, \dot{y} = -g$ and we can solve for $x(x_0, y_0, t), y(x_0, y_0, t)$ in terms of the initial data $(x_0, y_0)$ at $t = 0$. Show that in this case the resulting perturbations $\Delta x, \Delta y$ essentially grow linearly with time $t$ along the trajectory when we make perturbations $\Delta x_0, \Delta y_0$ in the initial data. That is to say the

distance along the trajectory $d \equiv \sqrt{(\Delta x)^2 + (\Delta y)^2} \sim \kappa t$ when $t$ is large and $\kappa$ is a suitable constant.

18. For the Galton board of Fig. 13.25 we may arrange things so that each piece of lead shot has an equal chance of rebounding just to the left or to the right at each direct encounter with a scattering pin at each level. Show that the probabilities of each piece of shot passing between the pins along a particular *row* $n$ are then given by $\binom{n}{r}(\frac{1}{2})^n$ where the binomial coefficient $\binom{n}{r} = n!/[(n-r)!r!]$ and $r = 0, 1, \ldots, n$. Use the result $\binom{n+1}{r+1} = \binom{n}{r} + \binom{n}{r+1}$ to generate the probability distribution for row $n = 16$. (For large numbers of pieces of shot and large $n$ the distribution of shot in the collection compartments approximates the standard normal error curve $y = k \exp(-x^2/2s^2)$ where $k, s$ are constants.)

# Chapter 14

# Order and Chaos in Hamiltonian Systems

In this chapter we consider the geometrical aspects of the particular class of dynamical systems which have the Hamiltonian structure described in Chapter 12. The dynamics of these systems depends crucially on the number of symmetries which they possess, leading to conserved quantities. It will become apparent that the larger the number of symmetries the greater is the restriction in the freedom of a trajectory in the corresponding phase space and the more ordered is the motion. In contrast a deficit in the number of such symmetries allows the development of irregularity, or chaos, in the evolution of states along the system trajectories.

## 14.1  Integrability

In §12.6 we discussed the direct relation between symmetries and conservation laws, indicating that conserved quantities (constants of motion) may be used to reduce progressively the effective order of a system. If there are enough symmetries and independent conserved quantities, then the reduction may be complete, in that it leads to the full solution of the system of equations ('integrability'). For a system with $n$ degrees of freedom the Hamiltonian $H$ may depend on the co-ordinates $q_1, \ldots, q_n$, momenta $p_1, \ldots, p_n$ and on time $t$. While there is no unique definition of *integrability*, it turns out that, for the complete solution to be obtainable *via* a sequence of integrations, we need $n$ independent functions $F_i$ $(i = 1, \ldots, n)$ of the same variables, which are themselves constants of the motion. That is to say, we require the same number of symmetries as there are degrees of freedom in the system and the system is then said to be '*integrable* (in the sense of Liouville)'. If there is not such a full set of $n$ conserved quantities then the system is said to be *non-integrable*.

Recalling (12.33),

$$\frac{\mathrm{d}F_i}{\mathrm{d}t} = \frac{\partial F_i}{\partial t} + [F_i, H] = 0. \tag{14.1}$$

We will confine our discussion here to autonomous systems, for which $H$ and other $F_i$ have no explicit dependence on $t$, and we then have $n$ Poisson-bracket relations $[F_i, H] = 0$ from (14.1). Of course $H$ itself can be taken as one of the functions $F_i$ and the invariance of $H$ corresponds to symmetry with respect to time translation.

An additional technical condition required of the $F_i$ is that for each pair of them we require

$$[F_i, F_j] = 0 \qquad \text{for} \qquad i, j = 1, \ldots n. \tag{14.2}$$

The functions $F_i$ are then said to be *in involution*.

Each of the $n$ functions $F_i = $ constant confines a particular trajectory to a $(2n-1)$-dimensional subspace of the full $2n$-dimensional phase space of the system. So the complete set of these restrictions confines a trajectory to the $n$-dimensional intersection $M$ of these subspaces. That this is a strong dimensional restriction on the way that a trajectory can move in the full phase space is indicated by Table 14.1.

Table 14.1

| Number of degrees of freedom | | 1 | 2 | 3 | $n$ |
|---|---|---|---|---|---|
| | phase space | 2 | 4 | 6 | $2n$ |
| Dimension of: | surface $F_i = $ constant | 1 | 3 | 5 | $2n - 1$ |
| | $M$ | 1 | 2 | 3 | $n$ |

Naturally the values of co-ordinates and momenta at a given initial time determine the $n$ values of the $F_i$ constants and hence pick out the particular $M$ to which the trajectory having these initial values is then confined for all time.

Now at each point on a particular $(2n - 1)$-dimensional surface $F_i = $ constant, the normal to the surface is in the direction given by the $2n$-component gradient vector $(\nabla_q F_i, \nabla_p F_i)$. (See §A.5.)

The Poisson-bracket relation (14.2) is a scalar-product relation and it expresses the orthogonality conditions between members of a set of $n$ vector fields $v_j = (\nabla_p F_j, -\nabla_q F_j)$, $j = 1, \ldots, n$ and the above gradient vector. It is apparent therefore that the vector fields $v_j$ at a particular point in the phase space are all parallel to the surface $M$ which contains that point. The fields $v_j$ allow us to define a co-ordinate grid continuously over the surface $M$ and without any singularity. The topology of $M$ is then necessarily that of a *torus* — an $n$-dimensional 'ring doughnut' embedded within the full $2n$-dimensional phase space.

For a system of one degree of freedom each $M$ is a 1-torus $F_1 \equiv H =$ constant, *i.e.* a one dimensional loop in the two-dimensional phase space. (See Fig. 14.1(a).) Evidently $v_1$ here is tangent to the curve at each point of $M$. For different values of $F_1$ we obtain different loops $M$, members of a set of 'nested' tori, as for example in Fig. 13.4.

For a system of two degrees of freedom each $M$ is a 2-torus, *i.e.* a two-dimensional 'ring doughnut' surface embedded in the four-dimensional phase space. (See Fig. 14.1(b).) In this case we can take $v_1, v_2$ everywhere parallel to distinct loops which can parametrize the surface. It is the case that $M$ cannot have, for example, the topology of a sphere, because all $v_1, v_2$ parametrizations of such a surface are necessarily singular, at least somewhere. In Fig. 14.1(c) the attempt to use a grid of lines of latitude and longitude is singular at the poles. That all such attempts for a sphere are bound to be singular is a consequence of the 'hairy ball theorem' of topology; this theorem also implies, among other things, the existence of at least one crown parting in the combing of a head of hair, and that a spherical magnetic bottle *must* leak!

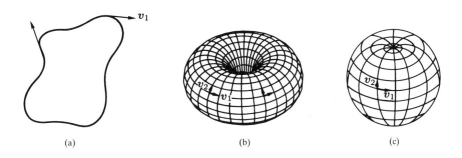

(a)  (b)  (c)

Fig. 14.1

The $n$-tori for an integrable system of $n$ degrees of freedom in its $2n$-dimensional phase space are often called *invariant tori* because a trajectory (orbit) which is initially on one of them ($M$) remains on it for ever. Different initial conditions then lead to tori which are nested within the phase space.

On a particular torus $M$ the separate vector fields correspond to $n$ independent types of linking circuit (see Fig. 14.1(a),(b)), and the integrable motion is then *exactly* equivalent to the combination of $n$ separate corresponding periodic motions, one for each type of linking circuit. Each of these periodic motions of an integrable system is a *nonlinear normal mode*, similar to those detailed in §11.3, but now with an analysis which is exact — no approximation has been made for small oscillations. This makes *integrability* a very special property.

The separate normal modes have associated natural *frequencies* $\omega_i$, $i = 1, \ldots, n$ on a particular torus $M$ and any trajectory on $M$ has $n$ associated natural *periods* $2\pi/\omega_i$. The trajectory is then said to be *multiply periodic*.

If the trajectory on $M$ is *closed* (see Fig. 14.2(a)) then it does not fill $M$ and is exactly *periodic*. For closure it is required that the frequencies of the the modes are rationally related, *i.e.* $\omega_i/\omega_j$ is rational for all $i, j$. This *degeneracy* is the exception rather than the usual case, so that closure is a special property. If the trajectory on $M$ is not closed, because of irrational $\omega_i/\omega_j$, then typically the trajectory covers $M$ densely eventually (see Fig. 14.2(b)) — we have *quasiperiodicity* and the trajectory is said to be *ergodic* on $M$.

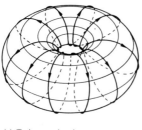
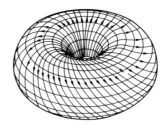

(a) Trajectory closed

(b) Trajectory eventually covers the surface densely

Fig. 14.2

Two important examples are given by:

### *Example:* Double harmonic oscillator and central force motion

Find suitable independent constants $F_1, F_2$ for
(a) the double harmonic oscillator,
(b) central force motion.

Are trajectories closed?

(a) For the *double harmonic oscillator* we have

$$H = \frac{1}{2m}(p_1^2 + p_2^2) + \tfrac{1}{2}m(\omega_1^2 q_1^2 + \omega_2^2 q_2^2) \equiv H_1 + H_2 \qquad (14.3)$$

with $H_1, H_2$ the Hamiltonians for the two separate oscillators. Here we may take $F_1, F_2$ to be *any pair* of $H, H_1, H_2$, since then $[F_i, H] = 0$. The frequencies are $\omega_1, \omega_2$ and this evidently leads to closed trajectories (periodicity) only when $\omega_1/\omega_2$ is rational, leading to the familiar Lissajous figures.

(b) *Central-force motion* was discussed in Chapter 4. The Hamiltonian takes the form (12.8) and we can take $F_1 = H, F_2 = p_\theta$, so that $[F_i, H] = 0$. For the isotropic harmonic oscillator of §4.1 we can take $\omega_1 = \omega_2$ in (14.3) above, so that all trajectories are closed. The oscillator describes an ellipse with centre at the origin and the frequencies of radial and transverse motions are related by $\omega_r = 2\omega_\theta$. For the inverse square law attractive force of §4.3 all bounded trajectories are closed. They are ellipses in general, with the centre of force at a focus and $\omega_r = \omega_\theta$. That the isotropic harmonic oscillator and the inverse square law of attraction are the only such force laws for which all bounded orbits are closed is known as Bertrand's theorem.

## 14.2 Surfaces of Section

If we consider two dynamical states which are initially close in the phase space of an integrable system, they do, in general, have different values of $F_1, F_2$ and so they reside on different tori. The evolution of these states along their respective trajectories takes place with slightly different frequencies, so that the distance between the states increases essentially linearly with time $t$. So, just as for the discussion of the libration motions of a simple pendulum in §13.7, there is for an integrable system no exponential divergence of states with time and no sensitivity to initial conditions — the motion is *ordered*.

When a full complement of constants of the motion does not exist, so that we have a system which is not integrable, then trajectories have much less restriction on their room to manoeuvre and sensitivity to initial

conditions is certain in at least some regions of the phase space — there is
then inevitably *chaotic* behaviour, as indicated in §13.7.

A method of detecting these ordered and chaotic behaviours was put
forward by Poincaré and involves the concept of a *surface of section*. If
we consider an autonomous system with two degrees of freedom, so that
we have a four-dimensional phase space, with three-dimensional surfaces
of constant 'energy' ($H =$ constant), then we can consider taking a 'slice'
through the space, and this may or may not be planar according to convenience.

If we look at a particular trajectory of the system and identify the points
at which this intersects the section (in the same sense) then the dynamics
induces a map of the section to itself. In Fig. 14.3 we have $H(q_1, q_2, p_1, p_2) =$
constant and we have chosen a section $q_2 = 0$ (with $p_2 > 0$) to demonstrate
how a trajectory maps points in the $(q_1, p_1)$ plane successively.

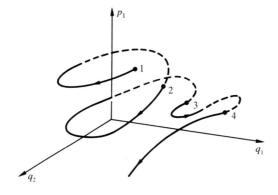

Fig. 14.3

When we map different $(q_1, p_1)$ points using different trajectories we
develop a *Poincaré return map* of the section to itself.

For integrable systems we may take a section through the nested tori
of §14.1. For two degrees of freedom the section through a 2-torus (ring
doughnut) is a closed curve $C$ (see Fig. 14.4) and a particular point $P_0$ on
this curve is mapped successively by the trajectory through it to $P_1$, then
$P_2$, and so on. If the trajectory is closed (see §14.1) then we have a periodic
motion and the map has the property that after $N$ steps, for some integer
$N$, it carries $P_0$ back to $P_0$. The successive points $P$ are twisted around
the curve $C$. If the motion is only quasiperiodic, however, there is no such

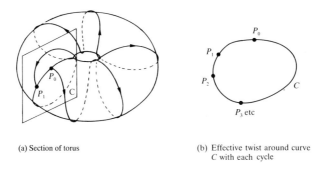

(a) Section of torus        (b) Effective twist around curve
                                              $C$ with each cycle

Fig. 14.4

exact repetition and the successive points $P$ eventually fill the whole curve $C$. *Integrable systems* are characterized by a *finite set of discrete points* or by *curve-filling* in the surface of section.

For systems which are not integrable the full torus structure does not exist. For two degrees of freedom we expect the system to explore the full three-dimensional 'energy' surface. The section return map is then generally two-dimensional and is not confined to points and lines. *Systems* which are *non-integrable* are characterized by *area-filling* in the surface of section.

It should be noted that the Poincaré return map for a Hamiltonian system can be shown to be *area-preserving*. This property is related to Liouville's theorem, treated in §12.5, although it is not derivable from it. Despite the equal-area property, the map induced by a particular dynamical system typically involves *stretching* and *squeezing* of finite and infinitesimal area elements, associated with, respectively, positive and negative Lyapunov exponents (see §13.7) whose effects balance. When these effects are combined with a *folding* of trajectories, necessitated by confinement to a bounded region of phase space, it is clear that the divergence of neighbouring trajectories leading to the *sensitivity to initial conditions* of chaos can then typically occur.

For systems which are not Hamiltonian the concept of a Poincaré return map for a surface of section is still useful. When the dynamics involves dissipation the equal-area property is replaced by area reduction at each iteration of the map — this would be the case *e.g.* for the Lorenz system of §13.6 — and here the effect of a negative Lyapunov exponent overcomes the effect of one which is positive. (There is some reference to the properties of discrete maps in Appendix D.)

## 14.3   Action/Angle Variables

Quite often the initial choice of variables $q_i, p_i$ $(i = 1, \ldots, n)$ used to describe a system of $n$ degrees of freedom may not be the best.

For integrable systems there is a natural set of co-ordinates and momenta which is particularly convenient and useful. We recall that these systems have $n$ distinct constants of the motion $(F_i = \text{constant})$ and we can transform to a new set of co-ordinates $\phi_i$ and momenta $I_i$, in such a way that

- the Hamiltonian form of the equations of motion is preserved,
- the new momenta $I_i$ are functions of the $F_i$,
- the new co-ordinates $\phi_i$ are all ignorable (as in §12.3).

For a suitable choice the $\phi_i$ are *angle* variables and the $I_i$ are the corresponding canonically conjugate *action* variables (see Table 14.2).

Table 14.2

| Old description | New description |
|---|---|
| $(q_i, p_i)$ $(i = 1, \ldots, n)$ | $(\phi_j, I_j)$ $(j = 1, \ldots, n)$ |
| $H(q_i, p_i)$ | $K(I_j) \equiv H$ |
| $\dot{q}_i = \dfrac{\partial H}{\partial p_i}$ | $\dot{\phi}_j = \dfrac{\partial K}{\partial I_j} \equiv \omega_j(I_i)$ |
| $\dot{p}_i = -\dfrac{\partial H}{\partial q_i}$ | $\dot{I}_j = -\dfrac{\partial K}{\partial \phi_j} = 0$ |

It is, of course, apparent that the new momenta $I_j$ are all constants of the motion and that the evolution of each $\phi_j$ is given by $\phi_j = \omega_j t + \beta_j$, where $\omega_j, \beta_j$ are constants. Evidently the angle variables evolve at a *uniform rate*.

Although these are general results, let us consider how the change of variables may be effected for a system of one degree of freedom $(n = 1)$. For such an autonomous system we have $H = \text{constant}$ ('energy') closed trajectories in the phase plane, representing libration motions. (See Fig. 14.5(a).) We need to choose the action $I$ to be a suitable function of $H$, such that the phase plane in the new description is as given in Fig. 14.5(b).

For Hamilton's equations to be preserved in form it is necessary that the change of variables $(q, p) \rightarrow (\phi, I)$ preserves areas, so that a suitable choice for $I$ is derived from the area within the corresponding trajectory in

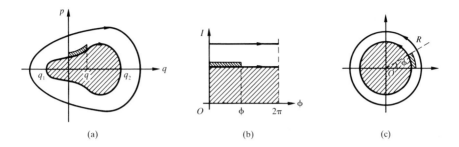

Fig. 14.5

the $(q, p)$ phase plane by

$$I = \frac{1}{2\pi} \oint p \, dq, \tag{14.4}$$

as indicated by the equality of the shaded areas $////$ in Fig. 14.5(a),(b). Naturally $I$ depends directly on $H$, and the natural frequency of the system for the corresponding trajectory is then given by (see Table 14.2)

$$\omega(I) = \frac{\partial K}{\partial I} \equiv \frac{\partial H}{\partial I}. \tag{14.5}$$

It also follows from the preservation of area that the defining relation for $\phi$ is given by

$$\phi = \frac{\partial}{\partial I} \int_0^q p \, dq \equiv \omega \frac{\partial}{\partial H} \int_0^q p \, dq, \tag{14.6}$$

again as indicated by the equality of the shaded areas $\backslash\backslash\backslash\backslash$ in Fig. 14.5(a),(b) for small changes $\Delta H, \Delta I$ in $H, I$ respectively. The 'angle' nature of the $\phi$ variable can be emphasized by a further change of variables to polars $(R, \phi)$, where $R = \sqrt{2I}$ in order to maintain the preservation of area property. (See Fig. 14.5(c).)

For the oscillator problems of §2.1 and §13.3 we have

$$H(q, p) = \frac{p^2}{2m} + V(q) = E, \tag{14.7}$$

so that $p = \pm\sqrt{2m[H - V(q)]}$ and

$$I = \frac{1}{2\pi} \oint p \, dq = \frac{2\sqrt{2m}}{2\pi} \int_{q_1}^{q_2} \sqrt{H - V(q)} \, dq. \tag{14.8}$$

*Example:* **Simple harmonic oscillator**

Find action/angle variables for the simple harmonic oscillator.

Here $V(q) = \frac{1}{2}kq^2$, so that

$$I = H/\omega_0, \quad \text{where} \quad \omega_0 = \sqrt{k/m}, \tag{14.9}$$

since $\oint p\,dq$ is the area within an ellipse with semi-axes $\sqrt{2H/k}$ and $\sqrt{2mH}$. (See Fig. 14.6.)

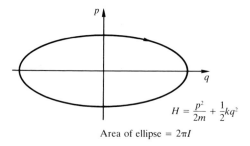

$$H = \frac{p^2}{2m} + \frac{1}{2}kq^2$$

Area of ellipse $= 2\pi I$

Fig. 14.6

For the simple harmonic oscillator we have action $\equiv$ energy/frequency. Evidently

$$\omega = \frac{\partial H}{\partial I} = \omega_0 \quad \text{(here)} \tag{14.10}$$

and action

$$I = \frac{p^2 + kmq^2}{2m\omega_0}, \tag{14.11}$$

with the angle variable

$$\phi = \frac{\partial}{\partial I} \int_0^q p\,dq = \frac{\partial}{\partial I} \int_0^q \sqrt{2m\omega_0 I - kmq^2}\,dq$$

$$= m\omega_0 \int_0^q \frac{dq}{\sqrt{2m\omega_0 I - kmq^2}}$$

$$= \arcsin\left(q\sqrt{\frac{m\omega_0}{2I}}\right)$$

$$\equiv \omega_0 t + \beta. \tag{14.12}$$

Of course, the original variables $q, p$ can be found in terms of $\phi, I$ in the form

$$q = \sqrt{\frac{2I}{m\omega_0}} \sin\phi, \qquad p = \sqrt{2m\omega_0 I} \cos\phi. \qquad (14.13)$$

These relations, together with the linear evolution of $\phi$ with time give the evolution of the original $q, p$ variables.

### The elastic bouncer

For a perfectly elastic ball bouncing normally between two walls (see Fig. 14.7) the Hamiltonian is $H = p^2/2m \equiv E = \frac{1}{2}mv^2$. Here the action $I = \sqrt{2mH}\,L/\pi$ so that $\omega = \pi^2 I/mL^2 \equiv \pi v/L$ and the period of this 'oscillation' is $\tau = 2\pi/\omega = 2L/v$ (of course).

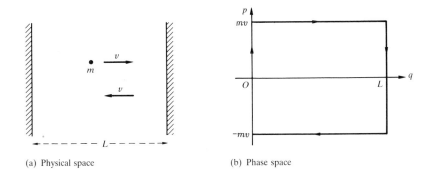

(a) Physical space                    (b) Phase space

Fig. 14.7

The results are hardly very surprising for these two cases, but the method provides a powerful means of obtaining the frequency (and so the period) of a general nonlinear libration oscillation.

The method may be extended to deal with rotations, which we will not consider further here, and to deal with systems which have two or more degrees of freedom. So we have the following:

### Central force motion

(See Chapter 4.)

We may write the Hamiltonian in plane polars in the form

$$H = \frac{p_r^2}{2m} + \frac{p_\theta^2}{2mr^2} + V(r) \equiv E, \tag{14.14}$$

with constants of the motion $F_1 = H, F_2 = p_\theta$. We can then calculate the action variables as

$$I_2 = \frac{1}{2\pi} \oint p_\theta \, d\theta = p_\theta,$$

$$I_1 = \frac{1}{2\pi} \oint p_r \, dr = \frac{\sqrt{2m}}{2\pi} \oint \sqrt{H - \frac{p_\theta^2}{2mr^2} - V(r)} \, dr. \tag{14.15}$$

### Example: Inverse square law

For the inverse square attractive force law $V(r) = -|k|/r$ (see §4.3 and Problem 9) we obtain

$$I_1 = -I_2 + \frac{|k|}{2} \sqrt{\frac{2m}{-H}}. \tag{14.16}$$

Show that in this case the nonlinear normal mode frequencies are the same, so that the orbits are closed.

Evidently

$$H = E = -\frac{mk^2}{2(I_1 + I_2)^2}. \tag{14.17}$$

The Hamiltonian depends only on the action variables, and the corresponding angle variables are ignorable, by design. The natural frequencies for this system are then given by

$$\omega_1 = \omega_2 = \frac{mk^2}{(I_1 + I_2)^3} \equiv \frac{m}{|k|} \left(-\frac{2E}{m}\right)^{3/2} \equiv \frac{2\pi}{\tau}, \tag{14.18}$$

where $\tau$ is the period. In fact this is just Kepler's third law of planetary motion (4.32), since the semi-major axis $a$ of the elliptical orbit of negative energy $E$ is given by $a = -|k|/2E$ (as in (4.30)).

Note that important results for this system have been obtained without explicit calculation of the form of the angle variables:

- The equality of the natural frequencies of the radial and transverse modes of oscillation, which leads to orbital closure. (See §14.1.)

- The relation between the period of oscillation and the energy $E$ of a bound orbit.

A further important property of the action variables will be indicated later in §14.5.

## 14.4  Some Hamiltonian Systems which Exhibit Chaos

We have seen in Chapter 13 and in previous sections of this current chapter that a system must have at least two degrees of freedom in order to have a phase space of large enough dimension to exhibit any chaotic behaviour. Again the system must also then have fewer symmetries (leading to conservation laws) than are required to give integrability and order.

### Orbital motion — n-body problem

In §§14.1, 14.2 we indicated that the central force problem of Chapter 4, with a fixed centre of force and two degrees of freedom, is integrable, so that it cannot exhibit the property of sensitivity to initial conditions characteristic of chaos.

This is also true for the two-body problem in the centre-of-mass frame. (See §7.2.) Less intuitively apparent is that the motion of a particle in a plane under the inverse square law of attraction to two fixed centres may also be integrated.

However the three-body problem is, in general, not integrable, even when it is restricted (as in Chapter 10, Problem 15); this problem, together with the extension to more bodies, was the strong motivation for Poincaré in his general considerations for dynamical systems.

The three-body problem has nine degrees of freedom and so gives rise to a system of differential equations of order eighteen. We may reduce this to order twelve by making use of the conservation of linear momentum, *i.e.* by going to centre-of-mass co-ordinates. By making use of conservation of energy and conservation of angular momentum we can reduce the order to eight. By further device and restricting our consideration to motions which are planar, the order can be reduced to four. This is the best that can be done in general. Even after all these reductions the problem is still extremely complicated and has kept mathematicians busy for several hundred years.

As an indication we can consider the equations of the *planar circular restricted three-body problem* (Chapter 10, Problem 15). In the frame of reference rotating with the primary masses $M_1, M_2$ about their centre of mass (see Fig. 14.8), the equations for the third (small mass $m$) body can

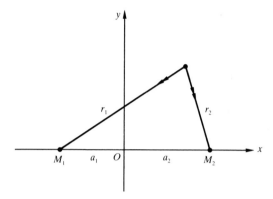

Fig. 14.8

be written in the form

$$\ddot{x} - 2\omega\dot{y} = -\frac{\partial V}{\partial x}$$

$$\ddot{y} + 2\omega\dot{x} = -\frac{\partial V}{\partial y}.$$

(14.19)

with

$$V = -\tfrac{1}{2}\omega^2(x^2 + y^2) - \frac{GM_1}{r_1} - \frac{GM_2}{r_2},$$

and

$$r_1 = \sqrt{(x + a_1)^2 + y^2}, \qquad r_2 = \sqrt{(x - a_2)^2 + y^2}.$$

This system can be expressed in Hamiltonian form with two degrees of freedom and there is only a single conserved quantity, *i.e.* the Hamiltonian itself, so that $\tfrac{1}{2}(\dot{x}^2 + \dot{y}^2) + V$ is a constant, which is known as the Jacobi integral. While there are some stable equilibria for this system (depending on the mass ratio $M_1/M_2$, see Chapter 12, Problem 12) there are also motions which are highly chaotic. However, the motion with given initial conditions is restricted by the *curves of zero velocity* at which $\dot{x} = 0 = \dot{y}$. (See Problem 12 of this Chapter.)

Interest in the restricted problem reflects the hierarchical nature of our Solar System and an increasing interest in space missions. The primaries are variously Sun–Jupiter (Chapter 12, Problem 13), Earth–Moon (Problem 12) and so on. The five relative equilibria referred to in Chapter 10, Problem 16 are apparent in these applications — as sites for orbiting observatories and as staging points for more extended missions by NASA and ESA.

The Lagrangian equilateral triangle rotating in its plane about its centre of mass is actually a relative equilibrium for the three-body problem with general masses.

### *Example:* **Three-body problem**

Show that the three-body problem with three general masses $m_1, m_2, m_3$ has a relative equilibrium (with the masses *not* in a straight line) if and only if the triangle formed by the masses is equilateral and find the rotation rate $\Omega$.

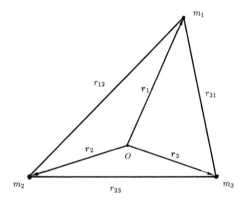

Fig. 14.9

Note that the origin in this figure is at the centre of mass. It is apparent that the masses remain in a plane. Then we have

$$m_1\boldsymbol{r}_1 + m_2\boldsymbol{r}_2 + m_3\boldsymbol{r}_3 = \boldsymbol{0}, \tag{14.20}$$

together with

$$\frac{Gm_1m_2(\boldsymbol{r}_2 - \boldsymbol{r}_1)}{r_{12}^3} + \frac{Gm_1m_3(\boldsymbol{r}_3 - \boldsymbol{r}_1)}{r_{31}^3} = -m_1\Omega^2\boldsymbol{r}_1, \tag{14.21}$$

and two similar equations for $m_2, m_3$. From (14.20) and (14.21) we obtain a relation $\alpha \boldsymbol{r}_2 + \beta \boldsymbol{r}_3 = \boldsymbol{0}$ where $\alpha, \beta$ are scalars. Since $\boldsymbol{r}_2, \boldsymbol{r}_3$ are by hypothesis not parallel, we must have $\alpha = \beta = 0$, which then leads to $r_{12} = r_{31} = a$ (say). Of course the similar equations give $r_{23} = a$ also, and by substitution we obtain $\Omega^2 = G(m_1 + m_2 + m_3)/a^3$.

Naturally, stability of this configuration is quite another matter — in general it is surely unstable and, typically, the system will separate.

The search for exact periodic solutions of the three-body problem has resulted in an astonishing solution for the case of three equal masses — the masses follow one another round a 'figure of eight' in a plane, visiting in turn each of the Euler three-body straight-line equilibria with one mass at the mid-point of the line joining the other two (see Chenciner and Montgomery, *Annals of Mathematics*, **152**, 881–901, 2000). While this solution is extremely special, it is *exact* and extensive searches have uncovered further exotic $n$-body solutions.

In general the vast majority of $n$-body systems are chaotic, particularly on account of close two-body encounters in their evolution.

### Charged particle in a magnetic field

In §5.5 as an example of rotating frames of reference we considered the effect of a magnetic field $\boldsymbol{B}$ on a charged particle moving in an orbit around a fixed charge. We assumed there that the magnetic field was sufficiently weak for the quadratic term in $\boldsymbol{B}$ to be neglected, leading to the Larmor precession of the Kepler elliptical orbit around the magnetic field vector. If the weak magnetic field approximation is not made, then the equation of motion of the moving charge can be written relative to the frame which rotates about the magnetic field with the Larmor frequency in terms of the Hamiltonian

$$H = \frac{\boldsymbol{p}^2}{2m} - \frac{|k_1|}{r} + \tfrac{1}{2}|k_2|\rho^2. \qquad (14.22)$$

Here the constant $\boldsymbol{B}$ field defines the $z$ axis and $\rho^2 = r^2 - z^2$, so that the Kepler inverse square law force of attraction towards the origin (having full rotational symmetry about all three axes) is supplemented by a Hooke attraction towards the $\boldsymbol{B}$ field axis (having rotational symmetry only about this axis). As a result the Hamiltonian system is not integrable — it is, in effect, a 'many-body' Hamiltonian and exhibits chaos. This *broken symme-*

*try* of *e.g.* the classical hydrogen atom is complex, but very important in its implications for the quantum mechanical treatment of the same problem (see *e.g.* Friedrich and Wintgen, *Physics Reports*, **183**, 37–79, 1989).

### Tops

In §10.3 and in §§12.3,12.4 we indicated the Lagrangian and Hamiltonian descriptions of the symmetric top, pivoted at a point on its axis of symmetry. The problem has three degrees of freedom and there are three independent conserved quantities $H, p_\varphi, p_\psi$, so that the system is integrable and the motions detailed are regular and ordered.

More complicated systems abound when the equations of rotational motion of an asymmetric body are considered, or where the system involves rolling motion, so that it is non-holonomic. For example, various asymmetric 'tops' appear to exhibit ordered and chaotic motion regimes and this is common for systems which are modelled by the Euler equations of §9.8 with general torques $\boldsymbol{G}$.

### Billiard systems

After Berry (*European Journal of Physics*, **2**, 91–102, 1981) we can consider the motion of a point 'ball' moving within a closed boundary (see Fig. 14.10) and bouncing perfectly on the wall. We assume motion in a straight line between collisions with the wall, simple reflection at each bounce and no dissipation, *i.e.* we assume that the Hamiltonian is $H = \boldsymbol{p}^2/2m$ within the enclosure. The ball follows a path just like that of a light ray with a boundary wall which is a perfect mirror.

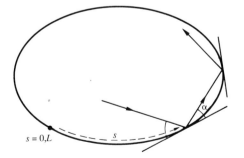
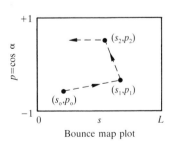

Bounce map plot

Fig. 14.10

Despite its apparent simplicity, this system turns out to be very rich and instructive, with dynamics depending particularly on the shape of the enclosure.

The dynamics can be examined using natural canonical variables, $s, p$, where $s$ is the distance measured along the boundary and $p = \cos\alpha$, where $\alpha$ is the angle of incidence (and reflection) at the wall. (Note that $p$ here is *not* the magnitude of the linear momentum, $|\boldsymbol{p}|$.) This leads to a *successive bounce map* $(s_n, p_n) \rightarrow (s_{n+1}, p_{n+1})$ which preserves area and leads to a Poincaré section for the dynamics — the slice through the overall dynamical space is the (non-planar) boundary shape.

For motion in a planar region of this kind the system has two degrees of freedom and $H$ is a constant of the motion. For some special enclosure shapes *e.g.* circular (see Fig. 14.11), elliptical (see Fig. 14.12) there is in each case another conserved quantity leading to integrability and order.

### *Example:* Circular billiard

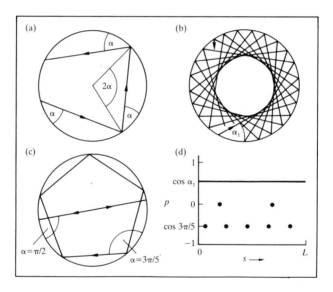

Fig. 14.11   Billiards in a circle: (a) basic orbit geometry; (b) typical orbit (never closing); (c) two closed orbits; (d) phase-space trajectories for orbits in (b) and (c).

Show that there is a second conserved quantity (independent of $H$) for the circular billiard.

Evidently the billiard trajectory for a circular enclosure is a succession of chords, each at the same perpendicular distance from the centre of the circle. So the angular momentum of the ball about the centre $(= a\sqrt{2mH}\,p$, where $a$ is the radius and $p = \cos\alpha)$ remains constant through successive bounces at the boundary. The angular momentum is the second conserved quantity and is evidently independent of the Hamiltonian $H$. Note that when a chord meets the circular boundary the angle $\alpha$ between it and the tangent to the circle is the same at each bounce — the bounce map for a particular trajectory therefore consists of a finite number of points or a line (each at $p = \cos\alpha$) according as the trajectory is closed or is not closed.

For the elliptical enclosure the second conserved quantity is the product of the angular momenta of the ball measured about the two foci of the ellipse. (See Problem 10.)

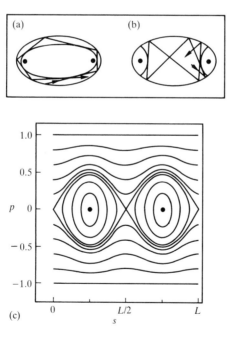

Fig. 14.12   Orbits in an ellipse: (a) repeatedly touching a confocal ellipse; (b) repeatedly touching confocal hyperbolae; (c) ellipse billiard mapping.

In these cases the bounce map (Poincaré section) is made up of points and lines characteristic of integrability. When the shape of the boundary is not so special, the bounce map has regions in which there is area filling, leading to chaos, as well as regions where ordered motion predominates. (See Fig. 14.13.)

Billiard systems with various different properties have been used to model other types of physical behaviour.

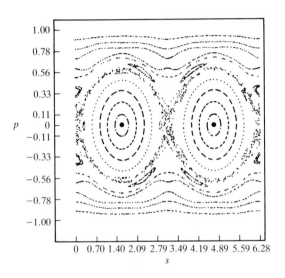

Fig. 14.13    Oval billiard mapping: 25 orbits followed through 200 bounces.

### Dice and coins

The motion when dice and coins are thrown is regular when there are no external torques, *i.e.* when moving and rotating through space, without any air resistance. However, there are impulsive forces which act at each bounce, so that for typical shapes the motion can be highly chaotic. Whereas billiards (see above) involve a regular object bouncing on an irregular boundary, dice (and coins) are a sort of 'dual' in that they are irregular objects bouncing on a regular surface. For each of these types of system the effects of small changes in physical geometry can be large in the statistical properties of the outcomes (see §13.7), whether any dissipation which may be present is very small or if it is such as to lead to rapid energy decay.

For example, when a regular cubical die is subject to a small change $\Delta$ in aspect ratio (away from 1 for the perfect cube), each 1% change in this geometric measure leads to a 3.4% change, approximately, in the probabilities associated with the faces displayed, a change modelled locally by $\Delta 2\sqrt{2}/\pi(4\sqrt{2/3} - 3)$ (Berkshire, *British Association for the Advancement of Science*, 1985).

### The swinging Atwood's machine

A simple physical system which exhibits a variety of dynamical behaviours is the swinging Atwood's machine, which consists of two masses $M, m$ connected by a light inextensible string which passes over a small pulley and through a small hole in a smooth vertical plane, so that the mass $m$ can rotate in this plane. The string is such that collisions of masses with pulley or hole can be discounted. (See Fig. 14.14.)

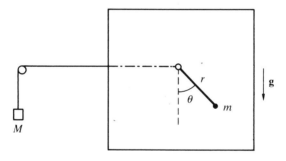

Fig. 14.14

This system was introduced first by Tufillaro, Abbott and Griffiths (*American Journal of Physics*, **52**, 895–903, 1984) and has been the subject of various subsequent papers.

The Hamiltonian for the system is

$$H = \frac{p_r^2}{2m(1 + \mu)} + \frac{p_\theta^2}{2mr^2} + mgr(\mu - \cos\theta), \quad \text{with} \quad \mu = M/m, \quad (14.23)$$

and the motion is not, in general, integrable, since $H$ is usually the only constant of the motion. In the case $\mu > 1$ we can show that the motion of $m$ is always bounded by a *curve of zero velocity* ($\boldsymbol{p} = \boldsymbol{0}$), which is an ellipse whose shape depends on the mass ratio $\mu$ and on the energy $H$. (See

Problem 11.) When $\mu \leq 1$ the motion is not, in general, bounded for any energy and eventually the mass $M$ passes over the pulley.

The system is integrable in the case $\mu = 3$ and apparently only for this particular value of the mass ratio. In that special case, there is a second conserved quantity given by

$$J = \frac{p_\theta}{4m}\left(p_r \cos\frac{\theta}{2} - \frac{2p_\theta}{r}\sin\frac{\theta}{2}\right) + mgr^2 \sin\frac{\theta}{2}\cos^2\frac{\theta}{2}. \qquad (14.24)$$

From Hamilton's equations, $\dot{r} = p_r/4m$, $\dot{p}_r = (p_\theta^2/mr^3) - mg(3 - \cos\theta)$, $\dot{\theta} = p_\theta/mr^2$, $\dot{p}_\theta = -mgr\sin\theta$, one can show by brute force that $\dot{J} = 0$. We note that while $J$ is also quadratic in the momenta $p_r, p_\theta$, it is independent of the Hamiltonian $H$.

It is certainly not yet clear what 'physical' property is represented by this 'hidden symmetry'. When $\mu = 3$ the motion is completely ordered. For all other mass ratio values there are at least pockets of highly irregular and chaotic motion.

### The Hénon–Heiles system

From the previous examples it should be apparent that a paradigm of chaotic dynamics is provided by a Hamiltonian system with two degrees of freedom which has only one conserved quantity, e.g. the Hamiltonian itself. Under these circumstances the system dynamics can show all the features of sensitivity to initial conditions and loss of predictability indicated in §13.7. As a final example consider the motion of a particle of unit mass in a two-dimensional asymmetric potential well (see Fig. 14.15) so that its Hamiltonian is

$$H = \frac{1}{2m}(p_x^2 + p_y^2) + V(x,y), \qquad (14.25)$$

where

$$V = \tfrac{1}{2}(x^2 + y^2) + x^2 y - \tfrac{1}{3}y^3$$
$$\equiv \tfrac{1}{2}r^2 + \tfrac{1}{3}r^3 \sin 3\theta \qquad \text{in plane polars.}$$

This system, with the analogy of a small ball rolling under gravity in an asymmetric bowl, was introduced by Hénon and Heiles (*Astronomical Journal*, **69**, 73–79, 1964) in order to model the motion of a star in a galaxy which has a simple smoothed-out gravitational potential to represent the attraction of the other stars.

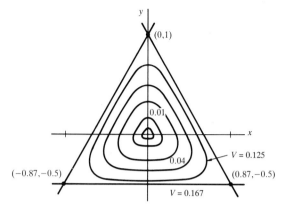

Fig. 14.15   Hénon–Heiles potential contours of $V$ near the local minimum at $(0,0)$.

For small energies the cubic terms in $V$ are insignificant when compared with the quadratic terms and in that limit the motion is integrable, consisting as it does of a combination of normal modes as obtained by the methods of Chapter 11. Of course the motion for small energies is just that of the isotropic oscillator of §4.1. For larger energies corresponding to larger displacements away from the potential minimum at the origin the motion becomes highly irregular.

There are very similar consequences for many other such systems, *e.g.* the double pendulum system of §11.1, when we now allow $\theta, \varphi$ not to be restricted to be small.

## 14.5   Slow Change of Parameters — Adiabatic Invariance

Let us suppose that in a cyclic system (librating or rotating) we allow some of the control parameters to change on a very long time scale, *i.e.* on a scale of many basic oscillation periods. Then it emerges that the change in an action $I$ is then also very slow, in the sense that we can change the parameters by as large an amount as we may want, while $I$ stays as near constant as we want, provided that the change of parameters is made slowly enough. The change of parameters is then said to be carried out *adiabatically* and the action variables are then called *adiabatic invariants*.

Now a general proof of this property is beyond our scope here, so let us consider by way of illustration the simple harmonic oscillator of §2.2, but now with a frequency which is slowly-varying. That is to say

$$\ddot{x} + [\omega(t)]^2 x = 0, \tag{14.26}$$

with $\omega = \sqrt{k/m}$, where $\dot{\omega}/\omega \ll \omega$. The variation in $\omega$ could, for example, be brought about by a spring constant $k$ which varies on a time scale which is very long compared with the natural period $\tau = 2\pi/\omega$. Over a single period (during which $\omega$ does not change appreciably) we may write

$$x(t) = a \cos[\omega(t)t - \theta], \tag{14.27}$$

as in (2.20).

The energy of the oscillation is

$$E = \tfrac{1}{2}m(\dot{x}^2 + \omega^2 x^2) \equiv \tfrac{1}{2}m\omega^2 a^2, \tag{14.28}$$

essentially, and this now changes with time at the slow rate

$$\dot{E} = m(\dot{x}\ddot{x} + \omega^2 x\dot{x} + \omega\dot{\omega}x^2) = m\omega\dot{\omega}x^2. \tag{14.29}$$

By averaging over a period of the oscillation, so that $e.g.$ $\langle E \rangle = \tfrac{1}{\tau}\int_0^\tau E \, d\tau = E$ (above) we obtain

$$\frac{d}{dt}\langle E \rangle = m\omega\dot{\omega}\langle x^2 \rangle = \tfrac{1}{2}m\omega\dot{\omega}a^2$$

$$\equiv \frac{\dot{\omega}}{\omega}\langle E \rangle. \tag{14.30}$$

We obtain immediately

$$\frac{d}{dt}\left[\frac{\langle E \rangle}{\omega}\right] = \frac{1}{\omega}\frac{d}{dt}\langle E \rangle - \frac{\dot{\omega}}{\omega^2}\langle E \rangle = 0, \tag{14.31}$$

so that $\langle E \rangle/\omega$ is essentially constant. This is, of course, our measure of the $action$ $I$ of this oscillator. (See (14.9).)

For more general systems the demonstration is carried out $via$ the definition (14.4) of the action variable. Particular physical applications are very common and valuable conclusions can be drawn about the variations of other variables on the assumption of the applicability of the $principle$ $of$ $adiabatic$ $invariance$.

### Example: Pendulum with slowly varying length

Find how the amplitude of (small) oscillation of a simple pendulum changes as the length is varied slowly.

We can consider the simple pendulum of §2.1, but now with length $l(t)$ changing slowly with time, and we can simplify considerably by confining ourselves to small oscillations. We then need only put $x = l\theta$ in the above analysis of the simple harmonic oscillator, together with $\omega = \sqrt{g/l}$. In this case the action $I$ is given by

$$
\begin{aligned}
I = E/\omega &= \tfrac{1}{2}m\omega a^2 \\
&= \tfrac{1}{2}m\sqrt{gl^3}(\theta_{max})^2,
\end{aligned}
\tag{14.32}
$$

since $a = l\theta_{max}$ here.

The assumption that $I$ is invariant leads to $\theta_{max} \propto l^{-3/4}$ (see Problem 13) and we can find from this how the energy, displacement, maximum acceleration of the pendulum bob change as $l$ is changed, possibly by orders of magnitude, on the assumption that the change in $l$ is effected sufficiently slowly.

A practical problem to contemplate when considering this matter is the consumption of a piece of spaghetti, although this is not usually a *simple* pendulum and politeness may well rule out the necessary long time scale for consumption! However, the general principle is the same, particularly with regard to the dangers of spaghetti sauce!

### The elastic bouncer

In §14.3 we considered a perfectly elastic ball bouncing normally between two walls and the action $I$ for that case was found to be $\sqrt{2mH}\,L/\pi$ where $m, H$ are, respectively, the mass and energy of the ball and $L$ is the distance between the walls. If the walls are now moved either towards or away from one another very slowly ('adiabatically'), then the principle of adiabatic invariance tells us that $L\sqrt{H}$ should be essentially constant, so that $v \propto 1/L$ and energy $H \propto 1/L^2$ (see Problem 14). Again it should be noted that $v, H$ can change by orders of magnitude, while the action $I$ remains essentially constant. We can note here that we can model a monatomic gas with elastic molecules bouncing in a cubical box and, under reasonable assumptions, obtain the relationship that pressure $\propto$ (density)$^{5/3}$ for adiabatic change.

### Attractive central force problem with varying strength

Evidently the force of attraction, *e.g.* of the Sun, would diminish as its mass decreases over what is, for us, a long time. It has also been suggested by Dirac that the gravitational 'constant' may, in fact, change on a long time scale. In either case we can look to the action/angle formulation in §14.3, where $I_2$ is an exact invariant and $I_1$ can be taken now as an adiabatic invariant. As a consequence we can show, in particular, that, as the strength $|k|$ of the attracting force decreases, the period and the semi-major axis of the Kepler elliptical orbit of a planet both increase (as we might expect), but that (rather unexpectedly) the eccentricity of the orbit remains constant and the orbit retains its shape. (See Problem 16.)

The adiabatic invariance of the action variables has enormous importance in the transition to quantum mechanics. The pendulum with slow variation in length was a discussion problem for Lorentz and Einstein in 1911, when the foundations of quantum mechanics were being laid. The effective constancy of the action variables by the principle of adiabatic invariance, in the classical formulation as control parameters are changed very slowly, makes the action variables the prime candidates for quantization.

## 14.6   Near-integrable Systems

When we have a system which is Hamiltonian with $H = H_0(\boldsymbol{q}, \boldsymbol{p})$ and integrable, so that its dynamics are regular and ordered, we can ask what happens for a system which is not itself integrable, but which is close by. That is to say this latter system has Hamiltonian $H = H_0 + \epsilon H_1$ where $\epsilon$ is very small.

When there is only one degree of freedom ($n = 1$) there is a well-developed perturbation theory, which allows us to find out how the action/angle variables for the $H_0$ system are changed by the perturbation $\epsilon H_1$. We won't detail this here, but it turns out that the only real difficulty with the perturbation expansions (as power series in $\epsilon$) is associated with attempting to expand near where the basic frequency $\omega_0$ of the unperturbed system $H_0$ is zero. An example of this is provided by the simple pendulum and the libration/rotation orbits which are situated close to the separatrices (at which $\omega_0 = 0$). The perturbation expansions break down in this neighbourhood, although they generally work well elsewhere. A simple remedy would seem to suggest itself strongly — avoid separatrices!

For systems with at least two degrees of freedom ($n \geq 2$) the perturbation analysis seems to be plagued by a similar difficulty, but much worse in implication. The expansions for the $n = 1$ case contain denominators $\omega_0$, so that the expansions become non-uniformly valid near $\omega_0 \cong 0$. For the $n \geq 2$ case we find in the expansions denominators like $\omega_0 \cdot \boldsymbol{N}$ where the lattice vector $\boldsymbol{N}$ has integer components and $\omega_0$ is the vector of normal frequencies of the particular torus (see §14.1) about which we are trying to expand our series solution.

Now when this denominator is actually zero we have a *degeneracy* (or *resonance*), in that some of the normal frequencies have rational ratios. This is bad enough, but the trouble is that, even for $\omega_0$ not exactly resonant in this manner, we can apparently make choices of $\boldsymbol{N}$ components in order to make $\omega_0 \cdot \boldsymbol{N}$ as close as we might (not) like to zero — thus destroying the usefulness of our expansion. This is usually called the *problem of small denominators* and it seems to imply that, only if the perturbation $\epsilon H_1$ to our Hamiltonian $H_0$ is such that $H = H_0 + \epsilon H_1$ is still integrable through the preservation of a sufficient number of conserved quantities, is the phase space torus structure itself preserved, although distorted, and with convergent perturbation expansions.

Very elaborate methods have been developed in an attempt to patch up the expansions in the general case. Especially noteworthy is a tour de force by Delaunay for the Sun–Earth–Moon system ('lunar theory') undertaken in the middle of the nineteenth century, which involved about twenty years of endeavour, but all such efforts seemed doomed to failure. It seemed that the slightest perturbation, however small, would cause the torus structure of an integrable system to collapse irretrievably. That this is not in fact the case was demonstrated by Kolmogorov, Arnol'd and Moser (KAM) in work originating between 1954 and 1963. This work has led to an explosion of research, results and applications. KAM showed that, as long as the perturbation introduced to a smooth system is itself smooth enough, then the torus destruction with a perturbed system is progressive as the size of the perturbation increases.

Key roles are played in the theory, and its consequences, by the following:

- *Rational approximation.* This means the closeness by which real numbers can be approximated by rationals. This is important, because the torus breakdown grows with the perturbation around the resonant ($\omega_0$ with rational ratios) tori.
- *Fractals.* These are 'self-similar' structures in *maps*, *i.e.* in Poincaré sections (like *e.g.* Fig. 14.13) and in full phase space. Properties of maps are explored briefly in Appendix D.

- *Divergence of trajectories.* Breakdown of the torus structure gives freedom to system trajectories to diverge strongly from their neighbours, leading to 'extreme sensitivity to initial conditions' (see §13.7) and to chaotic behaviour.
- *Lack of predictability.* We have effectively stochastic output resulting even from strictly deterministic systems, so that we lose knowledge of the connection between cause and effect.

To the big question 'Is the Solar System stable?', which has motivated a lot of work in Mechanics, the theory provides the answer 'a (definite) maybe'. The Solar System is an $n$-body problem, which is markedly hierarchical, so that the effect of *e.g.* planet/planet interactions is of lower order than the planet/Sun interactions which produce Kepler ellipses. For this reason the broad structure of the system is robust (fortunately for the human race!).

For large configurations, stable and unstable regimes may well be closely interwoven, so that in an idealized system we might not be able to tell from (necessarily) imprecise observations quite what regime we are in! However we should find the hierarchical nature of our system, with a strong central Sun, to be reassuring apart from the possibility of asteroid impact — in practice the fate of the whole Solar System is actually rather likely to be determined largely by dissipative and other non-gravitational forces.

## 14.7  Summary

In this chapter the importance of the number of symmetries of a Hamiltonian system (leading to conservation laws) has been shown in the restrictions they impose on the freedom of the system trajectories in the appropriate phase space.

Integrability of the system leads to a natural formulation in terms of action/angle variables, which allow us to detail the nonlinear normal modes of such a system, *i.e.* modes which are exact and involve no approximations for small-amplitude motions.

Examples have been given of simple/important Hamiltonian systems which have regimes which exhibit chaos and a resulting breakdown of predictability.

The action variables are effectively constant when system control parameters are changed very slowly. This principle of adiabatic invariance is a general result, which has non-intuitive and useful consequences for the way in which other quantities of physical interest may change, possibly by orders of magnitude on these long time scales.

The adiabatic invariance of action variables identifies them as prime candidates to be quantized in the transition from classical to quantum mechanics.

There are other routes to chaos. Indeed *e.g.* the limit cycles of Chapter 13, each with a single-point Poincaré section, can go through a period-doubling cascade to chaos (see Appendix D) as parameters are changed. When trajectories are stretched, squeezed and folded then chaos and 'mixing' typically result.

It is fitting that this book should end with the mention of results (KAM) which point the way to a whole host of future developments, by no means the least of which is the question of how to quantize a system which is classically chaotic!

As we said in the Preface, Classical Mechanics is a very old subject. However it is very much alive!

---

## Problems

1. A particle of mass $m$ is projected outward radially from the surface $(q = R)$ of a spherical planet. Show that the Hamiltonian is given by $H = p^2/2m - |k|/q$ (with $k$ constant), so that $H$ is a constant of the motion ($\equiv$ energy $E$). Sketch the phase portrait in the $(q, p)$ phase plane for $q \geq R$, distinguishing between trajectories which correspond to the particle returning and not returning to the planet's surface. When the particle does return show that the time taken to do this is

$$t_0 = \sqrt{2m} \int_R^h \frac{dq}{\sqrt{E + |k|/q}}, \quad \text{where} \quad h = \frac{|k|}{|E|}.$$

Evaluate this integral to find $t_0$ in terms of $h, R, |k|/m$. (*Hint*: the substitution $q = h\sin^2\theta$ is helpful!) By considering the limit $R/h \to 0$ show that the result is in accord with Kepler's third law (4.32).

2. For an autonomous Hamiltonian system with two degrees of freedom the Hamiltonian $H$ is a constant of the motion and such a system is integrable if another constant $F$ (independent of $H$) can be found — *i.e.* such that $[F, H] = 0$.

   (a) For the double harmonic oscillator (14.3) show that we may take $F = p_1^2/2m + \frac{1}{2}m\omega_1^2 q_1^2$ as the second constant.

   (b) For central force motion (14.14) show that we may take $F = p_\theta$ as the second constant.

(c) *When the potential $V(r) = -|k|/r$, show that the components of the vector

$$A = \frac{p \wedge J}{m|k|} - \frac{r}{r}$$

with $p, J$ respectively linear, angular momentum) are constants of the motion.

($A$ is the Laplace–Runge–Lenz vector. It can be shown using the results in §4.4 that it has magnitude given by the eccentricity $e$ of the conic-section orbit and is directed from the centre of force along the major axis towards the position of closest approach ('perihelion').)

3. A string of length $l$ with a mass $m$ at each end passes through a hole in a frictionless horizontal plane. One mass moves horizontally on the plane and the other mass hangs vertically downwards. Show that a suitable Hamiltonian for the system is

$$H = \frac{p_r^2}{4m} + \frac{p_\theta^2}{2mr^2} - mg(l - r),$$

where $(r, \theta)$ are the polar co-ordinates of the particle on the plane and $p_r, p_\theta$ are the corresponding momenta. Identify two constants of the motion. Show that a steady motion with $r = r_0$ is possible (for any $r_0 > 0$), if $p_\theta$ is chosen suitably, and that the period of small oscillations about this motion is $2\pi\sqrt{2r_0/3g}$.

4. Consider again the particle sliding on the inside of a smooth cone, as in Chapter 12, Problem 1. Show that when a stable circular motion is disturbed the resulting small oscillations are only closed if $\sqrt{3}\sin\alpha$ is a rational number.

5. For central force motion with an inverse square law force of attraction the Hamiltonian is (14.14), *i.e.*

$$H = \frac{p_r^2}{2m} + \frac{p_\theta^2}{2mr^2} - \frac{|k|}{r} \quad (\equiv \text{ energy } E).$$

If we fix $p_\theta$, show that $r$ is bounded ($r_1 \leq r \leq r_2$) only when $-k^2m/2p_\theta^2 \leq E < 0$ and that the energy minimum corresponds to an orbit in physical space which is a circle. Sketch the curves $H = $ constant in the $(r, p_r)$ projection of the full four-dimensional phase space for this system. Consider this projection in the light of the discussion of surfaces of section in §14.2.

6. A particle of mass $m$ is constrained to move under the action of gravity in the vertical $(x, z)$ plane on a smooth cycloid curve given paramet-

rically by $x = l(\theta + \sin\theta)$, $z = l(1 - \cos\theta)$. Show that a suitable Hamiltonian is

$$H = \frac{p_\theta^2}{4ml^2(1 + \cos\theta)} + mgl(1 - \cos\theta).$$

Use action/angle variables to show that the frequency of oscillation of the particle is independent of its amplitude, *i.e.* it is the same for all initial conditions with $|\theta| < \pi$.

(The substitution $s = \sin\frac{1}{2}\theta$ is useful. This *tautochrone* property of the cycloid was known to Huygens in the seventeenth century and, in principle at least, it leads to some quite accurate clock mechanisms. Contrast the tautochrone property with the brachistochrone property of Chapter 3, Problem 15.)

7. *A particle of mass $m$ is attached to the origin by a light elastic string of natural length $l$, so that it is able to move freely along the $x$-axis if its distance from the origin is less than $l$, but otherwise moves in a potential $V(x) = \frac{1}{2}k(|x| - l)^2$ for $|x| > l$. If the particle always moves in a straight line, sketch the potential and the phase-plane trajectories for different values of the energy $E$. Show that $E = (\sqrt{I\Omega + \beta^2} - \beta)^2$, where $I$ is the action, $\beta = \sqrt{2k}\,l/\pi$, $\Omega = \sqrt{k/m}$. Explain briefly (without detailed calculation) how the angle variable $\phi$ conjugate to the action may be found in the form $\phi(x)$ and how $x$ may be found as a function of the time $t$. What happens in the (separate) limits of small and large energies $E$?

8. *A particle of mass $m$ moves in a one-dimensional potential $V(q) = \frac{1}{2}(kq^2 + \lambda/q^2)$, where $k, \lambda, q > 0$. Sketch the potential and the phase portrait. Show that the energy $E$ and action $I$ are related by

$$E = \sqrt{k\lambda} + 2I\sqrt{k/m}$$

and that the period is then independent of amplitude. Discuss how the dependence of $q$ on the angle variable $\phi$ may be found and then the dependence of $q$ on the time $t$.

(This one-dimensional problem models the purely radial part of the motion of the isotropic harmonic oscillator of §4.1. The integral

$$\int_{x_1}^{x_2} \sqrt{(x_2 - x)(x - x_1)}\,\frac{dx}{x} = \frac{1}{2}\pi(x_1 + x_2) - \pi\sqrt{x_1 x_2}$$

where $x_2 > x_1 > 0$ and $x = q^2$ will prove useful!)

9. *For central force motion with an inverse square law force of attraction (see Problem 5) confirm the result of (14.16) that the radial action $I_1 =$

$-I_2 + \frac{1}{2}|k|\sqrt{2m/(-E)}$, where $I_2 = p_\theta$, so that the natural frequencies of radial and transverse motions of the system are the same.

(In order to evaluate the integral in (14.15) the result given at the end of Problem 8 will prove similarly useful!)

10. For the billiard in an elliptical enclosure (Fig. 14.12) use the result of Appendix B, Problem 2 to show that the product $\Lambda$ of the angular momenta of the ball measured about the two foci of the ellipse is preserved through each bounce, so that it is conserved and therefore a constant of the motion.

    Using elliptical co-ordinates (see Chapter 3, Problem 24), show that $\Lambda = (\cosh^2 \lambda - \cos^2 \theta)^{-1}(\sinh^2 \lambda\, p_\theta^2 - \sin^2 \theta\, p_\lambda^2)$ and that $H = [2mc^2(\cosh^2 \lambda - \cos^2 \theta)]^{-1}(p_\lambda^2 + p_\theta^2)$. Hence show that the reflected trajectory from the boundary $x^2/a^2 + y^2/b^2 = 1$ is necessarily tangent (when $\Lambda > 0$) to the ellipse $\lambda = \operatorname{arcsinh}\sqrt{\Lambda/2mc^2H}$.

    [Closure for such a trajectory implies closure for all trajectories tangent to the same inner ellipse — an example of a general result due to Poncelet (1822).]

11. *Consider the motion of the swinging Atwood's machine (see Fig. 14.14) for various different values of the mass ratio $\mu = M/m$.

    (a) Show that when $\mu > 1$ the motion of $m$ is always bounded by a zero-velocity curve which is an ellipse, whose shape depends on $\mu$ and on the constant energy $E$.

    (b) Show that when $\mu \leq 1$ the motion is not in general bounded for any $E$ by sketching the zero-velocity curves for mass $m$ in the $r, \theta$ plane (in various cases).

    (Hint: The results of §B.2 will prove helpful in identifying the zero-velocity curves.)

12. *Show that the equations of motion (14.19) for the restricted three-body problem may be put into Hamiltonian form using the substitution $p_x = \dot{x} - \omega y$, $p_y = \dot{y} + \omega x$. Show that, since the system is autonomous, $\frac{1}{2}(\dot{x}^2 + \dot{y}^2) + V = C$, where $C$ is a constant of the motion (Jacobi). Hence sketch the regions $V \leq C$ of possible motions in the $x, y$ plane, for various values of $C$ and of the mass ratio $M_1/M_2$, considering particularly those values of $C$ corresponding to the three 'equilibria' on the $x$ axis and to the two Lagrangian equilateral triangle 'equilibria' — see Chapter 10, Problem 16. (Note that for the Earth/Moon system with $M_1/M_2 = 81.3$ there is in consequence a value of $C$ below which an Earth–Moon transfer is not possible.)

13. For the small oscillations of a pendulum whose length $l$ is varying very slowly show that the maximum angular displacement $\theta_{max} \propto l^{-3/4}$. Hence show that the maximum sideways displacement from the vertical is $\propto l^{1/4}$ and that the maximum acceleration is $\propto l^{-3/4}$. (Note that this last result implies, as $l$ decreases slowly, an increasing risk for spaghetti eaters from sauce detachment!)

14. For the elastic bouncer (see Fig. 14.7) between two walls whose separation is being changed slowly ('adiabatically') show that the bouncer speed $v \propto 1/L$. We can model (crudely) a monatomic gas, confined within a cubical box of side $L$, by assuming that each molecule has the same speed $v$ and that one third of the molecules move normally to each pair of faces of the box! If intermolecular collisions do not affect these assumptions then $L$ can be changed very slowly. Assuming that the kinetic energy of the molecules gives a measure of temperature and that the gas is 'ideal', so satisfying the ideal gas law: (pressure)×(volume) $\propto$ (temperature), obtain the relationship (pressure) $\propto$ (volume)$^{5/3}$ for adiabatic change of a monatomic gas.

15. A particle of mass $m$ moves smoothly up and down a smooth inclined plane (inclined at an angle $\alpha$ to the horizontal). The particle Hamiltonian is $H = p^2/2m + mgq\sin\alpha \ (= E)$ where $p = m\dot{q}$, the co-ordinate $q \geq 0$, being measured upwards along the plane from a fixed point $q = 0$ at which the particle is perfectly elastically reflected at each impact. Show that the energy $E$ and action $I$ for this oscillator are related by $E = [(9\pi^2/8)g^2m\sin^2\alpha]^{1/3}I^{2/3}$ and find the frequency $\omega$ of small oscillation in terms of $g$, $\alpha$ and $q_0$ (the amplitude of the motion). Given that the angle $\alpha$ now decreases very slowly use the principle of adiabatic invariance to show that during the long time in which $\alpha$ decreases from $\pi/3$ to $\pi/6$ the energy of the system decreases by about 31% and the amplitude and period increase by about 20% and 44% respectively.

16. *For central force motions with an inverse square law force of attraction the relation between energy $E$ and actions $I_1$, $I_2$ is (14.17), demonstrated in Problem 9. If the strength of the force (*i.e.* $|k|$) decreases slowly ('tired sun'), use the principle of adiabatic invariance to show that the period $\tau$ of a bounded orbit (an ellipse, see Problem 5) varies so that $\tau \propto k^{-2}$ (*i.e.* $\tau$ increases). Find how the semi-major axis of the ellipse varies with $k$ and show that the eccentricity of the ellipse remains constant.

# Appendix A

# Vectors

In this appendix, we give a summary of the properties of vectors which are used in the text.

## A.1   Definitions and Elementary Properties

A *vector* $a$ is an entity specified by a magnitude, written $a$ or $|a|$, and a direction in space. It is to be contrasted with a *scalar*, which is specified by a magnitude alone. The vector $a$ may be represented geometrically by an arrow of length $a$ drawn from any point in the appropriate direction. In particular, the position of a point $P$ with respect to a given origin $O$ may be specified by the *position vector* $r$ drawn from $O$ to $P$ as in Fig. A.1.

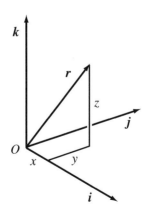

Fig. A.1

Any vector can be specified, with respect to a given set of Cartesian axes, by three components. If $x, y, z$ are the Cartesian co-ordinates of $P$, with $O$ as origin, then we write $r = (x, y, z)$, and say that $x, y, z$ are the *components* of $r$. (See Fig. A.1.) We often speak of $P$ as 'the point $r$'. When $P$ coincides with $O$, its position vector is the *zero vector* $\mathbf{0} = (0, 0, 0)$ of length 0 and indeterminate direction. For a general vector, we write $a = (a_x, a_y, a_z)$, where $a_x, a_y, a_z$ are its components.

The product of a vector $a$ and a scalar $c$ is $ca = (ca_x, ca_y, ca_z)$. If $c > 0$, it is a vector in the same direction as $a$, and of length $ca$; if $c < 0$, it is in the opposite direction, and of length $|c|a$. In particular, if $c = 1/a$, we obtain the *unit vector* in the direction of $a$, $\hat{a} = a/a$.

Addition of two vectors $a$ and $b$ may be defined geometrically by drawing one vector from the head of the other, as in Fig. A.2. (This is the 'parallelogram law' for addition of forces — or vectors in general.) Subtraction

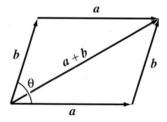

Fig. A.2

is defined similarly by Fig. A.3. In terms of components,

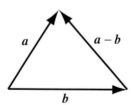

Fig. A.3

$$a + b = (a_x + b_x, a_y + b_y, a_z + b_z),$$
$$a - b = (a_x - b_x, a_y - b_y, a_z - b_z).$$

It is often useful to introduce three unit vectors $i, j, k$, pointing in the directions of the $x$-, $y$-, $z$-axes, respectively. They form what is known as an *orthonormal triad* — a set of three mutually perpendicular vectors of unit length. It is clear from Fig. A.1 that any vector $r$ can be written as a sum of three vectors along the three axes,

$$r = xi + yj + zk. \tag{A.1}$$

Mathematically, any set of three quantities may be grouped together and regarded as the components of a vector. It is important to realize, however, that when we say that some physical quantity is a vector we mean more than just that it needs three numbers to specify it. What we mean is that these three numbers must transform in the correct way under a change of axes.

For example, consider a new set of axes $i', j', k'$ related to $i, j, k$ by a rotation through an angle $\varphi$ about the $z$-axis (see fig. A.4):

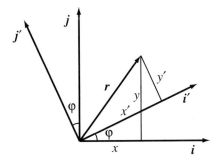

Fig. A.4

$$i' = i \cos \varphi + j \sin \varphi,$$
$$j' = -i \sin \varphi + j \cos \varphi, \tag{A.2}$$
$$k' = k.$$

The co-ordinates $x', y', z'$ of $P$ with respect to the new axes are defined by

$$r = x'i' + y'j' + z'k'.$$

Substituting (A.2) and comparing with (A.1), we see that $x = x' \cos \varphi - y' \sin \varphi$, *etc*, or equivalently $x' = x \cos \varphi + y \sin \varphi$, *etc*. Physically, then, a vector $a$ is an object represented with respect to any set of axes by three components $(a_x, a_y, a_z)$ *which transform under rotations in the same way as* $(x, y, z)$, *i.e.*, in matrix notation,

$$
\begin{bmatrix} a'_x \\ a'_y \\ a'_z \end{bmatrix} = \begin{bmatrix} \cos\varphi & \sin\varphi & 0 \\ -\sin\varphi & \cos\varphi & 0 \\ 0 & 0 & 1 \end{bmatrix} \begin{bmatrix} a_x \\ a_y \\ a_z \end{bmatrix}. \tag{A.3}
$$

## A.2   The Scalar Product

If $\theta$ is the angle between the vectors $\boldsymbol{a}$ and $\boldsymbol{b}$, then by elementary trigonometry the length of their sum is given by

$$|\boldsymbol{a} + \boldsymbol{b}|^2 = a^2 + b^2 + 2ab\cos\theta. \tag{A.4}$$

It is useful to define their *scalar product* $\boldsymbol{a} \cdot \boldsymbol{b}$ ('$\boldsymbol{a}$ *dot* $\boldsymbol{b}$') as

$$\boldsymbol{a} \cdot \boldsymbol{b} = ab\cos\theta. \tag{A.5}$$

Note that this is equal to the length of $\boldsymbol{a}$ multiplied by the projection of $\boldsymbol{b}$ on $\boldsymbol{a}$, or *vice versa*. (See Fig. A.5.)

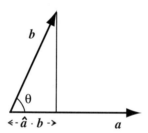

Fig. A.5

In particular, the *square of* $\boldsymbol{a}$ is

$$a^2 = \boldsymbol{a} \cdot \boldsymbol{a} = a^2.$$

Thus we can write (A.4) as

$$(\boldsymbol{a} + \boldsymbol{b})^2 = a^2 + b^2 + 2\boldsymbol{a} \cdot \boldsymbol{b},$$

and, similarly, the square of the difference is

$$(\boldsymbol{a} - \boldsymbol{b})^2 = a^2 + b^2 - 2\boldsymbol{a} \cdot \boldsymbol{b}.$$

All the ordinary rules of algebra are valid for the sums and scalar products of vectors, save one. (For example, the commutative law of addition, $a + b = b + a$ is obvious from Fig. A.2, and the other laws can be deduced from appropriate figures.) The one exception is the following: for two scalars, $ab = 0$ implies that either $a = 0$ or $b = 0$ (or, of course, both), but we can find two non-zero vectors $a$ and $b$ for which $a \cdot b = 0$. In fact, this is the case if $\theta = \pi/2$, that is, if the vectors are orthogonal:

$$a \cdot b = 0 \quad \text{if} \quad a \perp b.$$

The scalar products of the unit vectors $i, j, k$ are

$$i^2 = j^2 = k^2 = 1, \qquad i \cdot j = j \cdot k = k \cdot i = 0.$$

Thus, taking the scalar product of each in turn with (A.1), we find

$$i \cdot r = x, \qquad j \cdot r = y, \qquad k \cdot r = z.$$

These relations express the fact that the components of $r$ are equal to its projections on the three co-ordinate axes.

More generally, if we take the scalar product of two vectors $a$ and $b$, we find

$$a \cdot b = a_x b_x + a_y b_y + a_z b_z, \tag{A.6}$$

and in particular,

$$r^2 = r^2 = x^2 + y^2 + z^2. \tag{A.7}$$

## A.3 The Vector Product

Any two non-parallel vectors $a$ and $b$ drawn from $O$ define a unique axis through $O$ perpendicular to the plane containing $a$ and $b$. It is useful to define the *vector product* $a \wedge b$ ('$a$ *cross* $b$', sometimes also written $a \times b$) to be a vector along this axis whose magnitude is the area of the parallelogram with edges $a$ and $b$,

$$|a \wedge b| = ab \sin \theta. \tag{A.8}$$

(See Fig. A.6.) To distinguish between the two opposite directions along the axis, we introduce a convention: the direction of $a \wedge b$ is that in which a right-hand screw would move when turned from $a$ to $b$.

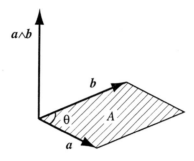

Fig. A.6

A vector whose sense is merely conventional, and would be reversed by changing from a right-hand to a left-hand convention, is called an *axial* vector, as opposed to an ordinary, or *polar*, vector. For example, velocity and force are polar vectors, but angular velocity is an axial vector (see §5.1). The vector product of two polar vectors is thus an axial vector.

The vector product has one very important, but unfamiliar, property. If we interchange $a$ and $b$, we reverse the sign of the vector product:

$$b \wedge a = -a \wedge b. \tag{A.9}$$

It is essential to remember this fact when manipulating any expression involving vector products. In particular, the vector product of a vector with itself is the zero vector,

$$a \wedge a = 0.$$

More generally, $a \wedge b$ vanishes if $\theta = 0$ or $\pi$:

$$a \wedge b = 0 \quad \text{if} \quad a \parallel b.$$

If we choose our co-ordinate axes to be right-handed, then the vector products of $i, j, k$ are

$$i \wedge i = j \wedge j = k \wedge k = 0,$$
$$i \wedge j = k, \qquad j \wedge i = -k,$$
$$j \wedge k = i, \qquad k \wedge j = -i,$$
$$k \wedge i = j, \qquad i \wedge k = -j. \tag{A.10}$$

Thus, when we form the vector product of two arbitrary vectors $\boldsymbol{a}$ and $\boldsymbol{b}$, we obtain

$$\boldsymbol{a} \wedge \boldsymbol{b} = \boldsymbol{i}(a_y b_z - a_z b_y) + \boldsymbol{j}(a_z b_x - a_x b_z) + \boldsymbol{k}(a_x b_y - a_y b_x).$$

This relation may conveniently be expressed in the form of a determinant

$$\boldsymbol{a} \wedge \boldsymbol{b} = \begin{vmatrix} \boldsymbol{i} & \boldsymbol{j} & \boldsymbol{k} \\ a_x & a_y & a_z \\ b_x & b_y & b_z \end{vmatrix}. \tag{A.11}$$

From any three vectors $\boldsymbol{a}, \boldsymbol{b}, \boldsymbol{c}$, we can form the *scalar triple product* $(\boldsymbol{a} \wedge \boldsymbol{b}) \cdot \boldsymbol{c}$. Geometrically, it represents the volume $V$ of the parallelepiped with adjacent edges $\boldsymbol{a}, \boldsymbol{b}, \boldsymbol{c}$ (see Fig. A.7). For, if $\varphi$ is the angle between $\boldsymbol{c}$

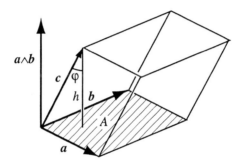

Fig. A.7

and $\boldsymbol{a} \wedge \boldsymbol{b}$, then

$$(\boldsymbol{a} \wedge \boldsymbol{b}) \cdot \boldsymbol{c} = |\boldsymbol{a} \wedge \boldsymbol{b}| c \cos \varphi = Ah = V,$$

where $A$ is the area of the base, and $h = c \cos \varphi$ is the height. The volume is reckoned positive if $\boldsymbol{a}, \boldsymbol{b}, \boldsymbol{c}$ form a right-handed triad, and negative if they form a left-handed triad. For example, $(\boldsymbol{i} \wedge \boldsymbol{j}) \cdot \boldsymbol{k} = 1$, but $(\boldsymbol{i} \wedge \boldsymbol{k}) \cdot \boldsymbol{j} = -1$.

In terms of components, we can evaluate the scalar triple product by taking the scalar product of $\boldsymbol{c}$ with (A.11). We find

$$(\boldsymbol{a} \wedge \boldsymbol{b}) \cdot \boldsymbol{c} = \begin{vmatrix} a_x & a_y & a_z \\ b_x & b_y & b_z \\ c_x & c_y & c_z \end{vmatrix}. \tag{A.12}$$

Either from this formula, or from its geometric interpretation, we see that the scalar triple product is unchanged by any cyclic permutation of

$a, b, c$, but changes sign if any pair is interchanged:

$$(a \wedge b) \cdot c = (b \wedge c) \cdot a = (c \wedge a) \cdot b$$
$$= -(b \wedge a) \cdot c = -(c \wedge b) \cdot a = -(a \wedge c) \cdot b. \qquad \text{(A.13)}$$

Moreover, we may interchange the dot and the cross:

$$(a \wedge b) \cdot c = a \cdot (b \wedge c). \qquad \text{(A.14)}$$

(For this reason, a more symmetrical notation, $[a, b, c]$, is sometimes used.)

Note that the scalar triple product vanishes if any two vectors are equal, or parallel. More generally, it vanishes if $a, b, c$ are coplanar.

From three vectors we can also form the *vector triple product* $(a \wedge b) \wedge c$. Since this vector is perpendicular to $a \wedge b$, it must lie in the plane of $a$ and $b$, and must therefore be a linear combination of these two vectors. It is not hard to show by writing out the components, that

$$(a \wedge b) \wedge c = (a \cdot c)b - (b \cdot c)a. \qquad \text{(A.15)}$$

Similarly,

$$a \wedge (b \wedge c) = (a \cdot c)b - (a \cdot b)c. \qquad \text{(A.16)}$$

Note that these vectors are unequal, so that we cannot omit the brackets in a vector triple product. It is useful to note that in both of these formulae the term with the positive sign is the middle vector times the scalar product of the other two.

## A.4   Differentiation and Integration of Vectors

We are often concerned with vectors which are functions of some scalar parameter, for example the position vector of a particle as a function of time, $r(t)$. The vector distance travelled by the particle in a short time interval $\Delta t$ is

$$\Delta r = r(t + \Delta t) - r(t).$$

(See Fig. A.8.) The velocity, or derivative of $r$ with respect to $t$, is defined just as for scalars, as the limit of a ratio,

$$\dot{r} = \frac{\mathrm{d}r}{\mathrm{d}t} = \lim_{\Delta t \to 0} \frac{\Delta r}{\Delta t}. \qquad \text{(A.17)}$$

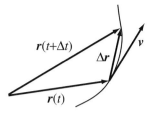

Fig. A.8

In the limit, the direction of this vector is that of the tangent to the path of the particle, and its magnitude is the speed in the usual sense. In terms of co-ordinates,

$$\dot{\boldsymbol{r}} = (\dot{x}, \dot{y}, \dot{z}).$$

Derivatives of other vectors are defined similarly. In particular, we can differentiate again to form the acceleration vector $\ddot{\boldsymbol{r}} = \mathrm{d}^2\boldsymbol{r}/\mathrm{d}t^2$.

It is easy to show that all the usual rules for differentiating sums and products apply also to vectors. For example,

$$\frac{\mathrm{d}}{\mathrm{d}t}(\boldsymbol{a} \wedge \boldsymbol{b}) = \frac{\mathrm{d}\boldsymbol{a}}{\mathrm{d}t} \wedge \boldsymbol{b} + \boldsymbol{a} \wedge \frac{\mathrm{d}\boldsymbol{b}}{\mathrm{d}t},$$

though in this particular case one must be careful to preserve the order of the two factors, because of the antisymmetry of the vector product.

Note that the derivative of the magnitude of $\boldsymbol{r}$, $\mathrm{d}r/\mathrm{d}t$, is *not* the same thing as the magnitude of the derivative, $|\mathrm{d}\boldsymbol{r}/\mathrm{d}t|$. For example, if the particle is moving in a circle, $r$ is constant, so that $\dot{r} = 0$, but clearly $|\dot{\boldsymbol{r}}|$ is not zero. In fact, applying the rule for differentiating a scalar product to $\boldsymbol{r}^2$, we obtain

$$2r\dot{r} = \frac{\mathrm{d}}{\mathrm{d}t}(r^2) = \frac{\mathrm{d}}{\mathrm{d}t}(\boldsymbol{r}^2) = 2\boldsymbol{r} \cdot \dot{\boldsymbol{r}},$$

which may also be written

$$\dot{r} = \hat{\boldsymbol{r}} \cdot \dot{\boldsymbol{r}}. \tag{A.18}$$

Thus the rate of change of the distance $r$ from the origin is equal to the radial component of the velocity vector.

We can also define the integral of a vector. If $v = \mathrm{d}r/\mathrm{d}t$, then we also write

$$r = \int v\,\mathrm{d}t,$$

and say that $r$ is the *integral* of $v$. If we are given $v(t)$ as a function of time, and the initial value of $r$, $r(t_0)$, then the position at any later time is given by the definite integral

$$r(t_1) = r(t_0) + \int_{t_0}^{t_1} v(t)\,\mathrm{d}t. \qquad (A.19)$$

This is equivalent to three scalar equations for the components, for example,

$$x(t_1) = x(t_0) + \int_{t_0}^{t_1} v_x(t)\,\mathrm{d}t.$$

One can show, exactly as for scalars, that the integral in (A.19) may be expressed as the limit of a sum.

## A.5    Gradient, Divergence and Curl

There are many quantities in physics which are functions of position in space; for example, temperature, gravitational potential, or electric field. Such quantities are known as *fields*. A *scalar field* is a scalar function $\phi(x, y, z)$ of position in space; a *vector field* is a vector function $A(x, y, z)$. We can also indicate the position in space by the position vector $r$ and write $\phi(r)$ or $A(r)$.

Now let us consider the three partial derivatives of a scalar field, $\partial\phi/\partial x, \partial\phi/\partial y, \partial\phi/\partial z$. They form the components of a vector field, known as the *gradient* of $\phi$, and written $\operatorname{grad}\phi$, or $\nabla\phi$ ('*del* $\phi$', or occasionally '*nabla* $\phi$'). To show that they really are the components of a *vector*, we have to show that it is defined in a manner which is independent of the choice of axes. We note that if $r$ and $r + \mathrm{d}r$ are two neighbouring points, then the difference between the values of $\phi$ at these points is

$$\mathrm{d}\phi = \phi(r + \mathrm{d}r) - \phi(r) = \frac{\partial\phi}{\partial x}\,\mathrm{d}x + \frac{\partial\phi}{\partial y}\,\mathrm{d}y + \frac{\partial\phi}{\partial z}\,\mathrm{d}z = \mathrm{d}r \cdot \nabla\phi. \qquad (A.20)$$

Now, if the distance $|\mathrm{d}r|$ is fixed, then this scalar product takes on its maximum value when $\mathrm{d}r$ is in the direction of $\nabla\phi$. Hence we conclude that the direction of $\nabla\phi$ is the direction in which $\phi$ increases most rapidly.

Moreover, its magnitude is the rate of increase of $\phi$ with distance in this direction. (This is the reason for the name 'gradient'.) Clearly, therefore, we could *define* $\nabla\phi$ by these properties, which are independent of any choice of axes.

We are often interested in the value of the scalar field $\phi$ evaluated at the position of a moving particle, $\phi(r(t))$. From (A.20) it follows that the rate of change of $\phi$ is

$$\frac{\mathrm{d}\phi(r(t))}{\mathrm{d}t} = \dot{r} \cdot \nabla\phi. \tag{A.21}$$

The symbol $\nabla$ may be regarded as a vector which is also a differential operator (like $\mathrm{d}/\mathrm{d}x$), given by

$$\nabla = i\frac{\partial}{\partial x} + j\frac{\partial}{\partial y} + k\frac{\partial}{\partial z}.$$

We can also apply it to a vector field $A$. The *divergence* of $A$ is defined to be the scalar field

$$\mathrm{div}\,A = \nabla \cdot A = \frac{\partial A_x}{\partial x} + \frac{\partial A_y}{\partial y} + \frac{\partial A_z}{\partial z}, \tag{A.22}$$

and the *curl* of $A$ to be the vector field

$$\mathrm{curl}\,A = \nabla \wedge A = \begin{vmatrix} i & j & k \\ \dfrac{\partial}{\partial x} & \dfrac{\partial}{\partial y} & \dfrac{\partial}{\partial z} \\ A_x & A_y & A_z \end{vmatrix}. \tag{A.23}$$

This latter expression is an abbreviation for the expanded form

$$\nabla \wedge A = i\left(\frac{\partial A_z}{\partial y} - \frac{\partial A_y}{\partial z}\right) + j\left(\frac{\partial A_x}{\partial z} - \frac{\partial A_z}{\partial x}\right) + k\left(\frac{\partial A_y}{\partial x} - \frac{\partial A_x}{\partial y}\right).$$

(Instead of curl $A$, the alternative notation rot $A$ is sometimes used, particularly in non-English-speaking countries.)

To understand the physical significance of these operations, it is helpful to think of the velocity field in a fluid: $v(r)$ is the fluid velocity at the point $r$.

Let us consider a small volume of fluid, $\delta V = \delta x\,\delta y\,\delta z$, and try to find its rate of change as it moves with the fluid. Consider first the length $\delta x$. To a first approximation, over a short time interval $\mathrm{d}t$, the velocity components in the $y$ and $z$ directions are irrelevant; the length $\delta x$ changes because the $x$ components of velocity, $v_x$, at its two ends are slightly different, by an amount $(\partial v_x/\partial x)\,\delta x$. Thus in a time $dt$, the change in $\delta x$ is $d\,\delta x = (\partial v_x/\partial x)\,\delta x\,dt$, whence

$$\frac{d\,\delta x}{dt} = \frac{\partial v_x}{\partial x}\,\delta x.$$

Taking account of similar changes in $\delta y$ and $\delta z$, we have

$$\frac{d}{dt}(\delta x\,\delta y\,\delta z) = \left(\frac{\partial v_x}{\partial x} + \frac{\partial v_y}{\partial y} + \frac{\partial v_z}{\partial z}\right)\delta x\,\delta y\,\delta z,$$

or, equivalently,

$$\frac{d\,\delta V}{dt} = (\boldsymbol{\nabla}\cdot\boldsymbol{v})\,\delta V. \tag{A.24}$$

Thus $\boldsymbol{\nabla}\cdot\boldsymbol{v}$ represents the proportional rate of increase of volume: positive $\boldsymbol{\nabla}\cdot\boldsymbol{v}$ means expansion, negative $\boldsymbol{\nabla}\cdot\boldsymbol{v}$ compression. In particular, if the fluid is *incompressible*, then $\boldsymbol{\nabla}\cdot\boldsymbol{v} = 0$.

It is possible to show in a similar way that a non-zero $\boldsymbol{\nabla}\wedge\boldsymbol{v}$ means that locally the fluid is rotating. This vector, called the *vorticity*, represents the local angular velocity of rotation (times 2; see Problem 10).

The rule for differentiating products can also be applied to expressions involving $\boldsymbol{\nabla}$. For example, $\boldsymbol{\nabla}\cdot(\boldsymbol{A}\wedge\boldsymbol{B})$ is a sum of two terms, in one of which $\boldsymbol{\nabla}$ acts on $\boldsymbol{A}$ only and in the other on $\boldsymbol{B}$ only. The gradient of a product of scalar fields can be written

$$\boldsymbol{\nabla}(\phi\psi) = \psi\boldsymbol{\nabla}\phi + \phi\boldsymbol{\nabla}\psi,$$

and similarly

$$\boldsymbol{\nabla}\cdot(\phi\boldsymbol{A}) = \boldsymbol{A}\cdot\boldsymbol{\nabla}\phi + \phi\boldsymbol{\nabla}\cdot\boldsymbol{A}.$$

But, when vector products are involved, we have to remember that the order of the factors as a product of vectors cannot be changed without affecting the signs. Thus we have

$$\boldsymbol{\nabla}\cdot(\boldsymbol{A}\wedge\boldsymbol{B}) = \boldsymbol{B}\cdot(\boldsymbol{\nabla}\wedge\boldsymbol{A}) - \boldsymbol{A}\cdot(\boldsymbol{\nabla}\wedge\boldsymbol{B}),$$

and, similarly,

$$\boldsymbol{\nabla}\wedge(\phi\boldsymbol{A}) = \phi(\boldsymbol{\nabla}\wedge\boldsymbol{A}) - \boldsymbol{A}\wedge(\boldsymbol{\nabla}\phi).$$

We may apply the vector differential operator $\nabla$ twice. The divergence of the gradient of a scalar field $\phi$ is called the *Laplacian* of $\phi$,

$$\nabla^2 \phi = \nabla \cdot \nabla \phi = \frac{\partial^2 \phi}{\partial x^2} + \frac{\partial^2 \phi}{\partial y^2} + \frac{\partial^2 \phi}{\partial z^2}. \tag{A.25}$$

Some operations always give zero. Just as $\boldsymbol{a} \wedge \boldsymbol{a} = \boldsymbol{0}$, we find that the curl of a gradient vanishes,

$$\nabla \wedge \nabla \phi = \boldsymbol{0}. \tag{A.26}$$

For example, its $z$ component is

$$\frac{\partial}{\partial x}\left(\frac{\partial \phi}{\partial y}\right) - \frac{\partial}{\partial y}\left(\frac{\partial \phi}{\partial x}\right) = 0.$$

Similarly, one can show that the divergence of a curl vanishes:

$$\nabla \cdot (\nabla \wedge \boldsymbol{A}) = 0. \tag{A.27}$$

An important identity, analogous to the expansion of the vector triple product (A.16), gives the curl of a curl,

$$\nabla \wedge (\nabla \wedge \boldsymbol{A}) = \nabla(\nabla \cdot \boldsymbol{A}) - \nabla^2 \boldsymbol{A}, \tag{A.28}$$

where of course

$$\nabla^2 \boldsymbol{A} = \frac{\partial^2 \boldsymbol{A}}{\partial x^2} + \frac{\partial^2 \boldsymbol{A}}{\partial y^2} + \frac{\partial^2 \boldsymbol{A}}{\partial z^2}.$$

It may easily be proved by inserting the expressions in terms of components.

## A.6    Integral Theorems

There are three important theorems for vectors which are generalizations of the fundamental theorem of the calculus,

$$\int_{x_0}^{x_1} \frac{\mathrm{d}f}{\mathrm{d}x}\,\mathrm{d}x = f(x_1) - f(x_0).$$

First, consider a curve $C$ in space, running from $\boldsymbol{r}_0$ to $\boldsymbol{r}_1$ (see Fig. A.9). Let the directed element of length along $C$ be $\mathrm{d}\boldsymbol{r}$. If $\phi$ is a scalar field, then according to (A.20), the change in $\phi$ along this element of length is

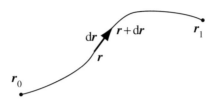

Fig. A.9

$d\phi = d\boldsymbol{r} \cdot \boldsymbol{\nabla}\phi$. Thus, integrating from $\boldsymbol{r}_0$ to $\boldsymbol{r}_1$, we obtain the first of the integral theorems,

$$\int_{r_0}^{r_1} d\boldsymbol{r} \cdot \boldsymbol{\nabla}\phi = \phi(\boldsymbol{r}_1) - \phi(\boldsymbol{r}_0). \tag{A.29}$$

The integral on the left is called the *line integral* of $\boldsymbol{\nabla}\phi$ along $C$. (Note that, as here, it is often more convenient to place the differential symbol $d\boldsymbol{r}$ to the *left* of the integrand.)

This theorem may be used to relate the potential energy function $V(\boldsymbol{r})$ for a conservative force to the work done in going from some fixed point $\boldsymbol{r}_0$, where $V$ is chosen to vanish, to $\boldsymbol{r}$. Thus, if $F = -\boldsymbol{\nabla}V$, then

$$V(\boldsymbol{r}) = -\int_{r_0}^{r} d\boldsymbol{r} \cdot \boldsymbol{F}. \tag{A.30}$$

When $\boldsymbol{F}$ is conservative, this integral depends only on its end-points, and not on the path $C$ chosen between them. Conversely, if this condition is satisfied, we can define $V$ by (A.30), and the force must be conservative. The condition that two line integrals of the form (A.30) should be equal whenever their end-points coincide may be restated by saying the the line integral round any *closed* path should vanish. Physically, this means that no work is done in taking a particle round a loop which returns to its starting point. The integral round a closed loop is usually denoted by the symbol $\oint_C$. Thus we require

$$\oint_C d\boldsymbol{r} \cdot \boldsymbol{F} = 0, \tag{A.31}$$

for all closed loops $C$.

This condition may be simplified by using the second of the integral theorems — *Stokes' theorem*. Consider a curved surface $S$, bounded by the closed curve $C$. If one side of $S$ is chosen to be the 'positive' side, then the positive direction round $C$ may be defined by the right-hand-screw convention (see Fig. A.10). Take a small element of the surface, of area $\mathrm{d}S$, and let

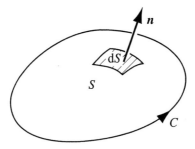

Fig. A.10

$n$ be a unit vector normal to the element, and directed towards its positive side. Then the *directed* element of area is defined to be $\mathrm{d}\boldsymbol{S} = \boldsymbol{n}\,\mathrm{d}S$. Stokes' theorem states that if $\boldsymbol{A}$ is any vector field, then

$$\iint_S \mathrm{d}\boldsymbol{S} \cdot (\boldsymbol{\nabla} \wedge \boldsymbol{A}) = \oint_C \mathrm{d}\boldsymbol{r} \cdot \boldsymbol{A}. \tag{A.32}$$

The application of this theorem to (A.31) is immediate. If the line integral round $C$ is required to vanish for all closed curves $C$, then the surface integral must vanish for all surfaces $S$. But this is only possible if the integrand vanishes identically. So the condition for a force to be conservative is

$$\boldsymbol{\nabla} \wedge \boldsymbol{F} = \boldsymbol{0}. \tag{A.33}$$

We shall not prove Stokes' theorem. However, it is easy to verify for a small rectangular surface. (The proof proceeds by splitting up the surface into small sub-regions.) Suppose $S$ is a rectangle in the $xy$-plane, of area $\mathrm{d}x\,\mathrm{d}y$. Then $\mathrm{d}\boldsymbol{S} = \boldsymbol{k}\,\mathrm{d}x\,\mathrm{d}y$, so the surface integral is

$$\boldsymbol{k} \cdot (\boldsymbol{\nabla} \wedge \boldsymbol{A})\,\mathrm{d}x\,\mathrm{d}y = \left(\frac{\partial A_y}{\partial x} - \frac{\partial A_x}{\partial y}\right)\mathrm{d}x\,\mathrm{d}y. \tag{A.34}$$

The line integral consists of four terms, one from each edge. The two terms arising from the edges parallel to the $x$-axis involve the $x$ component of $\boldsymbol{A}$ evaluated for different values of $y$. They therefore contribute

$$A_x(y)\,\mathrm{d}x - A_x(y + \mathrm{d}y)\,\mathrm{d}x = -\frac{\partial A_x}{\partial y}\,\mathrm{d}x\,\mathrm{d}y.$$

Similarly, the other pair of edges yields the first term of (A.34).

One can also find a necessary and sufficient condition for a field $\boldsymbol{B}(\boldsymbol{r})$ to have the form of a curl,

$$\boldsymbol{B} = \boldsymbol{\nabla} \wedge \boldsymbol{A}.$$

By (A.27), such a field must satisfy

$$\boldsymbol{\nabla} \cdot \boldsymbol{B} = 0. \tag{A.35}$$

The proof that this is also a sufficient condition (which we shall not give in detail) follows much the same lines as before. One can show that it is sufficient that the surface integral of $\boldsymbol{B}$ over any *closed* surface should vanish:

$$\iint_S \mathrm{d}\boldsymbol{S} \cdot \boldsymbol{B} = 0, \qquad (S \text{ closed})$$

and then use the third of the integral theorems, *Gauss' theorem.* This states that if $V$ is a volume in space bounded by the closed surface $S$, then for any vector field $\boldsymbol{B}$,

$$\iiint_V \mathrm{d}V \, \boldsymbol{\nabla} \cdot \boldsymbol{B} = \iint_S \mathrm{d}\boldsymbol{S} \cdot \boldsymbol{B}, \tag{A.36}$$

where $\mathrm{d}V$ denotes the volume element $\mathrm{d}V = \mathrm{d}x \, \mathrm{d}y \, \mathrm{d}z$, and the positive side of $S$ is taken to be the outside.

It is again easy to verify Gauss' theorem for a small rectangular volume, $\mathrm{d}V = \mathrm{d}x \, \mathrm{d}y \, \mathrm{d}z$. The volume integral is

$$\left( \frac{\partial B_x}{\partial x} + \frac{\partial B_y}{\partial y} + \frac{\partial B_z}{\partial z} \right) \mathrm{d}x \, \mathrm{d}y \, \mathrm{d}z. \tag{A.37}$$

The surface integral consists of six terms, one for each face. Consider the faces parallel to the $xy$-plane, with directed surface elements $\boldsymbol{k} \, \mathrm{d}x \, \mathrm{d}y$ and $-\boldsymbol{k} \, \mathrm{d}x \, \mathrm{d}y$. Their contributions involve $\boldsymbol{k} \cdot \boldsymbol{B} = B_z$, evaluated for different values of $z$. Thus they contribute

$$B_z(z + \mathrm{d}z) \, \mathrm{d}x \, \mathrm{d}y - B_z(z) \, \mathrm{d}x \, \mathrm{d}y = \frac{\partial B_z}{\partial z} \, \mathrm{d}x \, \mathrm{d}y \, \mathrm{d}z.$$

Similarly, the other terms of (A.37) come from the other pairs of faces.

## A.7 Electromagnetic Potentials

An important application of these theorems is to the electromagnetic field.

The basic equations of electromagnetic theory are *Maxwell's equations.* In the absence of dielectric or magnetic media, they may be expressed in terms of two fields, the electric field $\boldsymbol{E}$ and the magnetic field $\boldsymbol{B}$. There is one pair of homogeneous equations,

$$\boldsymbol{\nabla} \wedge \boldsymbol{E} + \frac{\partial \boldsymbol{B}}{\partial t} = \boldsymbol{0}, \qquad \boldsymbol{\nabla} \cdot \boldsymbol{B} = 0, \qquad (A.38)$$

and a second pair involving also the electric charge density $\rho$ and current density $\boldsymbol{j}$,

$$\mu_0^{-1} \boldsymbol{\nabla} \wedge \boldsymbol{B} - \epsilon_0 \frac{\partial \boldsymbol{E}}{\partial t} = \boldsymbol{j}, \qquad \epsilon_0 \boldsymbol{\nabla} \cdot \boldsymbol{E} = \rho, \qquad (A.39)$$

in which $\mu_0$ and $\epsilon_0$ are universal constants.

The second equation in (A.38) is just the condition (A.35). It follows that there must exist a *vector potential* $\boldsymbol{A}$ such that

$$\boldsymbol{B} = \boldsymbol{\nabla} \wedge \boldsymbol{A}. \qquad (A.40)$$

Then, substituting in the first of the equations (A.38), we find that $\boldsymbol{\nabla} \wedge (\boldsymbol{E} + \partial \boldsymbol{A}/\partial t) = \boldsymbol{0}$. It follows that there must exist a *scalar potential* $\phi$ such that

$$\boldsymbol{E} = -\boldsymbol{\nabla}\phi - \frac{\partial \boldsymbol{A}}{\partial t}. \qquad (A.41)$$

These potentials are not unique. If $\Lambda$ is any scalar field, then the potentials

$$\phi' = \phi + \frac{\partial \Lambda}{\partial t}, \qquad \boldsymbol{A}' = \boldsymbol{A} - \boldsymbol{\nabla}\Lambda \qquad (A.42)$$

define the same fields $\boldsymbol{E}$ and $\boldsymbol{B}$ as do $\phi$ and $\boldsymbol{A}$. This is called a *gauge transformation.* We may eliminate this arbitrariness by imposing an extra condition, for example the *radiation gauge* (or *Coulomb gauge*) condition

$$\boldsymbol{\nabla} \cdot \boldsymbol{A} = 0. \qquad (A.43)$$

In the static case, where all the fields are time-*in*dependent, Maxwell's equations separate into a pair of electrostatic equations, and a magnetostatic pair. Then $\phi$ becomes the ordinary electrostatic potential, satisfying

Poisson's equation (6.48). The vector potential, by (A.39) and (A.40) satisfies

$$\boldsymbol{\nabla} \wedge (\boldsymbol{\nabla} \wedge \boldsymbol{A}) = \mu_0 \boldsymbol{j}.$$

Using (A.28), and imposing the radiation gauge condition (A.43), we find

$$\nabla^2 \boldsymbol{A} = -\mu_0 \boldsymbol{j}. \tag{A.44}$$

This is the analogue of Poisson's equation. The solution is of the same form as (6.15), namely

$$\boldsymbol{A}(\boldsymbol{r}) = \frac{\mu_0}{4\pi} \iiint \frac{\boldsymbol{j}(\boldsymbol{r}')}{|\boldsymbol{r} - \boldsymbol{r}'|} \, \mathrm{d}^3 \boldsymbol{r}'. \tag{A.45}$$

Thus, given any static distribution of charges and currents, we may calculate the potentials $\phi$ and $\boldsymbol{A}$, and hence the fields $\boldsymbol{E}$ and $\boldsymbol{B}$.

## A.8   Curvilinear Co-ordinates

Another use of the integral theorems is to provide expressions for the gradient, divergence and curl in terms of curvilinear co-ordinates.

Consider a set of orthogonal curvilinear co-ordinates (see §3.5) $q_1, q_2, q_3$. Let us denote the elements of length along the three co-ordinate curves by $h_1 \, \mathrm{d}q_1, h_2 \, \mathrm{d}q_2, h_3 \, \mathrm{d}q_3$. For example, in cylindrical polars

$$h_\rho = 1, \qquad h_\varphi = \rho, \qquad h_z = 1, \tag{A.46}$$

while in spherical polars

$$h_r = 1, \qquad h_\theta = r, \qquad h_\varphi = r \sin \theta. \tag{A.47}$$

Now consider a scalar field $\psi$, and two neighbouring points $(q_1, q_2, q_3)$ and $(q_1 + \mathrm{d}q_1, q_2, q_3)$. Then the difference between the values of $\psi$ at these points is

$$\frac{\partial \psi}{\partial q_1} \, \mathrm{d}q_1 = \mathrm{d}\psi = \mathrm{d}\boldsymbol{r} \cdot \boldsymbol{\nabla}\psi = h_1 \, \mathrm{d}q_1 \, (\boldsymbol{\nabla}\psi)_1,$$

where $(\boldsymbol{\nabla}\psi)_1$ is the component of $\boldsymbol{\nabla}\psi$ in the direction of increasing $q_1$. Hence we find

$$(\boldsymbol{\nabla}\psi)_1 = \frac{1}{h_1} \frac{\partial \psi}{\partial q_1}, \tag{A.48}$$

with similar expressions for the other components. Thus in cylindrical and spherical polars, we have

$$\nabla\psi = \left(\frac{\partial\psi}{\partial\rho}, \frac{1}{\rho}\frac{\partial\psi}{\partial\varphi}, \frac{\partial\psi}{\partial z}\right), \tag{A.49}$$

and

$$\nabla\psi = \left(\frac{\partial\psi}{\partial r}, \frac{1}{r}\frac{\partial\psi}{\partial\theta}, \frac{1}{r\sin\theta}\frac{\partial\psi}{\partial\varphi}\right). \tag{A.50}$$

To find an expression for the divergence, we use Gauss' theorem, applied to a small volume bounded by the co-ordinate surfaces. The volume integral is

$$(\nabla\cdot A)\, h_1\, dq_1\, h_2\, dq_2\, h_3\, dq_3.$$

In the surface integral, the terms arising from the faces which are surfaces of constant $q_3$ are of the form $A_3 h_1\, dq_1\, h_2\, dq_2$, evaluated for two different values of $q_3$. They therefore contribute

$$\frac{\partial}{\partial q_3}(h_1 h_2 A_3)\, dq_1\, dq_2\, dq_3.$$

Adding the terms from all three pairs of faces, and comparing with the volume integral, we obtain

$$\nabla\cdot A = \frac{1}{h_1 h_2 h_3}\left(\frac{\partial(h_2 h_3 A_1)}{\partial q_1} + \frac{\partial(h_3 h_1 A_2)}{\partial q_2} + \frac{\partial(h_1 h_2 A_3)}{\partial q_3}\right). \tag{A.51}$$

In particular, in cylindrical and spherical polars,

$$\nabla\cdot A = \frac{1}{\rho}\frac{\partial(\rho A_\rho)}{\partial\rho} + \frac{1}{\rho}\frac{\partial A_\varphi}{\partial\varphi} + \frac{\partial A_z}{\partial z}, \tag{A.52}$$

and

$$\nabla\cdot A = \frac{1}{r^2}\frac{\partial(r^2 A_r)}{\partial r} + \frac{1}{r\sin\theta}\frac{\partial(\sin\theta\, A_\theta)}{\partial\theta} + \frac{1}{r\sin\theta}\frac{\partial A_\varphi}{\partial\varphi}. \tag{A.53}$$

To find the curl, we use Stokes' theorem in a similar way. Let us consider a small element of a surface $q_3 = $ constant, bounded by curves of constant $q_1$ and of $q_2$. Then the surface integral is

$$(\nabla\wedge A)_3\, h_1\, dq_1\, h_2\, dq_2.$$

In the line integral around the boundary, the two edges of constant $q_2$ involve $A_1 h_1 \, dq_1$ evaluated for different values of $q_2$, and so contribute

$$-\frac{\partial}{\partial q_2}(h_1 A_1) \, dq_1 \, dq_2.$$

Hence, adding the contribution from the other pair of edges, we obtain

$$(\boldsymbol{\nabla} \wedge \boldsymbol{A})_3 = \frac{1}{h_1 h_2}\left(\frac{\partial(h_2 A_2)}{\partial q_1} - \frac{\partial(h_1 A_1)}{\partial q_2}\right), \qquad (A.54)$$

with similar expressions for the other components. Thus, in particular, in cylindrical and spherical polars

$$\boldsymbol{\nabla} \wedge \boldsymbol{A} = \left(\frac{1}{\rho}\frac{\partial A_z}{\partial \varphi} - \frac{\partial A_\varphi}{\partial z}, \frac{\partial A_\rho}{\partial z} - \frac{\partial A_z}{\partial \rho}, \frac{1}{\rho}\left[\frac{\partial(\rho A_\varphi)}{\partial \rho} - \frac{\partial A_\rho}{\partial \varphi}\right]\right), \qquad (A.55)$$

and

$$\boldsymbol{\nabla} \wedge \boldsymbol{A} = \left(\frac{1}{r \sin\theta}\left[\frac{\partial(\sin\theta\, A_\varphi)}{\partial \theta} - \frac{\partial A_\theta}{\partial \varphi}\right],\right.$$
$$\left.\frac{1}{r \sin\theta}\frac{\partial A_r}{\partial \varphi} - \frac{1}{r}\frac{\partial(r A_\varphi)}{\partial r}, \frac{1}{r}\left[\frac{\partial(r A_\theta)}{\partial r} - \frac{\partial A_r}{\partial \theta}\right]\right). \qquad (A.56)$$

Finally, combining the expressions for the divergence and gradient, we can find the Laplacian of a scalar field. It is

$$\nabla^2 \psi = \frac{1}{h_1 h_2 h_3}\left[\frac{\partial}{\partial q_1}\left(\frac{h_2 h_3}{h_1}\frac{\partial \psi}{\partial q_1}\right) + \frac{\partial}{\partial q_2}\left(\frac{h_3 h_1}{h_2}\frac{\partial \psi}{\partial q_2}\right)\right.$$
$$\left. + \frac{\partial}{\partial q_3}\left(\frac{h_1 h_2}{h_3}\frac{\partial \psi}{\partial q_3}\right)\right]. \qquad (A.57)$$

In cylindrical polars,

$$\nabla^2 \psi = \frac{1}{\rho}\frac{\partial}{\partial \rho}\left(\rho \frac{\partial \psi}{\partial \rho}\right) + \frac{1}{\rho^2}\frac{\partial^2 \psi}{\partial \varphi^2} + \frac{\partial^2 \psi}{\partial z^2}, \qquad (A.58)$$

and, in spherical polars,

$$\nabla^2 \psi = \frac{1}{r^2}\frac{\partial}{\partial r}\left(r^2 \frac{\partial \psi}{\partial r}\right) + \frac{1}{r^2 \sin\theta}\frac{\partial}{\partial \theta}\left(\sin\theta \frac{\partial \psi}{\partial \theta}\right) + \frac{1}{r^2 \sin^2\theta}\frac{\partial^2 \psi}{\partial \varphi^2}. \qquad (A.59)$$

## A.9 Tensors

Scalars and vectors are the first two members of a family of objects known collectively as *tensors*, and described by $1, 3, 9, 27, \ldots$ components. Scalars and vectors are called tensors of *valence* 0 and *valence* 1, respectively. (Sometimes the word *rank* is used instead of 'valence', but there is then a possibility of confusion with a different usage of the same word in matrix theory.)

In this section, we shall be concerned with the next member of the family, the tensors of valence 2, often called *dyadics*. We shall use the word 'tensor' in this restricted sense, to mean a tensor of valence 2.

Tensors occur most frequently when one vector $b$ is given as a linear function of another vector $a$, according to the matrix equation

$$\begin{bmatrix} b_x \\ b_y \\ b_z \end{bmatrix} = \begin{bmatrix} T_{xx} & T_{xy} & T_{xz} \\ T_{yx} & T_{yy} & T_{yz} \\ T_{zx} & T_{zy} & T_{zz} \end{bmatrix} \begin{bmatrix} a_x \\ a_y \\ a_z \end{bmatrix}. \tag{A.60}$$

An example is the relation (9.17) between the angular momentum $J$ and angular velocity $\omega$ of a rigid body.

The nine elements of the $3 \times 3$ matrix in (A.60) are the components of a tensor, which we shall denote by the sans-serif capital $\mathsf{T}$. By an obvious extension of the dot product notation for the scalar product of two vectors, we may write (A.60) as

$$b = \mathsf{T} \cdot a. \tag{A.61}$$

For example, (9.17) may be written $J = \mathsf{I} \cdot \omega$, where $\mathsf{I}$ is the *inertia tensor*.

We can go on to form the scalar product of (A.61) with another vector, $c$, obtaining a scalar, $c \cdot \mathsf{T} \cdot a$. Note that in general this is not the same as $a \cdot \mathsf{T} \cdot c$. In fact,

$$a \cdot \mathsf{T} \cdot c = c \cdot \tilde{\mathsf{T}} \cdot a, \tag{A.62}$$

where $\tilde{\mathsf{T}}$ is the *transposed* tensor of $\mathsf{T}$, obtained by reflecting in the leading diagonal, *e.g.*, $\tilde{T}_{xy} = T_{yx}$.

The tensor $\mathsf{T}$ is called *symmetric* if $\tilde{\mathsf{T}} = \mathsf{T}$, *i.e.*, if $T_{ji} = T_{ij}$ for all $i, j$. It is *antisymmetric* if $\tilde{\mathsf{T}} = -\mathsf{T}$, or $T_{ji} = -T_{ij}$ for all $i, j$.

An interesting example of an antisymmetric tensor is provided by the relation (5.2) giving the velocity $v$ as a function of position $r$ in a body rotating with angular velocity $\omega$. It is a linear relation and so may be written in the form (A.60), specifically as

$$
\begin{bmatrix} v_x \\ v_y \\ v_z \end{bmatrix} = \begin{bmatrix} 0 & -\omega_z & \omega_y \\ \omega_z & 0 & -\omega_x \\ -\omega_y & \omega_x & 0 \end{bmatrix} \begin{bmatrix} x \\ y \\ z \end{bmatrix}.
$$

There is an important special tensor,

$$
\mathbf{1} = \begin{bmatrix} 1 & 0 & 0 \\ 0 & 1 & 0 \\ 0 & 0 & 1 \end{bmatrix},
$$

called the *unit tensor* or *identity tensor*, with the property that $\mathbf{1} \cdot \boldsymbol{a} = \boldsymbol{a}$ for all vectors $\boldsymbol{a}$.

From any two vectors $\boldsymbol{a}$ and $\boldsymbol{b}$, we can form a tensor $\mathbf{T}$ by multiplying their elements together (without adding), *i.e.*, $T_{ij} = a_i b_j$. This is the *tensor product* (or *dyadic product* or *outer product*) of $\boldsymbol{a}$ and $\boldsymbol{b}$, written $\mathbf{T} = \boldsymbol{ab}$, with no dot or cross. Note that $\mathbf{T} \cdot \boldsymbol{c} = (\boldsymbol{ab}) \cdot \boldsymbol{c} = \boldsymbol{a}(\boldsymbol{b} \cdot \boldsymbol{c})$, so the brackets are in fact unnecessary. In matrix notation, $\boldsymbol{ab}$ is the product of the column vector $\boldsymbol{a}$ and the row vector $\boldsymbol{b}$, while the scalar product (or *inner product*) $\boldsymbol{a} \cdot \boldsymbol{b}$ is the row $\boldsymbol{a}$ times the column $\boldsymbol{b}$.

We can deduce the correct transformation law of a tensor under a rotation of axes: its components transform just like the products of components of two vectors. If we symbolize (A.2) formally as $\boldsymbol{a}' = \mathbf{R} \cdot \boldsymbol{a}$, then the correct transformation law of a tensor is $\mathbf{T}' = \mathbf{R} \cdot \mathbf{T} \cdot \tilde{\mathbf{R}}$. (This denotes a product of three $3 \times 3$ matrices.)

The use of the tensor product allows us to write some old results in a new way. For example, for any vector $\boldsymbol{a}$,

$$
\mathbf{1} \cdot \boldsymbol{a} = \boldsymbol{a} = \boldsymbol{i}(\boldsymbol{i} \cdot \boldsymbol{a}) + \boldsymbol{j}(\boldsymbol{j} \cdot \boldsymbol{a}) + \boldsymbol{k}(\boldsymbol{k} \cdot \boldsymbol{a}) = (\boldsymbol{ii} + \boldsymbol{jj} + \boldsymbol{kk}) \cdot \boldsymbol{a},
$$

whence

$$
\boldsymbol{ii} + \boldsymbol{jj} + \boldsymbol{kk} = \mathbf{1}, \tag{A.63}
$$

as may easily be verified by writing out the components.

Similarly, we may write the relation (9.16) between angular momentum and angular velocity in the form

$$
\boldsymbol{J} = \sum m(r^2 \boldsymbol{\omega} - \boldsymbol{rr} \cdot \boldsymbol{\omega}) = \mathbf{I} \cdot \boldsymbol{\omega},
$$

where the inertia tensor $\mathbf{I}$ is given explicitly by

$$
\mathbf{I} = \sum m(r^2 \mathbf{1} - \boldsymbol{rr}).
$$

Note the difference between the unit tensor **1** and the inertia tensor **I**. It is easy to check that the nine components of this equation reproduce the relations (9.15).

Note that if $\mathbf{T} = \boldsymbol{ab}$, then $\tilde{\mathbf{T}} = \boldsymbol{ba}$, whence in particular the inertia tensor **I** is symmetric.

## A.10   Eigenvalues; Diagonalization of a Symmetric Tensor

In this section, we discuss a theorem that has very wide applicability.

Let **T** be a symmetric tensor. A vector $\boldsymbol{a}$ is called an *eigenvector* of **T**, with *eigenvalue* $\lambda$, if

$$\mathbf{T} \cdot \boldsymbol{a} = \lambda \boldsymbol{a}, \tag{A.64}$$

or, equivalently $(\mathbf{T} - \lambda \mathbf{1}) \cdot \boldsymbol{a} = \mathbf{0}$. (Compare (11.17), which is also an eigenvalue equation.) The condition for the existence of a non-trivial solution is that the determinant of the coefficients vanishes,

$$\det(\mathbf{T} - \lambda \mathbf{1}) = \begin{vmatrix} T_{xx} - \lambda & T_{xy} & T_{xz} \\ T_{yx} & T_{yy} - \lambda & T_{yz} \\ T_{zx} & T_{zy} & T_{zz} - \lambda \end{vmatrix} = 0.$$

This is a cubic equation for $\lambda$. Its three roots are either all real, or else one real and one complex conjugate pair. However, for a symmetric tensor **T** with real elements the latter possibility can be ruled out.

To see this, suppose that $\lambda$ is a complex eigenvalue, and let $\boldsymbol{a}$ be the corresponding eigenvector, whose components may also be complex. Now, taking the complex conjugate of $\mathbf{T} \cdot \boldsymbol{a} = \lambda \boldsymbol{a}$, we obtain $\mathbf{T} \cdot \boldsymbol{a}^* = \lambda^* \boldsymbol{a}^*$, where $\lambda^*$ denotes the complex conjugate of $\lambda$, and $\boldsymbol{a}^* = (a_x^*, a_y^*, a_z^*)$. Multiplying these two equations by $\boldsymbol{a}^*$ and $\boldsymbol{a}$ respectively, we obtain

$$\boldsymbol{a}^* \cdot \mathbf{T} \cdot \boldsymbol{a} = \lambda \boldsymbol{a}^* \cdot \boldsymbol{a}, \qquad \text{and} \qquad \boldsymbol{a} \cdot \mathbf{T} \cdot \boldsymbol{a}^* = \lambda^* \boldsymbol{a} \cdot \boldsymbol{a}^*.$$

But since **T** is symmetric, the left-hand sides of these equations are equal, by (A.62). Hence the right-hand sides must be equal too. Since $\boldsymbol{a}^* \cdot \boldsymbol{a} = |a_x|^2 + |a_y|^2 + |a_z|^2 = \boldsymbol{a} \cdot \boldsymbol{a}^*$, this means that $\lambda^* = \lambda$, *i.e.*, $\lambda$ must be real.

Thus we have shown that there are three real eigenvalues, say $\lambda_1, \lambda_2, \lambda_3$, and three corresponding real eigenvectors, $\boldsymbol{a}_1, \boldsymbol{a}_2, \boldsymbol{a}_3$. (We consider the case where two eigenvalues are equal below.) Next, we show that the eigenvectors are orthogonal. For, if

$$\mathbf{T} \cdot \boldsymbol{a}_1 = \lambda_1 \boldsymbol{a}_1, \qquad \mathbf{T} \cdot \boldsymbol{a}_2 = \lambda_2 \boldsymbol{a}_2,$$

then, multiplying the first equation by $a_2$ and the second by $a_1$, and again using the symmetry of $\mathbf{T}$, we obtain

$$\lambda_1 a_2 \cdot a_1 = \lambda_2 a_1 \cdot a_2.$$

Thus if $\lambda_1 \neq \lambda_2$, then $a_1 \cdot a_2 = 0$.

If all three eigenvalues are distinct, then the three eigenvectors are orthogonal. Moreover, it is clear that if $a$ is an eigenvector, then so is any multiple of $a$, so that we may choose to normalize it, defining $e_1 = a_1/a_1$. Then the three normalized eigenvectors form an orthonormal triad, $e_1, e_2, e_3$. If we choose these as axes, then $\mathbf{T}$ must take the diagonal form

$$\mathbf{T} = \begin{bmatrix} \lambda_1 & 0 & 0 \\ 0 & \lambda_2 & 0 \\ 0 & 0 & \lambda_3 \end{bmatrix}. \tag{A.65}$$

For, since $e_1 = (1, 0, 0)$, $\mathbf{T} \cdot e_1$ is simply the first column of $\mathbf{T}$, and this must be $\lambda_1 e_1 = (\lambda_1, 0, 0)$. Similarly for the other columns.

This relationship between $\mathbf{T}$ and the eigenvectors may also be expressed, using the tensor-product notation, in a co-ordinate-independent form, namely

$$\mathbf{T} = \lambda_1 e_1 e_1 + \lambda_2 e_2 e_2 + \lambda_3 e_3 e_3. \tag{A.66}$$

Finally, we have to show that these results still hold if two or three eigenvalues coincide. The simplest way to do this is to add a small quantity $\epsilon$ to one of the diagonal components of $\mathbf{T}$, to make the eigenvalues slightly different. So long as $\epsilon \neq 0$, the tensor must have three orthonormal eigenvectors. By continuity, this must still be true in the limit $\epsilon \to 0$. (The symmetry of $\mathbf{T}$ is important here, because without the consequent orthogonality of eigenvectors we could not exclude the possibility that two eigenvectors that are distinct for $\epsilon \neq 0$ have the same limit as $\epsilon \to 0$. Indeed, this *does* happen for non-symmetric tensors, as will be seen in a different context in Appendix C.)

We have shown, therefore, that any symmetric tensor may be diagonalized by a suitable choice of axes. This was the result we used for the inertia tensor in Chapter 9. In that case, the eigenvectors are the principal axes, and the eigenvalues the principal moments of inertia. The procedure for finding normal co-ordinates for an oscillating system, discussed in §11.2, is essentially the same. In that case, it is the potential energy function that is brought to 'diagonal' form. Eigenvalue equations also appear in the analysis of dynamical systems in Chapter 13 and in many other branches of physics, in particular playing a big role in quantum mechanics.

---

## Problems

1. Given $a = (3, -1, 2), b = (0, 1, 1)$ and $c = (2, 2, -1)$, find:
   (a) $a \cdot b, a \cdot c$ and $a \cdot (b + c)$;
   (b) $a \wedge b, a \wedge c$ and $a \wedge (b + c)$;
   (c) $(a \wedge b) \cdot c$ and $(a \wedge c) \cdot b$;
   (d) $(a \wedge b) \wedge c$ and $(a \wedge c) \wedge b$;
   (e) $(a \cdot c)b - (b \cdot c)a$ and $(a \cdot b)c - (b \cdot c)a$.

2. Find the angles between the vectors $a \wedge b$ and $c$, and between $a \wedge c$ and $b$, where $a, b, c$ are as in Problem 1.

3. Show that $c = (ab + ba)/(a + b)$ bisects the angle between $a$ and $b$, where $a$ and $b$ are any two vectors.

4. Find $\nabla \phi$ if $\phi = x^3 - xyz$. Verify that $\nabla \wedge \nabla \phi = 0$, and evaluate $\nabla^2 \phi$.

5. (a) Find the gradients of $u = x + y^2/x$ and $v = y + x^2/y$, and show that they are always orthogonal.
   (b) Describe the contour curves of $u$ and $v$ in the $xy$-plane. What does (a) tell you about these curves?

6. Draw appropriate figures to give geometric proofs for the following laws of vector algebra:

$$(a + b) + c = a + (b + c);$$
$$\lambda(a + b) = \lambda a + \lambda b;$$
$$a \cdot (b + c) = a \cdot b + a \cdot c.$$

(Note that $a, b, c$ need not be coplanar.)

7. Show that $(a \wedge b) \cdot (c \wedge d) = a \cdot c \, b \cdot d - a \cdot d \, b \cdot c$. Hence show that $(a \wedge b)^2 = a^2 b^2 - (a \cdot b)^2$.

8. Express $\nabla \wedge (a \wedge b)$ in terms of scalar products.

9. If the vector field $v(r)$ is defined by $v = \omega k \wedge r$, verify that $\nabla \cdot v = 0$, and evaluate the vorticity $\nabla \wedge v$.

10. *Show that, if $u$ and $v$ are scalar fields, the maxima and minima of $u$ on the surface $v = 0$ are points where $\nabla u = \lambda \nabla v$ for some value of $\lambda$. Interpret this equation geometrically. (*Hint*: On $v = 0$ only two coordinates can vary independently. Thus $\delta z$ for example can usually be expressed in terms of $\delta x$ and $\delta y$. We require that $\delta u$ should vanish for

all infinitesimal variations satisfying this constraint.) Show that this problem is equivalent to finding the *unrestricted* maxima and minima of the function $w(\mathbf{r}, \lambda) = u - \lambda v$ as a function of the *four* independent variables $x, y, z$ and $\lambda$. Here $\lambda$ is called a *Lagrange multiplier*. What is the role of the equation $\partial w / \partial \lambda = 0$?

11. *Evaluate the components of $\nabla^2 \mathbf{A}$ in cylindrical polar co-ordinates by using the identity (A.28). Show that they are *not* the same as the scalar Laplacians of the components of $\mathbf{A}$.

12. *Find the radiation-gauge vector potential at large distances from a circular loop of radius $a$ carrying an electric current $I$. [*Hint*: Consider first a point $(x, 0, z)$, and expand the integrand in powers of $a/r$, keeping only the linear term. Then express your answer in spherical polars.] Hence find the magnetic field — the field of a *magnetic dipole*. Express the results in terms of the *magnetic moment* $\boldsymbol{\mu}$, a vector normal to the loop, of magnitude $\mu = \pi a^2 I$.

13. *Calculate the vector potential due to a short segment of wire of directed length $d\mathbf{s}$, carrying a current $I$, placed at the origin. Evaluate the corresponding magnetic field. Find the force on another segment, of length $d\mathbf{s}'$, carrying current $I'$, at $\mathbf{r}$. (To compute the force, treat the current element as a collection of moving charges.) Show that this force does not satisfy Newton's third law. (To preserve the law of conservation of momentum, one must assume that, while this force is acting, some momentum is transferred to the electromagnetic field.)

14. *Given $u = \cos\theta$ and $v = \ln r$, evaluate $\mathbf{A} = u\nabla v - v\nabla u$. Find the divergence and curl of $\mathbf{A}$, and verify that $\nabla \cdot \mathbf{A} = u\nabla^2 v - v\nabla^2 u$ and that $\nabla \wedge \mathbf{A} = 2\nabla u \wedge \nabla v$.

15. *Show that the rotation which takes the axes $\mathbf{i}, \mathbf{j}, \mathbf{k}$ into $\mathbf{i}', \mathbf{j}', \mathbf{k}'$ may be specified by $\mathbf{r} \to \mathbf{r}' = \mathbf{R} \cdot \mathbf{r}$, where the tensor $\mathbf{R}$ is $\mathbf{R} = \mathbf{i}'\mathbf{i} + \mathbf{j}'\mathbf{j} + \mathbf{k}'\mathbf{k}$. Write down the matrix of components of $\mathbf{R}$ if the rotation is through an angle $\theta$ about the $y$-axis. What is the tensor corresponding to the rotation which takes $\mathbf{i}', \mathbf{j}', \mathbf{k}'$ back into $\mathbf{i}, \mathbf{j}, \mathbf{k}$? Show that $\tilde{\mathbf{R}} \cdot \mathbf{R} = \mathbf{1}$. (Such tensors are said to be *orthogonal*.)

16. *The *trace* of a tensor $\mathbf{T}$ is the sum of its diagonal elements, $\mathrm{tr}(\mathbf{T}) = \sum_i T_{ii}$. Show that the trace is equal to the sum of the eigenvalues, and that the determinant $\det(\mathbf{T})$ is equal to the product of the eigenvalues.

17. *The *double dot* product of two tensors is defined as $\mathbf{S}:\mathbf{T} = \mathrm{tr}(\mathbf{S} \cdot \mathbf{T}) = \sum_i \sum_j S_{ij} T_{ji}$. Evaluate $\mathbf{1} : \mathbf{1}$ and $\mathbf{1} : \mathbf{rr}$. Show that

$$(3\mathbf{r}'\mathbf{r}' - r'^2\mathbf{1}) : (\mathbf{rr} - \tfrac{1}{3}r^2\mathbf{1}) = 3(\mathbf{r}' \cdot \mathbf{r})^2 - r'^2 r^2.$$

Hence show that the expansion (6.19) of the potential may be written

$$\phi(\mathbf{r}) = \frac{1}{4\pi\epsilon_0} \left( q\frac{1}{r} + \mathbf{d}\cdot\frac{\mathbf{r}}{r^3} + \tfrac{1}{2}\mathbf{Q}:\frac{\mathbf{rr} - \tfrac{1}{3}r^2\mathbf{1}}{r^5} + \cdots \right),$$

and write down an expression for the *quadrupole tensor* $\mathbf{Q}$. Show that $\mathrm{tr}(\mathbf{Q}) = 0$, and that in the axially symmetric case it has diagonal elements $-\tfrac{1}{2}Q, -\tfrac{1}{2}Q, Q$, where $Q$ is the quadrupole moment defined in Chapter 6. Show also that the gravitational quadrupole tensor is related to the inertia tensor $\mathbf{I}$ by $\mathbf{Q} = \mathrm{tr}(\mathbf{I})\mathbf{1} - 3\mathbf{I}$.

18. *In an elastic solid in equilibrium, the force across a small area may have both a normal component (of compression or tension) and transverse components (shearing stress). Denote the $i$th component of force per unit area across an area with normal in the $j$th direction by $T_{ij}$. These are the components of the *stress tensor* $\mathbf{T}$. By considering the equilibrium of a small volume, show that the force across area $A$ with normal in the direction of the unit vector $\mathbf{n}$ is $\mathbf{F} = \mathbf{T}\cdot\mathbf{n}A$. Show also by considering the equilibrium of a small rectangular volume that $\mathbf{T}$ is symmetric. What physical significance attaches to its eigenvectors?

# Appendix B

# Conics

Conic sections, or *conics* for short, are most simply defined as curves in a plane whose equation in Cartesian co-ordinates is quadratic in $x$ and $y$. The name derives from the fact that they can be obtained by making a plane section through a circular cone. They turn up in several physical applications, particularly in the theory of orbits under an inverse square law force. It may be useful to gather together the relevant mathematical information.

## B.1 Cartesian Form

The most general conic would have an equation of the form

$$Ax^2 + 2Bxy + Cy^2 + 2Dx + 2Ey + F = 0,$$

where $A, B, \ldots, F$ are real constants, but by choosing the axes appropriately we can reduce this to a simpler form.

First, we look at the quadratic part, $Ax^2 + 2Bxy + Cy^2$. It is always possible by rotating the axes to eliminate the constant $B$. This is another example of the diagonalization process described in §11.3 and §A.10. The quadratic part of the equation is then reduced to a sum of squares, $A'x'^2 + C'y'^2$. We then forget about the original co-ordinates, and drop the primes. The nature of the curve is largely determined by the ratio $A/C$ of the new constants.

Let us assume for the moment that $A$ and $C$ are both non-zero (we will come back later to the special case where that isn't true). Then we can choose to shift the origin (adding constants to $x$ and $y$, e.g., $x' = x + D/A$) so as to remove $D$ and $E$. If $F$ is also non-zero, we can move it to the other side of the equation, and divide by $-F$, to get the standard form of the equation,

$$Ax^2 + Cy^2 = 1. \tag{B.1}$$

$F = 0$ is a degenerate case: if $A$ and $C$ have the same sign, the only solution is $x = y = 0$; if they are of opposite sign, the equation factorizes, and so represents a pair of straight lines, $y = \pm\sqrt{-A/C}\,x$.

We cannot allow both $A$ and $C$ in (B.1) to be negative; the equation would then have no solutions at all. So we can distinguish two cases:

1. Both $A$ and $C$ are positive. Without loss of generality we can assume that $A \leq C$. (If $A > C$, we simply interchange the $x$ and $y$ axes.) Defining new positive constants $a$ and $b$ by $A = 1/a^2$ and $C = 1/b^2$, we finally arrive at the canonical form of the equation,

$$\frac{x^2}{a^2} + \frac{y^2}{b^2} = 1. \tag{B.2}$$

This is the equation of an *ellipse* (see Fig. B.1). Here $a \geq b$; $a$ is the *semi-major axis* and $b$ is the *semi-minor axis*. (In the special case

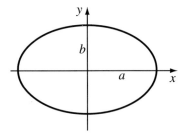

Fig. B.1

$a = b$, we have a *circle* of radius $a$.)

2. $A$ and $C$ have opposite signs. Again, without loss of generality, we can assume that $A > 0$ and $C < 0$. So, defining $A = 1/a^2$ and $C = -1/b^2$, we get

$$\frac{x^2}{a^2} - \frac{y^2}{b^2} = 1, \tag{B.3}$$

the equation of a *hyperbola* (see Fig. B.2); $a$ and $b$ are still called the *semi-major axis* and the *semi-minor axis* respectively, although it is no longer necessarily true that $a$ is the larger. Note that this curve has two

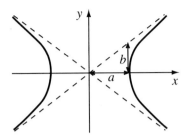

Fig. B.2

separate branches on opposite sides of the origin. At large distances, it asymptotically approaches the two straight lines $y = \pm(b/a)x$ shown on the figure.

We still have to consider the special case where one of the constants $A$ and $C$ vanishes. (They cannot both vanish, otherwise we have simply a linear equation, representing a straight line.) Without loss of generality, we may assume that $A = 0$ and $C \neq 0$. As before, we can shift the origin in the $y$ direction to eliminate $E$. On the other hand, $D$ cannot be zero (otherwise $x$ doesn't appear at all in the equation). This time, we can choose the origin

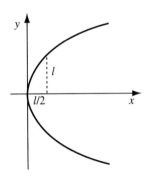

Fig. B.3

in the $x$ direction to make $F = 0$ (by setting $x' = x + F/2D$). Finally, defining $l = -D/C$, we arrive at the canonical form

$$y^2 = 2lx, \tag{B.4}$$

which is the equation of a *parabola* (see Fig. B.3).

Areas are easy to compute using the Cartesian form of the equation. For example, we can solve (B.2) for $y$ and integrate to find the area of the ellipse; the result, which generalizes the familiar $\pi r^2$ for a circle, is $\pi ab$. (In fact, the ellipse may be thought of as a circle of radius $a$ which has been squashed uniformly in the $y$ direction in the ratio $b/a$.)

## B.2   Polar Form

When we are looking for orbits under the action of a central force, it is usually convenient to use polar co-ordinates. The form of the equation that emerged from the discussion in §4.4 was

$$r(e\cos\theta \pm 1) = l, \tag{B.5}$$

where $e$ and $l$ are constants satisfying $e \geq 0$, $l > 0$ (the upper and lower signs refer to the attractive and repulsive cases, respectively).

It is interesting to see how this form is related to the Cartesian form above. If we rearrange (B.5) and square it, we obtain for both signs the equation

$$x^2 + y^2 = (l - ex)^2. \tag{B.6}$$

This can easily be put into one of the canonical forms above; which one depends on the value of $e$.

I. If $e < 1$, we can 'complete the square' in (B.6) and write it as

$$(1 - e^2)x^2 + 2elx + \frac{e^2 l^2}{1 - e^2} + y^2 = \frac{l^2}{1 - e^2}. \tag{B.7}$$

Dividing by $l^2/(1 - e^2)$, this reduces almost to the form (B.2), where

$$a = \frac{l}{1 - e^2}, \qquad b = \frac{l}{\sqrt{1 - e^2}}. \tag{B.8}$$

The only difference is that the origin is not at the centre of the ellipse: (B.7) is equivalent to

$$\frac{(x + ae)^2}{a^2} + \frac{y^2}{b^2} = 1, \tag{B.9}$$

an ellipse with centre at $(-ae, 0)$.

II. If $e > 1$, we complete the square in the same way and divide by $l^2/(e^2 - 1)$, obtaining

$$\frac{(x - ae)^2}{a^2} - \frac{y^2}{b^2} = 1, \tag{B.10}$$

where now

$$a = \frac{l}{e^2 - 1}, \qquad b = \frac{l}{\sqrt{e^2 - 1}}. \tag{B.11}$$

This is a hyperbola with centre at $(ae, 0)$. The left-hand branch, intersecting the $x$-axis at $(ae - a, 0)$, corresponds to an orbit under an attractive inverse square law force, while the right-hand one, meeting it at $(ae + a, 0)$, corresponds to the repulsive case.

III. Finally, if $e = 1$, the equation can be written

$$y^2 = l^2 - 2lx, \tag{B.12}$$

which is a parabola with its apex at $(l/2, 0)$, and oriented in the opposite direction to (B.4).

In all these cases, the position of the origin is one *focus* of the conic. In cases I and II there is a second focus symmetrically placed on the other side of the centre; for the parabola, the second focus is at infinity. (The plural of *focus* is *foci*.) The reason for the name is an intriguing geometric property (see Problem 2): if we have a perfect mirror in the shape of an ellipse light from a source at one focus will converge to the second focus. Similarly, a source at the focus of a parabolic mirror generates a parallel beam, which makes parabolic mirrors ideal for certain applications. For a hyperbolic mirror with a source at one focus, the reflected light will appear to come from a virtual image at the second focus.

---

## Problems

1. The equation (B.2) of an ellipse can be written in parametric form as $x = a \cos \psi$, $y = b \sin \psi$. Show [using the identity $b^2 = (1 - e^2)a^2)$] that the distances between the point labelled $\psi$ and the two foci, $(\pm ae, 0)$, are $a(1 \mp e \cos \psi)$, and hence that the sum of the two distances is a constant. (This result provides a commonly used method of drawing an ellipse, by tying a string between two pegs at the foci, stretching it round a pencil, and drawing a curve while keeping the string taut.)

2. *Using the parametrization of the previous question, show that the slope of the curve is given by $dy/dx = -(b/a)\cot\psi$. Hence show that the angles between the curve and the two lines joining it to the foci are equal. (One way is to find the scalar products between the unit vector tangent to the curve and the unit vectors from the two foci. This result provides a proof of the focussing property: light from one focus converges to the other.)

[The results stated in Problems 1, 2 imply that all radiation originating at one focus of an ellipse at a particular time is then reflected to the other focus with the same time of arrival — a consequence with many applications, both peaceful and otherwise.]

# Appendix C

# Phase Plane Analysis near Critical Points

In this appendix we give a summary of the types of behaviour exhibited by a general autonomous dynamical system near critical points in the phase plane ($n = 2$), as indicated in §13.3.

## C.1   Linear Systems and their Classification

We saw in §13.3 that, in the local expansion near a critical point $(x_0, y_0)$, the key to the local behaviour and to the stability of the equilibrium at the critical point is, normally, the behaviour of the linear system

$$\begin{bmatrix} \dot{\xi} \\ \dot{\eta} \end{bmatrix} \equiv \frac{\mathrm{d}}{\mathrm{d}t} \begin{bmatrix} \xi \\ \eta \end{bmatrix} = M \begin{bmatrix} \xi \\ \eta \end{bmatrix}, \tag{C.1}$$

which is obtained from (13.14) by neglecting higher-order terms in the expansion. The $2 \times 2$ Jacobian matrix $M$ [a tensor of valence 2 (§A.10)] has constant entries, which are found as derivatives of the functions $F(x, y), G(x, y)$ evaluated at the critical points $(x_0, y_0)$, as in (13.11), (13.13).

Consider

$$M = \begin{bmatrix} a & b \\ c & d \end{bmatrix},$$

where $a, b, c, d$ are real constants. For the critical point itself, at which $\xi = 0, \eta = 0$, to be an *isolated critical point* it is necessary that the determinant of $M$ is non-zero. That is to say $ad - bc \neq 0$ and $M$ then has an inverse. If this condition is not satisfied, so that $M$ is singular, then there is at least a *critical line* through $\xi = 0, \eta = 0$, rather than just the single point; we do not consider this case further here.

If we seek a solution to (C.1) in the form

$$\boldsymbol{\xi}(t) \equiv \begin{bmatrix} \xi \\ \eta \end{bmatrix} = \begin{bmatrix} \xi_0 \\ \eta_0 \end{bmatrix} e^{\lambda t} \equiv \boldsymbol{\xi}_0 e^{\lambda t}, \tag{C.2}$$

then we require

$$M\boldsymbol{\xi}_0 = \lambda \boldsymbol{\xi}_0 \tag{C.3}$$

and this is an eigenvalue/eigenvector problem. (See §A.10, although $M$ may not now be symmetric.)

Here the eigenvalues $\lambda_1, \lambda_2$ satisfy the quadratic equation

$$\lambda^2 - (a+d)\lambda + (ad - bc) = 0,$$

that is

$$\lambda^2 - (\mathrm{tr}M)\lambda + (\det M) \equiv (\lambda - \lambda_1)(\lambda - \lambda_2) = 0, \tag{C.4}$$

so that $\lambda_1 + \lambda_2 = \mathrm{tr}M$, the *trace* of $M$ and $\lambda_1\lambda_2 = \det M$, the *determinant* of $M$ (see Appendix A, Problem 16). The eigenvalues $\lambda_1, \lambda_2$ lead to corresponding eigenvectors $\boldsymbol{\xi}_{01}, \boldsymbol{\xi}_{02}$ respectively, in principle, but, since the matrix $M$ is not necessarily symmetric we have eigenvalues which may not be real and a set of eigenvectors which may not be orthogonal or even complete.

There are various cases depending on the nature of the eigenvalues and we can consider separately the cases $\lambda_1 \neq \lambda_2$ and $\lambda_1 = \lambda_2$.

1. $\lambda_1 \neq \lambda_2$. In this case, because of the linearity of the system, we can write

$$\boldsymbol{\xi}(t) = c_1 \boldsymbol{\xi}_{01} e^{\lambda_1 t} + c_2 \boldsymbol{\xi}_{02} e^{\lambda_2 t}, \tag{C.5}$$

with $c_1, c_2$ constants. The vectors $\boldsymbol{\xi}_{01}, \boldsymbol{\xi}_{02}$ are independent in this case and any vector can be expressed as a linear combination of them. In particular $\boldsymbol{\xi}_0 \equiv \boldsymbol{\xi}(0)$ leads to the unique values of $c_1, c_2$ corresponding to given initial conditions. In this situation we can carry out a linear change of variables

$$\begin{bmatrix} \xi \\ \eta \end{bmatrix} = S \begin{bmatrix} \bar{\xi} \\ \bar{\eta} \end{bmatrix},$$

similar to the change to normal co-ordinates in §11.4, in such a way that

$$\frac{d}{dt}\begin{bmatrix}\bar{\xi}\\\bar{\eta}\end{bmatrix} = \begin{bmatrix}\lambda_1 & 0\\0 & \lambda_2\end{bmatrix}\begin{bmatrix}\bar{\xi}\\\bar{\eta}\end{bmatrix}. \tag{C.6}$$

In this *similarity transformation* the $2 \times 2$ matrix $S = [\boldsymbol{\xi}_{01} \vdots \boldsymbol{\xi}_{02}]$ and the diagonal matrix in (C.6) above then takes the form $S^{-1}MS$. It should be noted here that $\lambda_i, \boldsymbol{\xi}_{0i}, c_i$ $(i = 1, 2)$ could be complex, but even then (C.5) is the formal expression of the solution for $\boldsymbol{\xi}$. In the case when $M$ is symmetric then the eigenvalues $\lambda_1, \lambda_2$ are real and the eigenvectors $\boldsymbol{\xi}_{01}, \boldsymbol{\xi}_{02}$ are orthogonal. If the eigenvectors are normalized to have unit length then $S^{-1} \equiv \tilde{S}$, *i.e.* $S$ is a *rotation matrix*.

2. $\lambda_1 = \lambda_2 (\equiv \lambda)$. In this case, $\lambda$ is necessarily real and we may find that the matrix reduction to diagonal form indicated above may, or may not, be possible:

   (a) If we *can* find two distinct eigenvectors corresponding to $\lambda$ then the above machinery will go through trivially, since the matrix $M = \lambda I$ in this case, where $I$ is the unit matrix, and *all* non-zero vectors are eigenvectors!

   (b) If there are *not* two distinct eigenvectors corresponding to $\lambda$ then the best that can be done by a linear transformation is to reduce the system to

   $$\frac{d}{dt}\begin{bmatrix}\bar{\xi}\\\bar{\eta}\end{bmatrix} = \begin{bmatrix}\lambda & 0\\1 & \lambda\end{bmatrix}\begin{bmatrix}\bar{\xi}\\\bar{\eta}\end{bmatrix}, \tag{C.7}$$

   since the diagonal form is not now achievable. The system (C.7) has the solution

   $$\begin{bmatrix}\bar{\xi}\\\bar{\eta}\end{bmatrix} = c_1\begin{bmatrix}0\\1\end{bmatrix}e^{\lambda t} + c_2\begin{bmatrix}1\\t\end{bmatrix}e^{\lambda t}, \tag{C.8}$$

   with $c_1, c_2$ constants.

Depending on the eigenvalues $\lambda_1, \lambda_2$ there are then various possible cases to consider. These are listed below together with sketches of typical patterns of local trajectories. The orientation and sense of rotation in these patterns depends on the system concerned. However, in each case the directions of the arrows indicate evolution with time $t$ along the trajectories.

## Case 1

$\lambda_1, \lambda_2$ real, unequal, *same sign* $\implies$ *(improper)* node, *e.g. negative* sign (Fig. C.1).

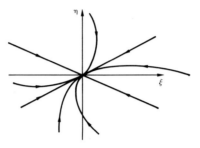

Fig. C.1

All trajectories except for one pair approach the critical point tangent to the same line. The critical point is *asymptotically stable*.

If the sign of $\lambda_1, \lambda_2$ is *positive* then the local structure is similar to that above, but with the sense of the arrows reversed. The critical point is then *unstable*.

## Case 2

$\lambda_1, \lambda_2$ real, equal or unequal magnitude, *opposite sign* $\implies$ *saddle* (or *hyperbolic point*) (Fig. C.2). This type of critical point is always *unstable*.

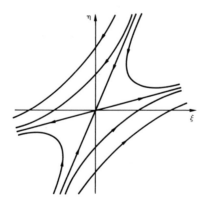

Fig. C.2

## Case 3

$\lambda_1 = \lambda_2 = \lambda$ (necessarily real).

1. When $M = \lambda I$ we have a (*proper*) *node, e.g.* $\lambda$ *negative* (Fig. C.3). This
   critical point is *asymptotically stable*.

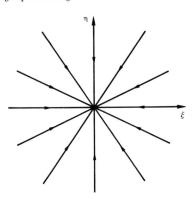

Fig. C.3

If $\lambda$ is *positive* then the local structure is similar to that above, but with
the sense of the arrows reversed. The critical point is then *unstable*.
(A *proper node* is sometimes called a *star, focus, source* or *sink* as appropriate.)

2. When $M$ may *not* be diagonalized, so that there is only a single eigenvector corresponding to $\lambda$, we have an *improper* (or *inflected*) *node, e.g.*
   $\lambda$ *negative* (Fig. C.4). This critical point is *asymptotically stable*.

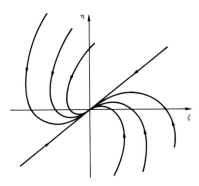

Fig. C.4

If $\lambda$ is *positive* then the local structure is similar to that above, but with the sense of the arrows reversed. The critical point is then *unstable*.

## Case 4

$\lambda_{1,2}$ a complex conjugate pair $\mu \pm i\nu$, with $\mu \neq 0 \implies$ *spiral* (*point*), e.g. $\mu$ *negative* (Fig. C.5).

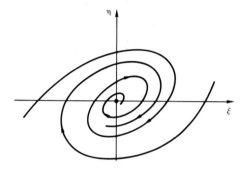

Fig. C.5

This critical point is *asymptotically stable*.

If $\mu$ is *positive* then the local structure is similar to that above, but with the sense of the arrows reversed. The critical point is then *unstable*.

(A *spiral* is sometimes called a *spiral source* or *spiral sink* as appropriate.)

## Case 5

$\lambda_{1,2}$ a pure imaginary conjugate pair $\pm i\nu \implies$ *centre* (or *elliptic point*) (Fig. C.6).

The sense of the arrows may be different, but this type of critical point is always *stable*.

We can solve equation (C.4) for the eigenvalues $\lambda_1, \lambda_2$ in terms of $\mathrm{tr}M$ and $\det M$ obtaining

$$\lambda_{1,2} = \tfrac{1}{2}(\mathrm{tr}M \pm \sqrt{\Delta}), \qquad (C.9)$$

where the discriminant $\Delta = (\mathrm{tr}M)^2 - 4(\det M)$. We may then represent the types of behaviour in the phase plane near a critical point schematically.

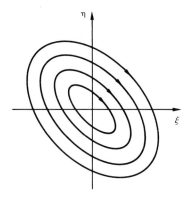

Fig. C.6

(See Fig. C.7.) Note that along the line $\det M = 0$ the critical point is not isolated.

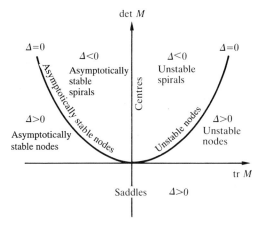

Fig. C.7

## C.2    Almost Linear Systems

We have seen that the local analysis near a typical critical point in the phase plane leads to (13.14) and this equation differs from the linear system (C.1) in that it includes some higher-order terms. For the almost linear system (13.14) the classification of the corresponding linear system (C.1) determines the local phase portrait and the type of stability in almost every case. Small changes produced by the higher-order terms are evidently going to be crucial, if at all, only in the particular cases:

## Case 3

The equal real eigenvalues $\lambda, \lambda$ (*node*) *could* split to give $\lambda \pm \epsilon$ (*node*) or $\lambda \pm i\epsilon$ (*spiral*), where $\epsilon$ is small. However, the *stability* of the critical point would still be just that predicted by the linear system analysis.

## Case 5

The pure imaginary conjugate pair of eigenvalues $\pm i\nu$ (*centre*) *could* become $\pm i(\nu + \epsilon)$ (*centre*) or $\epsilon \pm i\nu$ (*spiral*), where $\epsilon$ is small. Naturally a centre would still indicate that the critical point is *stable*. However, the spiral would be crucially dependent for its stability on the sign of the new real part $\epsilon$ of the eigenvalue pair. If $\epsilon > 0$ then the critical point is *unstable*, whereas if $\epsilon < 0$ then the critical point is *asymptotically stable*.

So, for the systems we are considering, it is only when the exactly linear analysis of §C.1 predicts that a critical point is a centre that we need to be suspicious of the predictions of the exactly linear analysis. Whether (13.14) has a *true centre* or an *unstable* or *asymptotically stable spiral* has to be resolved by a closer scrutiny of the particular system in hand.

It is the case, in fact, that the trajectories near a critical point in the phase plane have a topological equivalence in the *linear* and *almost linear* systems except when there is a zero eigenvalue (*i.e.* the critical point is not isolated) or when the eigenvalues are pure imaginary (*i.e.* a centre) — this is guaranteed by a *theorem* due to *Hartman and Grobman*. For example, we can examine the system

$$\frac{dx}{dt} = x,$$
$$\frac{dy}{dt} = -y + x^2,$$

(C.10)

which has only one critical point (at the origin $x = 0, y = 0$). For the linear system, in the expansion about the origin we have

$$M = \begin{bmatrix} 1 & 0 \\ 0 & -1 \end{bmatrix},$$

so that the eigenvalues are $\lambda_1 = 1, \lambda_2 = -1$ with eigenvectors

$$\begin{bmatrix} 1 \\ 0 \end{bmatrix}, \quad \begin{bmatrix} 0 \\ 1 \end{bmatrix}$$

respectively. The trajectories near the origin, which is a *saddle*, are indicated in Fig. C.8(a). For the exact nonlinear system (C.10) we can write

$$\frac{\mathrm{d}y}{\mathrm{d}x} = -\frac{y}{x} + x, \tag{C.11}$$

so that

$$y = \frac{x^2}{3} + \frac{c}{x}, \qquad \text{with } c \text{ constant,}$$

together with a second solution $x = 0$ (for all $y$).

The exact family of trajectories near the origin is indicated in Fig. C.8(b). It should be noted that the trajectories which go directly into and directly out of the critical point $O$ (respectively the stable and unstable manifolds) correspond directly at and near $O$ for the exactly linear and almost linear systems — a general result usually known as the *stable manifold theorem*.

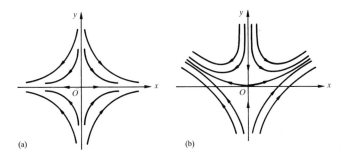

Fig. C.8

## C.3   Systems of Third (and Higher) Order

As we indicated in §13.6, higher-order systems can be analyzed in a similar fashion to that carried out in §13.3 and earlier in this appendix. That is to say the critical points are found and local analysis effected about each of

them by linearization. The resulting eigenvalue/eigenvector problems determine the local stability of the critical points and the *local* phase portrait
structures normally determine the *global* phase portrait for the complete
system.

For a third-order system, *e.g.* the Lorenz system of §13.6, the matrix $M$
for a particular critical point is $3 \times 3$ [again a tensor of valence 2 (§A.9,10)],
so that the three eigenvalues satisfy a cubic equation. This implies that at
least one of the eigenvalues must be real, with the others either both real
or a complex conjugate pair. We only note here that, if all the eigenvalues are negative real or have negative real part, then the critical point is
asymptotically stable. Even if only one of the eigenvalues is positive or if
the complex conjugate pair has positive real part, then the critical point is
unstable.

---

## Problems

1. Find the critical points of the following systems and classify them according to their local linear approximations:
   (a) $\dot{x} = -3x + y$, $\dot{y} = 4x - 2y$;
   (b) $\dot{x} = 3x + y$, $\dot{y} = 2x + 2y$;
   (c) $\dot{x} = -6x + 2xy - 8$, $\dot{y} = y^2 - x^2$;
   (d) $\dot{x} = -2x - y + 2$, $\dot{y} = xy$;
   (e) $\dot{x} = 4 - 4x^2 - y^2$, $\dot{y} = 3xy$;
   (f) $\dot{x} = \sin y$, $\dot{y} = x + x^3$;
   (g) $\dot{x} = y$, $\dot{y} = \left[\frac{\omega^2 - \alpha - y^2}{1 + x^2}\right] x$ in the cases $\omega^2 < \alpha$ and $\omega^2 > \alpha$.

2. For the nonlinear oscillator equation $\ddot{x} + x = x^3$, write $\dot{x} = y$ and show
   that there are two saddle points and one centre in the linear approximation about the critical points in the $(x, y)$ phase plane. Integrate the
   equations of the system to obtain an 'energy' equation and use this to
   show that
   (a) the centre is a *true* centre for the full system;
   (b) the equation of the separatrices through the saddles is

   $$2y^2 = x^2(x^2 - 2) + 1.$$

# Appendix D

# Discrete Dynamical Systems — Maps

In this appendix we consider discrete dynamical systems in which a space is effectively mapped onto itself repeatedly. We recognized in Chapter 13 that for some systems it is appropriate and useful to observe at discrete time intervals, which are not necessarily equal — this is, for example, often the case for biological systems.

Also we saw in §14.2 the concept of a Poincaré return map, where the evolution of a system through its dynamics induces a map of a Poincaré section onto itself. Examining properties of maps, in their own right, will give insight into mechanisms of chaotic breakdown in continuous systems as well.

## D.1   One-dimensional Maps

We consider a map given by

$$x_{n+1} = F(x_n), \tag{D.1}$$

for $n = 0, 1, 2, \ldots$ and with $F$ a known function, and consider possible behaviours of $x_n$ for suitable initial values $x_0$, as we *iterate* to find successively $x_1 = F(x_0)$, $x_2 = F(x_1) \equiv F\big(F(x_0)\big) \equiv F^{(2)}(x_0)$, etc. We can expect to find any *fixed points* $X$ as solutions of

$$X = F(X). \tag{D.2}$$

To examine the stability of the fixed point $X$, we may write $x_n + \epsilon_n = X$, for each $n$, and, when $\epsilon_n$ is small, we can expand $F(x_n)$ in (D.1) in the form

$$F(x_n) = F(X - \epsilon_n) = F(X) - \epsilon_n F'(X) + \tfrac{1}{2}\epsilon_n^2 F''(X) + \ldots, \tag{D.3}$$

where $F'(X) = [dF(x)/dx]_{x=X}$, etc.

A fixed point $X$ is asymptotically stable (and therefore an attractor) if $|F'(X)| < 1$. We may consider different cases (where $\epsilon_n \to 0$):

- $0 < |F'(X)| < 1 \implies \epsilon_{n+1} \simeq F'(X)\epsilon_n$ as $n \to \infty$, and we have *first-order convergence*.
- $F'(X) = 0$, $F''(X) \neq 0 \implies \epsilon_{n+1} \simeq -\frac{1}{2}F''(X)\epsilon_n^2$ as $n \to \infty$, and we have *second-order convergence*.

While this sequence may be continued, the key criterion is that stated above for $|F'(X)|$. We note that the case $|F'(X)| > 1$ leads to instability of $X$, and that the case $|F'(X)| = 1$ depends more specifically on the function $F(X)$.

A familiar example of what is normally second-order convergence is the *Newton–Raphson* iteration process to find roots of a single equation $f(x) = 0$. Here $x_{n+1} = x_n - f(x_n)/f'(x_n)$ and each root has a *basin of attraction*, so that we can find all the roots by judicious choices of $x_0$.

The very simplest map is the linear map:

$$x_{n+1} = rx_n, \tag{D.4}$$

with $r$ constant (and, say, non-negative), and it is evident that $x_n = r^n x_0$ in this case. Here there are various behaviours depending on $r$:

- $0 \le r < 1$: $x_n \to 0$ for all $x_0$ [asymptotic stability of $X = 0$].
- $r = 1$: $x_n = x_0$ for all $x_0$ [steady state].
- $r > 1$: $x_n = x_0 \exp(n \ln r)$ [exponential growth].

If this were a biological model, of *e.g.* a seasonal breeding population $x_n$, then the rate constant $r$ is crucial in determining the fate of any initial population $x_0$.

### The logistic map

A simple nonlinear map, derived from (D.4), is

$$x_{n+1} = rx_n - sx_n^2, \tag{D.5}$$

which is called the *logistic map* and has apparent similarity with the logistic differential equation (13.3). In a biological context the rate constant $r$, quantifying the ability of the population to reproduce, is balanced by the parameter $s$, which quantifies the effect of overcrowding. This model formed the centrepiece of what has become a very influential paper — 'Simple mathematical models with very complicated dynamics', May, *Nature*, **261**, 459–467, 1976.

A simple scaling $\bar{x}_n = sx_n/r$ leads to $\bar{x}_{n+1} = r\bar{x}_n(1 - \bar{x}_n)$, and it is evident that the overbar may then be dropped, in order to find the map

$$x_{n+1} = rx_n(1 - x_n), \tag{D.6}$$

which is the *logistic map in standard form*, with $r$ the single key parameter.

Naturally the primary physical/biological interest is in the case where the $x$ interval $[0, 1]$ is mapped by (D.6) onto $[0, 1]$, which requires $0 \leq r \leq 4$ — for other applications this $r$ restriction might well be absent.

Despite the apparent similarity with the continuous system (13.3) the maps (D.5), (D.6) have some very different and complex properties.

We see immediately that there are two fixed points of (D.6) — at $X = 0$, $X = 1 - 1/r$ — and, in each case, $F'(X) = r(1 - 2X)$.

It is helpful for maps to consider [see Fig. D.1(a)] a plot of $y = x$ and $y = F(x)$ so that by tracing the vertical and horizontal lines between the two we can follow the sequence $x_0 \to x_1 \to x_2 \to \ldots$, and thus see whether or not it might converge.

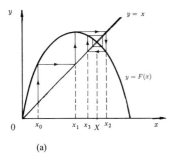

(a)

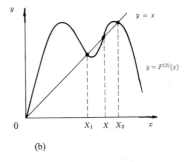

(b)

Fig. D.1

The behaviours of the map (D.6) for different values of $r$ can be summarized as follows:

- $0 \leq r < 1$: $X = 0$ is asymptotically stable and $X = 1 - 1/r$ is unstable. [$X = 0$ is a *point attractor* — if the corresponding *linear* model population cannot sustain itself then overcrowding makes it worse!]
- $1 < r < 3$: $X = 0$ is unstable and $X = 1 - 1/r$ is asymptotically stable — see Fig. D.1(a) for example. [$X = 1 - 1/r$ is a *point attractor* —

exponential growth is stabilized by overcrowding, in very similar fashion to the behaviour of the logistic differential equation (13.3).]
- $3 < r \leq 4$: $X = 0$ and $X = 1 - 1/r$ are now both unstable. As $r$ increases successive 'period-doubling bifurcations' occur as asymptotic stability is exchanged between lower- and higher-order cycles (termed a supercritical flip bifurcation):

  * $3 < r < 1 + \sqrt{6} = 3.449\,48\cdots$: $x_n \to$ an asymptotically stable 2-cycle (or period-2 solution). If we examine $x_{n+2} = F(x_{n+1}) = F^{(2)}(x_n)$, then we obtain an equation of degree 4 for the fixed points of this iterated mapping. Of course $X = 0$, $X = 1 - 1/r$ are two of the roots of this equation. The nontrivial solutions $X_1, X_2$, such that $F(X_1) = X_2$, $F(X_2) = X_1$ [see Fig. D.1(b)], are roots of the quadratic $r^2 X^2 - r(r+1)X + (r+1) = 0$. For asymptotic stability it is necessary (see Problem 1) that $|4 + 2r - r^2| < 1$ or equivalently, for positive $r$, $3 < r < 1 + \sqrt{6}$.
  * $3.449\,48\cdots < r < 3.544\,09\cdots$: $x_n \to$ an asymptotically stable 4-cycle found from an equation of degree 16; four more roots are the trivial solutions $0, 1 - 1/r, X_1, X_2$ (the other roots, when real, give 4-cycles arising through a different process).
  * $3.544\,09\cdots < r < 3.564\,40\cdots$: $x_n \to$ an asymptotically stable 8-cycle, and so on (each time in a shorter interval in $r$) ... until
  * $r = 3.569\,94\cdots$: Accumulation point of $2^\infty$-cycle.
  * $3.569\,94\cdots < r \leq 4$: For some values of $r$ there are asymptotically stable cycles of different lengths, but for others the $x_n$ values range seemingly over a whole continuous interval, in an apparently random fashion. An intriguing fact is that odd-period cycles only appear for $r > 3.678\,57\cdots$.

The logistic map attractors are shown in Fig. D.2. The numbers along the top are the cycle periods.

It should be noted that this figure has an approximate self-similarity at higher magnification, in that the period-doubling cascade is broadly repeated as other-period asymptotically stable cycles become unstable — e.g. the period-3 cycle in Fig. D.2.

That the period-doubling proceeds broadly in geometric fashion in the limit was discovered by Feigenbaum in 1975. Here the $r$-intervals $\Delta_i$ between bifurcations and the measures $d_i$ of width in $X$ of successive cycles

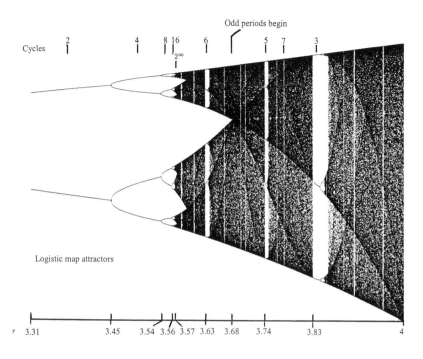

Fig. D.2

are such that (see Figs. D.2, D.3)

$$
\begin{aligned}
\Delta_i/\Delta_{i+1} &\to \delta = 4.669\,201\,6\ldots, \\
d_i/d_{i+1} &\to \alpha = 2.502\,907\,8\ldots,
\end{aligned}
\tag{D.7}
$$

in each case in the limit as $i \to \infty$.

Feigenbaum noted that essentially all 'humped' mapping functions $F(x)$ lead to intersections of $y = x$ with $y = F(x)$, $y = F^{(2)}(x)$, $y = F^{(4)}(x),\ldots$ which are similar up to a rescaling. In a process of '*renormalization*' the precise form of $F(x)$ is lost and in 1976 Feigenbaum discovered a universal function $g(x)$, which is self-reproducing under such rescaling and iteration, so describing this universal property:

$$
g(x) = -\alpha g\left[g\left(\frac{-x}{\alpha}\right)\right]
\tag{D.8}
$$

(see Feigenbaum, *Journal of Statistical Physics*, **21**, 669–706, 1979).

The universality leads to $\delta, \alpha$ as universal constants and $\delta$ is usually identified as the *Feigenbaum number*.

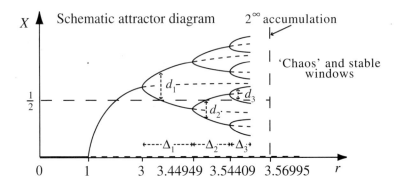

Fig. D.3

As to the onset of cycles with other periods, there is a theorem, due to *Sarkovskii* (1964), which is as follows:

If a mapping function $F(x)$ has a point $x_p$ which is cyclic of order $p$, then it must also have a point $x_q$ of period $q$, for every $q$ which precedes $p$ in the sequence:

$$1 \Leftarrow 2 \Leftarrow 4 \Leftarrow \quad \cdots \quad \Leftarrow 2^n \Leftarrow \quad \cdots$$

$$\cdots \Leftarrow 2^m.9 \Leftarrow 2^m.7 \Leftarrow 2^m.5 \Leftarrow 2^m.3$$

$$\cdots \Leftarrow 2^2.9 \Leftarrow 2^2.7 \Leftarrow 2^2.5 \Leftarrow 2^2.3$$
$$\cdots \Leftarrow \quad 2.9 \Leftarrow \quad 2.7 \Leftarrow \quad 2.5 \Leftarrow \quad 2.3$$
$$\cdots \Leftarrow \quad 9 \Leftarrow \quad 7 \Leftarrow \quad 5 \Leftarrow \quad 3.$$

For example, the existence of a 3-cycle implies the existence of cycles of all the other periods!

Naturally the odd periods cannot arise from period-doubling, but do so *via* a rather different process which may be examined by similar methods to those employed above.

This theorem can be proved using a continuity/intermediate value theorem argument. However it says nothing about stability of these cycles, or the ranges of $r$ (in our logistic example) for which they may be observed. The vast majority of these cycles are unstable when all are present, and it is this which leads to a brief summary statement in the form 'Period 3 implies chaos', as essentially random behaviour of iterates $x_n$ occurs (Li and Yorke, *American Mathematical Monthly*, **82**, 985–992, 1975, in which, incidentally, the term *chaos* was first introduced!).

What can be said about values of $r$ in the logistic map which lead to distributions of iterates, rather than to asymptotically stable cycles?

Analytically this is a tough problem. However, it happens that there are two positive values of $r$ for which a formal exact solution of (D.6) is known — $r = 2$ and $r = 4$.

The former $(r = 2)$ allows $(1 - 2x_{n+1}) = (1 - 2x_n)^2$ leading to $x_n = \frac{1}{2}[1 - (1 - 2x_0)^{2^n}]$. This is not especially interesting, since $x_n \to \frac{1}{2}$ as we should expect.

However the latter possibility $(r = 4)$ allows us to substitute $x_n = \frac{1}{2}[1 - \cos(2\pi\theta_n)] \equiv \sin^2(\pi\theta_n)$ and this leads us to $\theta_{n+1} = 2\theta_n$ and then $\theta_n = 2^n\theta_0$. Here $\theta$ is evidently periodic, with period 1, in that the same $x$ is generated by $\theta$ and by $\theta + 1$. Thus we may write any $\theta_0$ we choose in a binary representation just using negative powers of 2 — for example, $\theta_0 = \frac{1}{2} + \frac{1}{8} + \frac{1}{16} + \frac{1}{64} + \cdots = 0.101\,101\ldots$. Then $\theta_1 = 0.011\,01\ldots$, $\theta_2 = 0.110\,1\ldots$, etc., since the integer part may be cancelled at each stage on account of the periodic property.

For almost all choices of $\theta_0$ then the $\theta_n$ will be uniformly distributed on the interval $[0, 1]$, since each digit in the binary expansion of $\theta_0$ could be chosen with equal likelihood to be a 0 or a 1. As a consequence of $\theta$ being uniformly distributed, then $x$ is not. Indeed $x$ has a probability distribution given by $P(x) = 1/[\pi\sqrt{x(1 - x)}]$ (see Figs. D.4, D.2).

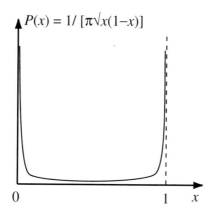

Fig. D.4

We note that $P(x)$ itself *is* the attractor here for $r = 4$, in the sense that for almost all choices of $x_0$ the distribution of $x_n$ will approach $P(x)$ for large $n$. It is called an *invariant probability distribution*.

For other values of $r$ there are theoretical results for the corresponding *probability distribution attractors* (via an equation for the distribution function — Perron–Frobenius), but no slick result like that above for $r = 4$.

The distribution attractors are also characterized by an exponential divergence of iterates, leading to the sensitivity to initial conditions characteristic of chaos. If we choose to examine $x_0$ and $x_0 + \epsilon_0$, with $\epsilon_0$ very small, then $\epsilon_n \simeq \epsilon_0 \exp(\lambda n)$ on average, and we have divergence or convergence of iterates according as $\lambda > 0$ or $\lambda < 0$. Here $\lambda$ is a *Lyapunov exponent* (see §13.7).

Since we have $F^{(n)}(x_0 + \epsilon_0) - F^{(n)}(x_0) \simeq \epsilon_0 e^{\lambda n}$, then we have

$$\lambda \simeq \lim_{n \to \infty} \left\{ \frac{1}{n} \ln \left( \left| \frac{\mathrm{d}}{\mathrm{d}x} F^{(n)}(x) \right| \right) \right\}$$

$$= \lim_{n \to \infty} \left\{ \frac{1}{n} \sum_{i=0}^{n-1} \ln |F'(x_i)| \right\}. \tag{D.9}$$

Values of $\lambda$ can be found by numerical computation. In measuring $\lambda$ empirically in a particular practical case, we allow $n$ to become large enough so that our estimate of $\lambda$ can settle down to a steady value. We also average over various different $x_0$, in order to avoid an atypical result through a single unfortunate choice. For the logistic map Fig. D.5 shows a plot of $\lambda$ as $r$ varies.

Here $F'(x) = r(1 - 2x)$, so that we expect:

- $0 < r < 1$: $\lambda = \ln r$,
- $1 < r < 3$: $\lambda = \ln |2 - r|$ (leading to an infinite spike when $r = 2$),
- $3 < r < 1 + \sqrt{6}$: $\lambda = \frac{1}{2} \ln |r(1 - 2X_1)| + \frac{1}{2} \ln |r(1 - 2X_2)|$ [leading to an infinite spike when $X_1 = \frac{1}{2}$ (requiring $r = 1 + \sqrt{5} = 3.236$)],
- $r = 4$: since $P(x) = 1/[\pi \sqrt{x(1 - x)}]$ so we have

$$\lambda = \int_0^1 (\ln |F'(x)|) P(x) \, \mathrm{d}x = \int_0^1 \frac{\ln |4(1 - 2x)| \, \mathrm{d}x}{\pi \sqrt{x(1 - x)}} = \ln 2 = 0.693\,147\ldots.$$

In any event, the regions of $r$ for which there are positive values of $\lambda$ indicate the sensitivity to initial conditions at these $r$ values. While there are other measures of the complexity/disorder of the iterates in these cases —

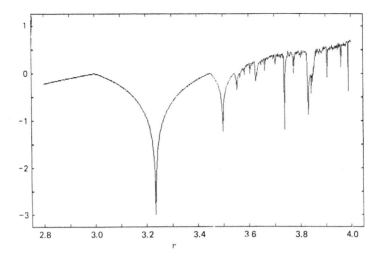

Fig. D.5  [Baker and Gollub (*Chaotic Dynamics*, 2nd ed., Cambridge University Press, 1996)]

*e.g.* the system 'entropy' — these will not be pursued further here. However, see Problem 4 for another (and simpler) example of a chaotic map.

The logistic map (D.6), together with other similar maps, involves *stretching* and *folding*, so that states diverging exponentially are still broadly confined to a bounded region. The loss of information about initial conditions, as the iteration process proceeds in a chaotic regime, is associated with the *non-invertibility* of the mapping function $F(x)$, *i.e.* while $x_{n+1}$ is uniquely determined from $x_n$, each $x_n$ can come from 2 possible $x_{n-1}$, 4 possible $x_{n-2}$, etc., and eventually from $2^n$ possible $x_0$. Hence system memory of initial conditions becomes blurred!

Many continuous systems — for example the Lorenz system of §13.6 — exhibit a similar period-doubling in their dynamics.

## D.2   Two-dimensional Maps

For two-dimensional maps any new universality has proved harder to find! However, there are interesting phenomena. We may write our generic map in the form

$$x_{n+1} = F(x_n, y_n),$$
$$y_{n+1} = G(x_n, y_n),$$

(D.10)

for integer $n$ and with $F, G$ known functions. Again we may seek fixed points $(X, Y)$ as solutions of

$$X = F(X, Y),$$
$$Y = G(X, Y).$$
(D.11)

By a Taylor expansion, near $X, Y$ and similar to that carried out in §13.3 and in Appendix C, we obtain

$$\begin{pmatrix} x_{n+1} - X \\ y_{n+1} - Y \end{pmatrix} \simeq M \begin{pmatrix} x_n - X \\ y_n - Y \end{pmatrix}, \text{ with } M = \begin{pmatrix} \frac{\partial F}{\partial x} & \frac{\partial F}{\partial y} \\ \frac{\partial G}{\partial x} & \frac{\partial G}{\partial y} \end{pmatrix}_{(x,y)=(X,Y)}.$$
(D.12)

It now follows that $(x_n, y_n) \to (X, Y)$ if and only if the eigenvalues of the matrix $M$ all have modulus less than 1; this is needed to force $M^n \to$ the zero matrix as $n \to \infty$.

When instability sets in as parameters are changed, we then typically have a 2-cycle to examine with $(X_1, Y_1) \rightleftarrows (X_2, Y_2)$ and where these points are the solutions of

$$X = F[F(X, Y), G(X, Y)],$$
$$Y = G[F(X, Y), G(X, Y)],$$
(D.13)

other than the fixed points of (D.10) found earlier and which satisfy (D.11). The asymptotic stability of this 2-cycle depends on the eigenvalues of the matrix product

$$M_1 M_2 = \begin{pmatrix} \frac{\partial F}{\partial x} & \frac{\partial F}{\partial y} \\ \frac{\partial G}{\partial x} & \frac{\partial G}{\partial y} \end{pmatrix}_{(x,y)=(X_1,Y_1)} \begin{pmatrix} \frac{\partial F}{\partial x} & \frac{\partial F}{\partial y} \\ \frac{\partial G}{\partial x} & \frac{\partial G}{\partial y} \end{pmatrix}_{(x,y)=(X_2,Y_2)},$$
(D.14)

and so on.

### The Hénon map

Probably the most celebrated example is the Hénon map (Hénon, *Communications in Mathematical Physics*, **50**, 69–77, 1976):

$$x_{n+1} = 1 - ax_n^2 + y_n,$$
$$y_{n+1} = bx_n,$$
(D.15)

with $a, b$ real parameters, and where the normal physical interest is in $|b| \leq$ 1. This map was constructed to exhibit some behaviours similar to those of the Lorenz system of §13.6. Geometrically the map may be considered to be a composition of three separate simple maps — an area-preserving fold, a contraction $|b|$ in the $x$ direction, an area-preserving rotation.

There are two fixed points given by

$$X = X_\pm \equiv [-(1-b) \pm \sqrt{(1-b)^2 + 4a}]/2a,$$
$$Y = Y_\pm \equiv bX_\pm, \qquad \text{(D.16)}$$

and these are real and distinct if and only if $a > a_0 = -\frac{1}{4}(1-b)^2$. One of the fixed points is then always unstable and the other is asymptotically stable if $a < a_1 = \frac{3}{4}(1-b)^2$ (see Problem 7). For $a > a_1$, both fixed points are unstable and we get period-doubling to a 2-cycle, 4-cycle, etc. The two-cycle stability is determined through (D.14) by the eigenvalues of the matrix product

$$M_1 M_2 = \begin{pmatrix} 4a^2 X_1 X_2 + b & -2aX_1 \\ -2abX_2 & b \end{pmatrix}.$$

This leads to asymptotic stability of the 2-cycle only when $a_1 < a < a_2$ with $a_2 = (1-b)^2 + \frac{1}{4}(1+b)^2$. The period-doubling cascade then continues (see Fig. D.6).

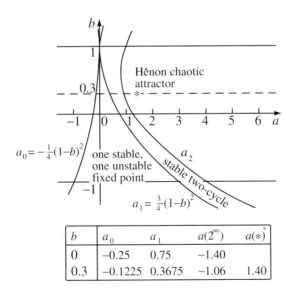

| $b$ | $a_0$ | $a_1$ | $a(2^\infty)$ | $a(*)$ |
|-----|-------|-------|---------------|--------|
| 0 | −0.25 | 0.75 | −1.40 | |
| 0.3 | −0.1225 | 0.3675 | −1.06 | 1.40 |

$b=0$ corresponds to the one-dimensional logistic map (see Problem 2)

Fig. D.6

Beyond the cascade, and embedded among other-period cycles, there are $(a, b)$ values where the attractor is very complex (see, *e.g.*, Fig. D.7, where $a = 1.4, b = 0.3$).

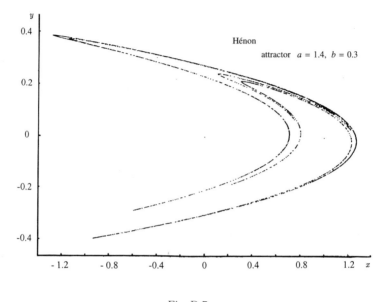

Fig. D.7

Here there is *stretching* along the strands of the attractor and *squeezing* across them — with associated Lyapunov exponents which are respectively positive (along) and negative (across) (see Problem 8).

This chaotic attractor — termed '*strange*' — has some similar features to the Lorenz attractor of Fig. 13.20. It also has a fractal character in that it has a broad similarity in features and relative scale at all magnifications.

The repetition of stretching, squeezing and folding onto the original region is characteristic of a technical construct — the Smale 'horseshoe' (1960) — which is now known to be a trademark of chaotic systems.

Smale 'stretched' the unit square in one co-ordinate direction and 'squeezed' it in the other direction, then placing the resulting strip (with one 'fold') over the original square — with inevitable overlap. Infinite repetition of this sequence of operations leads to the identification of an *attractor* consisting of all the points which remain within the original square indefinitely. This attractor has great topological complexity and, while it *guarantees* the 'sensitivity to initial conditions' of chaos, *proving* its existence for a particular system may be very tough.

A formal demonstration of *chaotic* dynamics for the Hénon map is contained in a paper by Benedicks and Carleson (*Annals of Mathematics*, **133**, 73–169, 1991).

In passing (and for reference in §D.3) we note that a map which preserves area would have $|\det M| = 1$ in (D.12) and then the stretching and squeezing exactly compensate each other (see Problem 11). For the Hénon map this is the case only when $|b| = 1$. Equal-area maps are of special interest since the Poincaré return map for a section through a Hamiltonian system (see §14.2) has the equal-area property. Some consequences are explored briefly in §D.3.

## D.3   Twist Maps and Torus Breakdown

In §14.1 we noted that each trajectory of an integrable Hamiltonian system with $n$ degrees of freedom is confined to the surface of an $n$-torus in the $2n$-dimensional phase space. The $n$-tori corresponding to the range of initial conditions are 'nested' in the phase space.

In §14.2 we introduced the concept of a Poincaré surface of section, as a slice through the dynamical structure. For $n = 2$ degrees of freedom this section of the nested torus structure is a continuum of closed curves, each one of which is intersected by one of its own torus trajectories in a sequence of points (see Fig. 14.4).

If we make use of action/angle variables, as described in §14.3 and particularly in the polar form of Fig. 14.5(c), then the closed curves of the Poincaré section can be taken as concentric circles. The intersection points of a trajectory with its own particular one of these circles (of radius $r$) will necessarily be twisted successively around the origin (the centre of the circle) through an angle $2\pi\alpha$. Here $\alpha$ is the *rotation number*, which is the ratio of normal frequencies characterizing the particular torus concerned — it therefore depends on $r$.

We can then introduce the notion of a *twist map*, $T$ [Moser (1973)]; in polars:

$$\begin{pmatrix} r_{n+1} \\ \theta_{n+1} \end{pmatrix} = \begin{pmatrix} r_n \\ \theta_n + 2\pi\alpha(r_n) \end{pmatrix} \equiv T\left[\begin{pmatrix} r_n \\ \theta_n \end{pmatrix}\right]. \tag{D.17}$$

When a perturbation Hamiltonian is introduced, as in §14.6, we can model the modified situation using a *perturbed twist map*, $T_\epsilon$:

$$\begin{pmatrix} r_{n+1} \\ \theta_{n+1} \end{pmatrix} = \begin{pmatrix} r_n + \epsilon f(r_n, \theta_n) \\ \theta_n + 2\pi\alpha(r_n) + \epsilon g(r_n, \theta_n) \end{pmatrix} \equiv T_\epsilon\left[\begin{pmatrix} r_n \\ \theta_n \end{pmatrix}\right], \qquad \text{(D.18)}$$

with $\epsilon$ small and positive and with $f, g$ known smooth functions.

We can now ask what happens to a circle of a particular radius, which is mapped to itself (with a twist) by $T$, as the perturbation quantified by $\epsilon$ in $T_\epsilon$ grows from zero — *i.e.* as the *integrable system* $T$ becomes a *near-integrable system* $T_\epsilon$, in such a way that *area is still preserved.*

In fact the answer to this question depends on the rotation number $\alpha$ corresponding to our particular circle chosen.

If $\alpha$ is an *irrational* number then, for $\epsilon$ small enough, the circle undergoes some distortion (perturbation) but is certainly not destroyed. This is also true for the corresponding torus in the full phase space. This result is in accord with the KAM theory referred to in §14.6.

As we shall see, it is circles for (D.17), and their tori, corresponding to *rational* $\alpha$, which break down under perturbation, leading to sensitivity to initial conditions and chaos.

The rational $\alpha$ are 'scanty, but dense' among the real numbers (see Problem 12) and these $\alpha$ correspond to resonances in the system. As in the discussion of the problem of small denominators in §14.6, the sensitivity to initial conditions is strongest for the rational $\alpha = k/s$ with small values of $s$. As the perturbation grows with $\epsilon$, the breakdown associated with each such rational $\alpha$ broadens, so that progressive overlap occurs, leading eventually to complete breakdown of the torus structure.

Let us examine the twist map (D.17) for three neighbouring circles $C_-, C, C_+$ corresponding respectively to $\alpha_- < k/s$, $\alpha = k/s$, $\alpha_+ > k/s$ (say), with $k, s$ positive integers and with $\alpha_-, \alpha_+$ irrational. (We have here chosen to take $\alpha$ to be an increasing function of radius $r$.)

Then applying the twist map $T$ successively $s$ times we find that the effect of $T^{(s)}$ (see Fig. D.8) is to map the circles to themselves, with $C$ invariant and with $C_-, C_+$ twisted 'rigidly' clockwise, anticlockwise respectively — this is so since $2\pi s\alpha_- < 2\pi k$ and $2\pi s\alpha_+ > 2\pi k$. All points of $C$ are fixed under the iterated mapping $T^{(s)}$.

Since $\alpha_-, \alpha_+$ are irrational the circles $C_-, C_+$ are only mildly distorted when we apply the map $T_\epsilon$ successively $s$ times instead — *i.e.* when we iterate to consider the effect of $T_\epsilon^{(s)}$. However, the inner and outer circles $C_-, C_+$ are still twisted clockwise, anticlockwise respectively — see Fig. D.9.

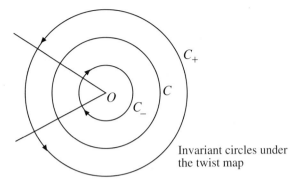

Invariant circles under
the twist map

Fig. D.8

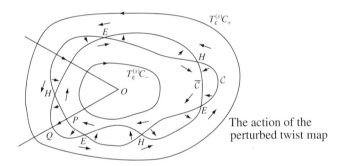

The action of the
perturbed twist map

Fig. D.9

Moving outwards radially from $O$ in any direction we can, by continuity, find a point on that radius arm which undergoes *no* net twist when the iterated map $T_\epsilon^{(s)}$ is applied — $P$ is mapped only *radially* (to $Q$) by this process. By considering all possible radius arms we construct the closed curve $\mathcal{C}$ consisting of all such points $P$ which are mapped only radially by $T_\epsilon^{(s)}$ — to $\bar{\mathcal{C}}$.

For the map $T_\epsilon$ to model a section of a Hamiltonian dynamical system (as, of course, does $T$ itself) the map $T_\epsilon$ — and hence $T_\epsilon^{(s)}$ — must be one of equal-area, as reflected in §14.2.

Since the areas contained within the curves $C$ and $\bar{C}$ must be *equal*, there is in general an even number of intersections of these curves, which correspond, of course, to fixed points of the iterated map $T_\epsilon^{(s)}$. These points are all that remains fixed from the original invariant circle $C$ of the unperturbed iterated map $T^{(s)}$. We now note that the fixed points are of alternating type as we move around $C$ (or $\bar{C}$) — see Fig. D.10, where the iterated mapping sense of flow is indicated.

(a) Elliptic point $(E)$   stable          (b) Hyperbolic point $(H)$   unstable

Fig. D.10

There are evidently in all $2ns$ such fixed points of the iterated map $T_\epsilon^{(s)}$, where $n$ is a positive integer (usually 1).

The statement of existence of these fixed points, of their multiplicity and their alternating stability is a result known as the Poincaré–Birkhoff Theorem (1927).

It turns out that the elliptic fixed points $E$ are themselves surrounded (at higher scales) by elliptic and hyperbolic fixed points corresponding to even higher-order frequency resonances.

For the hyperbolic fixed points $H$ the unstable and stable manifolds (*q.v.* also in §C.2) for neighbouring such points in the same family cross to form what are called *homoclinic intersections*. The resulting instability at all scales leads inevitably to sensitivity to initial conditions, in that trajectories of the system have to twist and turn, forming a *homoclinic tangle*, in order not to *self*-intersect, while maintaining the equal-area property of the return map. This results in the stretching, squeezing and folding associated with the *Smale horseshoe* referred to in §D.2 and this forces a dense interweaving of ordered and chaotic motions within Hamiltonian systems (Fig. D.11). Recognition by Poincaré of the existence of such tangles was the first mathematical realization of the presence of what we now call chaos.

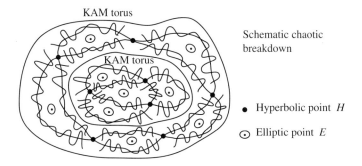

KAM torus

Schematic chaotic breakdown

• Hyperbolic point $H$

⊙ Elliptic point $E$

Fig. D.11

An example of a map in which some of the breakdown to chaos in a Poincaré section is apparent is that for the oval billiard map (Fig. 14.13).

For further exploration of breakdown to chaos under parameter change, see Problem 13.

---

## Problems

1. For the logistic map (D.6) with $r$ non-negative:

   (a) Show that there is a point attractor for $0 \leq r < 3$.
   (b) Show that there is a two-cycle attractor for $3 < r < 1 + \sqrt{6}$.
   (c) Show that $a, b, s$ can be found such that $y_n$ satisfies the logistic map with parameter $s$ ($\neq r$) and $x_n = a + b y_n$.
   (d) Hence determine the principal period-doubling bifurcation points for the logistic map on the range $-2 \leq r \leq 4$.

2. Show that the map $y_{n+1} = 1 - a y_n^2$ is just the logistic map (D.6) for $x_n$ with $x_n, y_n$ related linearly by $y_n = \alpha + \beta x_n$ and $a = \frac{1}{4} r(r - 2)$. [This is an example of the fact that all quadratic maps $y_{n+1} = A + B y_n + C y_n^2$ are essentially just the logistic map, in that (D.6) can be obtained by a suitably chosen linear relation between $y_n$ and $x_n$.]

3. *Allowing $r$ and $x_n$ to be complex in the logistic map (D.6), find regions of the complex $r$ plane for which the map has (a) a point attractor, (b) a 2-cycle attractor. (c) Sketch the corresponding regions when these results are expressed in terms of the complex $a$ plane for the $y_n$ map of Problem 2.

4. The tent map

$$x_{n+1} = \begin{cases} 2x_n, & (0 \le x_n \le \frac{1}{2}) \\ 2(1 - x_n), & (\frac{1}{2} < x_n \le 1) \end{cases}$$

has unstable fixed points. Show that this map exhibits extreme sensitivity to initial conditions, in that an uncertainty $\epsilon_0$ in $x_0$ is rapidly magnified. Estimate the number of iterations after which the range of uncertainty in the iterates is the complete interval $[0, 1]$.

5. For the cubic map $x_{n+1} = ax_n - x_n^3$, where $a$ is real, show that, when $|a| < 1$, there is an asymptotically stable fixed point $X = 0$ and that, when $1 < a < 2$ there are two such fixed points at $X = \pm\sqrt{a - 1}$. What happens when $a$ becomes $> 2$?

6. Explore the one-dimensional maps of Problems 1–5 using a programmable calculator or (better) a computer. Further interesting examples are $x_{n+1} = \exp[a(1 - x_n)]$, $x_{n+1} = a \sin x_n$.

7. For the Hénon map (D.15) show that:

   (a) when $-\frac{1}{4}(1 - b)^2 < a < \frac{3}{4}(1 - b)^2$ there are two real fixed points, one of which is asymptotically stable,

   (b) *when $\frac{3}{4}(1-b)^2 < a < (1-b)^2 + \frac{1}{4}(1+b)^2$ there is an asymptotically stable 2-cycle.

   [*Hint*: Since $M = \begin{pmatrix} -2aX & 1 \\ b & 0 \end{pmatrix}$ the eigenvalues of $M$ are $\lambda_1, \lambda_2$ with $\lambda_1\lambda_2 = -b$, $\lambda_1 + \lambda_2 = -2aX$. To determine stability, it is useful to consider a sketch of the function $f(\lambda) \equiv b/\lambda - \lambda$ and look for points where $f(\lambda) = 2aX$ in order to find $(a, b)$ such that both the eigenvalues satisfy $|\lambda_i| < 1$.]

8. *Show that a small circle of radius $\epsilon$ centred at any $(X, Y)$ becomes a small ellipse under a single iteration of the Hénon map (D.15). Explain how the semi-axes of the ellipse are related to Lyapunov exponents $\lambda_1, \lambda_2$ and show that $\lambda_1 + \lambda_2 = \ln |b|$ with $\lambda_1 > 0 > \lambda_2$, implying simultaneous 'stretch' and 'squeeze'.

9. *The Lozi map is (D.10) with $F(x, y) = 1 + y - a|x|$, $G(x, y) = bx$, where $a, b$ are real parameters.

   (a) When $|b| < 1$ and $|a| < 1 - b$, show that there is one asymptotically stable fixed point.

   (b) Find the 2-cycle when $|b| < 1$ and $a > 1 - b$ and determine its stability.

10. Explore the two-dimensional maps of Problems 7–9 using a computer.

11. By calculating Lyapunov exponents examine sensitivity to initial conditions of the equal-area maps of the unit square ($0 \leq x, y \leq 1$):

   (a) Arnold's cat map $x_{n+1} = x_n + y_n$, $y_{n+1} = x_n + 2y_n$ (each modulo 1).

   (b) The baker's transformation

$$(x_{n+1}, y_{n+1}) = \begin{cases} (2x_n, \frac{1}{2}y_n) & (0 \leq x_n < \frac{1}{2}), \\ (2x_n - 1, \frac{1}{2}(y_n + 1)) & (\frac{1}{2} \leq x_n \leq 1). \end{cases}$$

12. Consider a system of two degrees of freedom with two natural frequencies $\omega_1, \omega_2$ in the light of the discussion of periodicity and degeneracy in §§14.1,14.6 and the real ratio $\omega_1/\omega_2$. Show that for numbers on the real line:

   (a) between any two irrationals we can certainly find a rational;
   (b) between any two rationals we can certainly find an irrational.

   (Note that, despite (a), there are vastly more irrationals than rationals — unlike the former the latter are 'countable', so that the rationals are 'scanty, but dense'. This emphasizes that for most systems periodicity (closure) is relatively rare, although it is still the case that for any irrational $\omega_1/\omega_2$ there are rationals arbitrarily close by.)

13. *As an exercise on near-integrable systems explore the '*standard* map' (Chirikov–Taylor) analytically/computationally:

$$I_{n+1} = I_n + K \sin \phi_n,$$
$$\phi_{n+1} = \phi_n + I_{n+1}.$$

   (*Note*: This equal-area map models the twist around a Poincaré section by the dynamics (as in §14.2 and in more detail in §D.3), when the integrable case ($K = 0$) is expressed in action/angle variables ($I, \phi$). As $K$ increases there appear resonance zones, periodic orbits and bands of chaos as tori $I = $ constant undergo progressive breakdown. See Chirikov, *Physics Reports*, **52**, 263–379, 1979.)

# Answers to Problems

1. $m_A/m_B = 3$; $3\boldsymbol{v}/4$.
2. $m_1/m_2 = r_2/r_1$.
3. $2m\ddot{\boldsymbol{r}} = \boldsymbol{F} = \boldsymbol{F}_{21} + \boldsymbol{F}_{31}$; $\boldsymbol{r} = (m_2\boldsymbol{r}_2 + m_3\boldsymbol{r}_3)/(m_2 + m_3)$.
4. $0.12\,\mathrm{m}$.
5. $\boldsymbol{r}'_{ij} = \boldsymbol{r}_{ij}$, $\boldsymbol{p}'_i = \boldsymbol{p}_i - m_i\boldsymbol{v}$, $\boldsymbol{F}'_{ij} = \boldsymbol{F}_{ij}$.
6. $400\,\mathrm{N}$, $300\,\mathrm{N}$.
7. $\arcsin 0.135 = 7.76°$ E of N; $60.6\,\mathrm{min}$; $130\,\mathrm{km}$, $8.62°$ W of S.
8. $1.2\,\mathrm{m}$; $(\pm 0.4, 0.6, -1.2)$, $(0, -0.9, -1.2)$; $7\,\mathrm{N}$, $7\,\mathrm{N}$, $10\,\mathrm{N}$.
10. $(2.2 \times 10^{-3})° = 7.9''$.

CHAPTER 2

1. ($x$ in m, $t$ in s) $x = -3\cos 2t + 4\sin 2t = 5\cos(2t - 2.214) =$ Re$[(-3 - 4\mathrm{i})\mathrm{e}^{2\mathrm{i}t}]$; $t = 0.322\,\mathrm{s}$, $1.107\,\mathrm{s}$.
2. $z = -(mg/k)(1 - \cos\omega t)$, $\omega = \sqrt{k/m}$.
3. $0.447\,\mathrm{s}$, $14.2\,\mathrm{mm}$.
4. $15.7°\,\mathrm{s}^{-1}$, $\theta = 5°\cos\pi t - 8.66°\sin\pi t = 10°\cos(\pi t + \pi/3)$.
5. $V = -GMm/x$; $\sqrt{2GM(R^{-1} - a^{-1})}$; $8\,\mathrm{km\,s}^{-1}$.
6. $V = \frac{1}{4}cx^4$; $\sqrt{c/2m}\,a^2$; $x = \pm a$.
7. $V = \frac{1}{2}kx^2 - c\ln x$; $x = \sqrt{c/k}$; $\omega = \sqrt{2k/m}$.
8. $F = -mk$, $F = mk$; oscillation; $2\sqrt{2a/k}$.
9. $F = kx$, $|x| < a$; $F = 0$, $|x| > a$; oscillation between two turning points if $k < 0$ and $E < 0$, 1 turning point if $k > 0$ and $E < \frac{1}{2}ka^2$, otherwise no turning points.
10. Earlier by $(2a/v) - (2/\omega)\arctan(\omega a/v)$.
11. (a) no turning points, (b) 1 turning point, (c) 1 or 2 turning points.

12. $x = -a$; $2\pi\sqrt{2ma^3/c}$; (a) $|v| < \sqrt{c/ma}$, (b) $v < -\sqrt{c/ma}$ or
    $\sqrt{c/ma} < v < \sqrt{2c/ma}$, (c) $v > \sqrt{2c/ma}$.

13. $z = (g/\gamma^2)(1 - e^{-\gamma t}) - gt/\gamma$; $\dot{z} \to -g/\gamma$.

14. $8.05\,\text{s}$, $202\,\text{m}$.

15. $\sqrt{gk}\arctan(\sqrt{k/g}\,u)$, $(1/2k)\ln(1 + ku^2/g)$.

16. $\sqrt{g/k}$, $(gk)^{-1/2}\ln(e^{kh} - \sqrt{e^{2kh} - 1})$.

17. $\pm\sqrt{2(g/l)(\cos\theta - \cos\theta_0)}$; $2\pi\sqrt{l/g}$; $\theta = \theta_0\cos(\sqrt{g/l}\,t)$.

18. For $\theta = \pi - \alpha$, $\ddot{\alpha} = (g/l)\alpha$; $0.95\,\text{s}$; $2\pi\,\text{s}^{-1}$.

19. $\sqrt{c/m(a^2 - x^2)}$; $x = a\tanh(\sqrt{c/m}\,at)$.

20. $x = a$; $\pi/\omega$; $\sqrt{a^2 + v^2/4\omega^2} \pm v/2\omega$.

21. $z = (mg/k)[-1 + (1 + \gamma t)e^{-\gamma t}]$, $\gamma = \sqrt{k/m}$; $16\,\text{mm}$.

22. $1.006\,\text{s}$; $5.33°$, $1.17°$.

23. $x = (v/\omega)e^{-\gamma t}\sin\omega t \to vte^{-\gamma t}$.

25. $\omega_1 = \sqrt{\omega_0^2 + \gamma^2} \pm \gamma$.

26. $\bar{E} = \frac{1}{4}ma_1^2(\omega_1^2 + \omega_0^2)$, $W = 2\pi m\gamma\omega_1 a_1^2$.

27. $3$; final velocities: $-6\,\text{m\,s}^{-1}$, $7\,\text{m\,s}^{-1}$, $10\,\text{m\,s}^{-1}$; $T = 364.5\,\text{J}$.

28. $v_n = e^n\sqrt{2gh}$.

29. $a_n = c/m\omega^2 n(1 + n^2)$.

31. $\tau = 1.017 \times 2\pi\sqrt{l/g}$.

32. $G(t) = (e^{-\gamma_- t} - e^{-\gamma_+ t})/m(\gamma_+ - \gamma_-)$, $t > 0$;
$$x = \frac{c}{m}\left[\frac{1}{\gamma_+ - \gamma_-}\left(\frac{1}{\gamma_-^2}e^{-\gamma_- t} - \frac{1}{\gamma_+^2}e^{-\gamma_+ t}\right) - \frac{2\gamma}{\omega_0^4} + \frac{t}{\omega_0^2}\right].$$

## CHAPTER 3

1. (a) $V = -\frac{1}{2}ax^2 - ayz - bxy^2 - \frac{1}{3}bz^3$, (c) $V = -ar^2\sin\theta\sin\varphi$,
   (f) $V = -\frac{1}{2}(\boldsymbol{a}\cdot\boldsymbol{r})^2$.

2. $\frac{1}{2}a + b$.

3. (i) $0$, (ii) $\frac{1}{2}a$.

4. (a) $\pi a^2$, $\boldsymbol{F}$ not conservative; (b) $0$, $\boldsymbol{F}$ may be conservative.

5. $\boldsymbol{F} = c[3(\boldsymbol{k}\cdot\boldsymbol{r})\boldsymbol{r} - r^2\boldsymbol{k}]/r^5$; $F_r = 2c\cos\theta/r^3$, $F_\theta = c\sin\theta/r^3$, $F_\varphi = 0$.

6. $382\,\text{m}$, $883\,\text{m}$; $30°$; $17.7\,\text{s}$, $10.2\,\text{s}$.

7. $z = x\tan\alpha - gx^2/2v^2\cos^2\alpha$; $\alpha = \pi/4 + \beta/2$.

8. $z = wx/u - gx^2/2u^2 - \gamma gx^3/3u^3$; $42.3°$, $823\,\text{m}$.

9. $6.89\,\text{km}$, $7.35\,\text{km}$, $7.18\,\text{km}$.

10. $4\omega$; $m\omega^3 l^4/r^3$; $\Delta T = \frac{3}{2}m\omega^2 l^2$.

11. $v/2$, $v^2 = 4ka^2/3m$; $4ka/3m$, $-5ka/6m$.

12. $\ddot{\theta} = 2F/ma - (g/a)\sin\theta$, $\dot{\theta}^2 = 4F\theta/ma - (2g/a)(1 - \cos\theta)$;
    $F_0 = 0.362\,mg$.

13. $\ddot{\theta} = (1 - \sin\theta)g/3a$, $\dot{\theta}^2 = (\theta - 1 + \cos\theta)2g/3a$; $F = mg(1 + 2\sin\theta)/6$; $\theta = 7\pi/6$; thereafter the two bodies move independently until string tautens.

14. $\dot{u} = (M + m)g\sin\alpha/(M + m\sin^2\alpha)$, $\dot{v} = mg\sin\alpha\cos\alpha/(M + m\sin^2\alpha)$; $\alpha = \arcsin(2/3) = 41.8°$.

15. $z = c^{-2}\sin^2\theta$, $x = c^{-2}(\theta - \frac{1}{2}\sin 2\theta)$.

17. $\cot\theta = \cot\theta_0\cos(\varphi - \varphi_0)$, $(\theta_0, \varphi_0$ constants$)$.

18. $\frac{m}{4}\left[\frac{\xi+\eta}{\xi}\ddot{\xi} - \frac{1}{2}\eta\left(\frac{\dot{\xi}}{\xi} - \frac{\dot{\eta}}{\eta}\right)^2\right] = F_\xi$, $\frac{m}{4}\left[\frac{\xi+\eta}{\eta}\ddot{\eta} - \frac{1}{2}\xi\left(\frac{\dot{\xi}}{\xi} - \frac{\dot{\eta}}{\eta}\right)^2\right] = F_\eta$.

19. (a) and (b): $r = k/(c + g)$, $\theta = 0$, unstable; (a) only: $r = k/(c - g)$, $\theta = \pi$, stable.

20. $T = \frac{1}{2}m\sum_i h_i^2\dot{q}_i^2$, $p_i = mh_i^2\dot{q}_i$, $\boldsymbol{e}_i \cdot \boldsymbol{p} = mh_i\dot{q}_i$.

21. $\ddot{\boldsymbol{r}} = (\ddot{\rho} - \rho\dot{\varphi}^2)\boldsymbol{e}_\rho + (\rho\ddot{\varphi} + 2\dot{\rho}\dot{\varphi})\boldsymbol{e}_\varphi + \ddot{z}\boldsymbol{k} = (\ddot{r} - r\dot{\theta}^2 - r\sin^2\theta\,\dot{\varphi}^2)\boldsymbol{e}_r + (r\ddot{\theta} + 2\dot{r}\dot{\theta} - r\sin\theta\cos\theta\,\dot{\varphi}^2)\boldsymbol{e}_\theta + (r\sin\theta\,\ddot{\varphi} + 2r\cos\theta\,\dot{\theta}\dot{\varphi} + 2\dot{r}\sin\theta\,\dot{\varphi})\boldsymbol{e}_\varphi$.

23. $\partial\boldsymbol{e}_r/\partial\theta = \boldsymbol{e}_\theta$, $\partial\boldsymbol{e}_\theta/\partial\theta = -\boldsymbol{e}_r$, $\partial\boldsymbol{e}_r/\partial\varphi = \boldsymbol{e}_\varphi\sin\theta$, $\partial\boldsymbol{e}_\theta/\partial\varphi = \boldsymbol{e}_\varphi\cos\theta$, $\partial\boldsymbol{e}_\varphi/\partial\varphi = -(\boldsymbol{e}_r\sin\theta + \boldsymbol{e}_\theta\cos\theta)$, others zero.

24. $mc^2[(\cosh^2\lambda - \cos^2\theta)\ddot{\lambda} + \frac{1}{2}\sinh 2\lambda(\dot{\lambda}^2 - \dot{\theta}^2) + \sin 2\theta\,\dot{\lambda}\dot{\theta}] = F_\lambda$, $mc^2[(\cosh^2\lambda - \cos^2\theta)\ddot{\theta} - \frac{1}{2}\sin 2\theta(\dot{\lambda}^2 - \dot{\theta}^2) + \sinh 2\lambda\,\dot{\lambda}\dot{\theta}] = F_\theta$.

25. $y = \lambda + a\cosh[(x - b)/a]$, $a, b, \lambda$ constants.

26. A circle, $x^2 + y^2 - 2by = a^2$, $b$ constant.


## CHAPTER 4

1. $4.22 \times 10^4$ km.

2. $1.61 \times 10^5$ km, $0.176$ AU $= 2.64 \times 10^7$ km.

3. $11.9$ yrs, $13.1$ km s$^{-1}$.

4. $5.46$ yrs.

5. $38.6$ km s$^{-1}$, $7.4$ km s$^{-1}$.

6. $1.62$ m s$^{-2}$, $2.38$ km s$^{-1}$; $25.8$ m s$^{-2}$, $60.2$ km s$^{-1}$.

7. $84.4$ min, $108$ min, $173$ min.

8. $1.0 \times 10^{11}$ $M_S$ (assuming that the mass distribution is spherical — this is the mass inside the radius of the Sun's orbit).

9. $\sqrt{k/m}\,a$; $r^2 = \frac{1}{2}a^2(3 + 2\cos\alpha \pm \sqrt{5 + 4\cos\alpha})$; $r = 2a, a$; $r = a, 0$.

10. $U = J^2/2mr^2 + \frac{1}{2}k(r - a)^2$; $\omega = \sqrt{1/2}\,\omega_0$; $\omega' = \sqrt{5/2}\,\omega_0$; $2.24$ radial oscillations per orbit.

11. $(2/3\pi)$ yrs $= 77.5$ days.

12. Ratio is $1.013$, $2.33$, $134$.

13. $1/r^2 = mE/J^2 + \sqrt{(mE/J^2)^2 - mk/J^2}\cos 2(\theta - \theta_0)$.

14. Hyperbola with origin at the centre.
15. 7.77 days.
16. $8.8\,\mathrm{km\,s^{-1}}$, $5.7\,\mathrm{km\,s^{-1}}$; 97° ahead of Earth; 82° ahead of Jupiter.
17. $4.26R_E$, 18.4°; $38.3\,\mathrm{km\,s^{-1}}$, 4.87 yrs.
19. $GMm/2a$, $-GMm/a$.
20. $x = a(e - \cosh\psi)$, $y = b\sinh\psi$; $r = a(e\cosh\psi - 1)$,
    $t = (abm/J)(e\sinh\psi - \psi)$.
21. $5.7\,\mathrm{km\,s^{-1}}$, opposite to Jupiter's orbital motion; $5.7\,\mathrm{km\,s^{-1}}$;
    $3.9 \times 10^6\,\mathrm{km} = 56R_J$, $23R_J$.
22. $14.3\,\mathrm{km\,s^{-1}}$ at 23.5° to Jupiter's orbital direction; 9.2 AU, 16.2 yrs;
    3.6 AU.
23. $14.3\,\mathrm{km\,s^{-1}}$, in plane normal to Jupiter's orbit, at 23.5° to orbital
    direction; 7.8 AU, 16.2 yrs; 2.5 AU.
24. $\cos\theta = (1 - l/R)/\sqrt{1 - l/a}$; 60°, $6.45\,\mathrm{km\,s^{-1}}$.
25. With $n^2 = |1 + mk/J^2|$, $b^2 = J^2/2m|E|$:
    $J^2 + mk > 0, E > 0 : r\cos n(\theta - \theta_0) = b$;
    $J^2 + mk = 0, E > 0 : r(\theta - \theta_0) = \pm b$;
    $J^2 + mk < 0, E > 0 : r\sinh n(\theta - \theta_0) = \pm b$;
    $J^2 + mk < 0, E = 0 : re^{\pm n\theta} = r_0$;
    $J^2 + mk < 0, E < 0 : r\cosh n(\theta - \theta_0) = b$.
26. $d\sigma/d\Omega = k\pi^2(\pi - \theta)/mv^2\theta^2(2\pi - \theta)^2\sin\theta$.
27. $\omega = \sqrt{(-ka^2 - c)/ma^5}$, $\omega' = \sqrt{(-ka^2 + c)/ma^5}$.
28. 0.123 m, 2.44 m.
29. $1.13 \times 10^{-11}$ m, $8.1 \times 10^3\,\mathrm{s^{-1}}$.
30. $\dot{r} = eJ\sin\theta/ml$, $\ddot{r} = eJ^2\cos\theta/m^2lr^2$, rad. accel. $= -J^2/m^2lr^2$.

CHAPTER 5

1. $2.2\,\mathrm{m\,s^{-2}} = 0.086\,g_J$, $5.1 \times 10^{-3}\,\mathrm{m\,s^{-2}} = 1.9 \times 10^{-5}\,g_S$.
2. $15.3\,\mathrm{s^{-1}}$.
3. 0.20 mm, 78 mm.
4. $465\,\mathrm{m\,s^{-1}}$; (a) $542\,\mathrm{m\,s^{-1}}$, (b) $187\,\mathrm{m\,s^{-1}}$, (c) $743\,\mathrm{m\,s^{-1}}$.
5. (a) 99.88 t wt, (b) 100.29 t wt, (c) 99.46 t wt.
6. $2.53 \times 10^{-3}$ N to south, $1.46 \times 10^{-3}$ N up.
7. $0.013\,\mathrm{mbar\,km^{-1}}$.
8. 0.155°.
9. 12.0 s; 100 kg wt, 20 kg wt; decreasing weight and a Coriolis force of
   79 N.
10. 47.4 mm.

11. $\boldsymbol{F} = m\ddot{\boldsymbol{r}} = q(\boldsymbol{E} + \boldsymbol{v} \wedge \boldsymbol{B})$ with $\boldsymbol{E} = E\boldsymbol{k}$, $\boldsymbol{B} = B\boldsymbol{k}$;
$x = (mv/qB)\sin(qBt/m), y = (mv/qB)[\cos(qBt/m) - 1]$,
$z = qEt^2/2m$; $z = (2mE/qa^2B^2)y^2$; depends only on $m/q$.

12. $l = \pi mv/qB$; $E = 2 \times 10^6 \, \mathrm{V\,m^{-1}}$, $l = 0.089 \, \mathrm{m}$.

13. $\sim 10^5 \, \mathrm{T}$, $1.76 \times 10^{11} \, \mathrm{s^{-1}}$.

14. $5.1 \times 10^{16} \, \mathrm{s^{-1}}$, $-3.3 \times 10^{16} \, \mathrm{s^{-1}}$.

15. $124 \, \mathrm{m}$.

18. $m\ddot{\boldsymbol{r}} = \boldsymbol{F} - m\boldsymbol{a}$.

19. $\begin{bmatrix} x^* \\ y^* \\ z^* \end{bmatrix} = \begin{bmatrix} \cos\omega t & -\sin\omega t & 0 \\ \sin\omega t & \cos\omega t & 0 \\ 0 & 0 & 1 \end{bmatrix} \begin{bmatrix} x \\ y \\ z \end{bmatrix}.$

20. $T = \frac{1}{2}m(\dot{\boldsymbol{r}} + \boldsymbol{\omega} \wedge \boldsymbol{r})^2$.

## CHAPTER 6

1. $\phi = (q/2\pi\epsilon_0 a^2)(\sqrt{a^2 + z^2} - |z|)$, $\boldsymbol{E} = (q\boldsymbol{k}/2\pi\epsilon_0 a^2)\left(\frac{z}{|z|} - \frac{z}{\sqrt{a^2+z^2}}\right)$;
$\boldsymbol{E} \to \boldsymbol{k}(\sigma/2\epsilon_0)(z/|z|)$.

2. $Q = -\frac{1}{2}qa^2$; for $\theta = 0$, $\phi \approx (q/4\pi\epsilon_0)(1/z - a^2/4z^3)$.

3. When $\boldsymbol{d}$ is in same direction as $\boldsymbol{E}$.

4. $\boldsymbol{E} = (3\boldsymbol{d} \cdot \boldsymbol{r}\boldsymbol{r} - r^2\boldsymbol{d})/4\pi\epsilon_0 r^5$, $V = (r^2\boldsymbol{d} \cdot \boldsymbol{d}' - 3\boldsymbol{d} \cdot \boldsymbol{r}\,\boldsymbol{d}' \cdot \boldsymbol{r})/4\pi\epsilon_0 r^5$;
  (a) $\boldsymbol{F} = -\boldsymbol{F}' = -6\boldsymbol{k}(dd'/4\pi\epsilon_0 r^4)$, $\boldsymbol{G} = \boldsymbol{G}' = \boldsymbol{0}$;
  (b) $\boldsymbol{F} = -\boldsymbol{F}' = 3\boldsymbol{k}(dd'/4\pi\epsilon_0 r^4)$, $\boldsymbol{G} = \boldsymbol{G}' = \boldsymbol{0}$;
  (c) $\boldsymbol{F} = -\boldsymbol{F}' = 3\boldsymbol{i}(dd'/4\pi\epsilon_0 r^4)$, $\boldsymbol{G} = -\boldsymbol{j}(dd'/4\pi\epsilon_0 r^3)$, $\boldsymbol{G}' = 2\boldsymbol{G}$;
  (d) $\boldsymbol{F} = \boldsymbol{F}' = \boldsymbol{0}$, $\boldsymbol{G} = -\boldsymbol{G}' = -\boldsymbol{k}(dd'/4\pi\epsilon_0 r^3)$.

5. $V = \frac{1}{2}\sum_{i \neq j}(q_i q_j/4\pi\epsilon_0 r_{ij})$.

6. $\frac{3}{5}(q^2/4\pi\epsilon_0 a)$; $\frac{1}{2}(q^2/4\pi\epsilon_0 a)$.

7. $4.5 \times 10^5 \, \mathrm{C}$, $0.28 \, \mathrm{J\,m^{-2}}$.

8. $8\pi\sigma_0 a^4/5$, $2\pi\sigma_0^2 a^3/25\epsilon_0$.

9. $3qa^2(x^2 - y^2)/4\pi\epsilon_0 r^5$,
$\boldsymbol{E} = (3qa^2/4\pi\epsilon_0 r^7)\big((3x^2 - 7y^2 - 2z^2)x, (7x^2 - 3y^2 + 2z^2)y, 5(x^2 - y^2)z\big)$.

10. $-Gm/r + (Gma^2/r^3)(3\cos^2\theta - 1)$;
$-Gm/a + (Gm/4a^3)(2z^2 - x^2 - y^2)$, $\boldsymbol{g} = (Gm/2a^3)(x, y, -2z)$.

11. $-6Gm/r + (7Gma^4/4r^5)[3 - 5(x^4 + y^4 + z^4)/r^4]$.

12. $-\sqrt{(8\pi G\rho_0 a^3/3)}(r^{-1} - a^{-1})$; $6.7 \times 10^6$ yrs; $14.9$ mins, $29.5$ mins.

13. $4.0 \times 10^{40} \, \mathrm{J}$.

14. $1/9.5$.

15. $3.0$.

16. $79 \, \mathrm{m}$; $78 \, R_{\mathrm{E}}$.

17. $(M_{\mathrm{E}}/M_{\mathrm{M}})^2(R_{\mathrm{M}}/R_{\mathrm{E}})^4 = 35$; $2.8 \, \mathrm{km}$.

18. 12.5 m.

19. $(8\pi/5)\rho_0 d_0 r^4$; 1.13.

22. $q\mu^2 e^{-\mu r}/4\pi\epsilon_0 r$; $-q$.

23. $2\pi G\rho^2 R^2/3 = 1.7 \times 10^{11}$ Pa $= 1.7$ Mbar.

24. $\Phi = -k\rho$; $a = \sqrt{\pi k/4G}$.

25. $-1.43 \times 10^{-6}\,\mathrm{s}^{-1}$; 51 days.

26. 17.9 yrs (should be 18.6 yrs).

## CHAPTER 7

1. 258 days.

2. $7.4 \times 10^5$ km from centre of Sun, *i.e.*, just outside the Sun; $0.28°$.

3. $0.00125\,M_0$; $m_1 \geq 0.00125\,M_0$.

4. $z_1 = l + m_1 vt/M - \frac{1}{2}gt^2 + (m_2 v/M\omega)\sin\omega t$,
   $z_2 = m_1 vt/M - \frac{1}{2}gt^2 - (m_1 v/M\omega)\sin\omega t$, with $\omega = \sqrt{k/\mu}$ and $v < l\omega$.

5. $m_1/m_2 = 1$.

6. 12; 0.071.

7. $62.7°, 55.0°, 640$ keV.

8. $T_1^* = m_2 Q/M, T_2^* = m_1 Q/M$; 3.2 MeV, 0.8 MeV.

9. $\ln 10^6/\ln 2 \approx 20$.

10. $90°$; $45°, 45°$.

11. $2.41b$; $(0.65v, 0.15v, 0), (0.35v, -0.15v, 0)$.

12. $T^*/T = m_2/M$; $\to 1$ or 0.

13. $3 \times 10^{-6}, +450$ km, $+2.4$ min.

15. $a^2 \cos\theta(1/\sin^4\theta + 1/\cos^4\theta)$, where $a = e^2/2\pi\epsilon_0 mv^2$. (The second term comes from recoiling target particles.)

16. $1.8 \times 10^3\,\mathrm{s}^{-1}$, same for both.

17. $2m\ddot{\boldsymbol{R}} = \boldsymbol{0}$, $\frac{1}{2}m\ddot{\boldsymbol{r}} = q\boldsymbol{E} - (q^2/4\pi\epsilon_0 r^3)\boldsymbol{r}$; $z = 2qE/m\omega^2$.

## CHAPTER 8

1. $0.99\,\mathrm{km\,s}^{-1}$, 164 kg.

2. $(2.44 + 1.48 =) 3.91\,\mathrm{km\,s}^{-1}$, 143 kg.

3. $4.74\,\mathrm{km\,s}^{-1}$.

4. 3 stages, $1.48 \times 10^5$ kg.

5. $14.2\,\mathrm{km\,s}^{-1}$, $2.06\,\mathrm{km\,s}^{-1}$, $2.8 \times 10^6$ kg.

6. $\frac{1}{2}M_0 u^2(1 - e^{-v/u})$.

7. 44.6 km, 33.9 km.

8. $10.3\,\mathrm{km\,s}^{-1}$, $(3.07 - 0.07 =) 3.0\,\mathrm{km\,s}^{-1}$; 6.13 t.

9. If $\boldsymbol{u}_1 = (v, 0)$: $(-1,0)v/5, (3, \pm\sqrt{3})v/5, (|\boldsymbol{v}_2| = 2\sqrt{3}v/5)$.

10. $-\rho A v^2$, $A = \pi r^2$; because scattering is isotropic.
11. $\delta a = -2(I/m)\sqrt{(1+e)a^3/GM(1-e)}$.
12. $da/dt = dl/dt = -2\rho A v a/m$.
13. $-20.9\,\text{s}$, $-16.9\,\text{km}$; $-3.37\,\text{s}$, $-2.78\,\text{km}$.
14. (a) $6.2\,\text{h}$, $1.85\,\text{d}$; (b) $3.15\,\text{d}$, $40.4\,\text{d}$.

## CHAPTER 9

1. $4\sqrt{2}a/3$, $(3\sqrt{2g/4a})^{1/2}$.
2. $64\,\text{r.p.m.}$, $5.3 \times 10^{-6}\,\text{J}$ from work done by insect; dissipated to heat.
3. (a) $E, \boldsymbol{P}$; (b) $\boldsymbol{J}$ about leading edge; (c) $E$; $3v/8a, 5/8$;
   $[16(\sqrt{2}-1)ga/3]^{1/2}$.
4. (a) $1.011\,\text{s}$, (b) $1.031\,\text{s}$.
5. $4a/3$; $3bX/4Ma^2$, $3bX/4a$; $b = 4a/3$.
6. (a) $3Mg\cos\varphi$; (b) $(Mg/8)(-9\sin 2\varphi, 11 + 9\cos 2\varphi)$,
   $(3Mg/2)(-\sin 2\varphi, 1 + \cos 2\varphi)$.
7. $9 \times 10^{-6}\,\text{kg m}^2$, $16 \times 10^{-6}\,\text{kg m}^2$, $25 \times 10^{-6}\,\text{kg m}^2$;
   $(1.08, 1.44, 0) \times 10^{-4}\,\text{kg m}^2\,\text{s}^{-1}$; $6.3 \times 10^{-3}\,\text{N}$.
8. (a) $(8,8,2)Ma^2/3$; (b) $(11,11,2)Ma^2/3$.
9. $2M(a^5 - b^5)/5(a^3 - b^3)$.
10. $25.6\,\text{s}$, $1.097 \times 10^3\,\text{J}$.
11. $60°$.
12. $I_1 = I_2 = 3M(a^2 + 4h^2)/20$, $I_3 = 3Ma^2/10$; $1/2$; $Z = 3h/4$,
    $I_1^* = I_2^* = 51Ma^2/320$, $I_3^* = I_3$.
13. $1.55\,\text{s}$.
14. $112\,\text{s}$.
15. $0.244\,\text{s}^{-1}$.
17. $8.83\,\text{m}$.
18. (a) $2.64\,\text{Hz}$ $(\Omega = 16.6\,\text{s}^{-1})$; (b) $3.44\,\text{Hz}$ $(\Omega = 21.6\,\text{s}^{-1})$.
20. $2.50 \times 10^{-12}\,\text{s}^{-1} = 16.3''\,\text{yr}^{-1}$.
21. $22.3\,\text{s}$.

## CHAPTER 10

1. $\pm g/4$.
2. $\sqrt{4mgl/(M + 2m)a^2}$.
3. $Mmg/(M + 2m)$.
4. $M^2mg/k(M + 2m)^2$.
6. $g/7, 3g/7, -5g/7$.
7. $24mg/7, 12mg/7$.

8. $62.62\,\mathrm{s}^{-1}$, $4.347\,\mathrm{s}^{-1}$ ($cf.$ $4.065\,\mathrm{s}^{-1}$); $371.5\,\mathrm{s}^{-1}$ ($3548\,\mathrm{r.p.m.}$).

9. $(M + m\sin^2\theta)l\ddot{\theta} + ml\dot{\theta}^2\cos\theta\sin\theta + (M + m)g\sin\theta = 0$; $1.40\,\mathrm{s}^{-1}$.

10. $I_1\ddot{\varphi} = I_3\omega_3\Omega\sin\lambda\cos\varphi - I_1\Omega^2\sin^2\lambda\sin\varphi\cos\varphi$; $(I_1/I_3)\Omega\sin\lambda$; east and west.

11. $I_1, I_3$ are replaced by $I_1^* < I_1, I_3^* = I_3$; large $\Omega$ is bigger, small $\Omega$ is slightly smaller.

12. $\arcsin(1/\sqrt{3}) = 35.3°$.

14. (a) as at $t = 0$ except that for $l/2 - ct < x < l/2 + ct$, $y = a - 2act/l$; (b) $y = 0$; (c) $y(x, l/c) = -y(x, 0)$.

15. $\ddot{x} - 2\omega\dot{y} - \omega^2 x = -GM_1(x + a_1)/r_1^3 - GM_2(x - a_2)/r_2^3$,
$\ddot{y} + 2\omega\dot{x} - \omega^2 y = -GM_1 y/r_1^3 - GM_2 y/r_2^3$, with
$r_1^2 = (x + a_1)^2 + y^2, r_2^2 = (x - a_2)^2 + y^2, \omega^2 = GM/a^3$.

## CHAPTER 11

1. $x = a(\cos\omega_1 t\cos\omega_2 t + \frac{1}{2}\sqrt{2}\sin\omega_1 t\sin\omega_2 t)$,
$y = a(2\cos\omega_1 t\cos\omega_2 t + \frac{3}{2}\sqrt{2}\sin\omega_1 t\sin\omega_2 t)$,
where $\omega_{1,2} = \frac{1}{2}(\omega_+ \pm \omega_-)$ and $\omega_\pm = \sqrt{(2 \pm \sqrt{2})g/l}$.

2. $\omega^2 = g/l, g/l + k/M + k/m$; $A_X/A_x = 1, -m/M$; $2Ma/(M + m)$; no.

3. $x_0 = 2mg/k, y_0 = 3mg/k$; $\omega^2 = (3 \pm \sqrt{5})k/2m$.

4. $\omega^2 = \omega_0^2, \omega_0^2 + \omega_s^2, \omega_0^2 + 3\omega_s^2$;
$A_x : A_y : A_z = 1 : 1 : 1, \ 1 : 0 : -1, \ 1 : -2 : 1$.

5. $2a/3, a$.

6. (a) $\omega^2 = k/m, 3k/m$; (b) $\omega^2 = (1 - a/l)k/m, 3(1 - a/l)k/m$.

7. $\omega^2 = (M + m)g/ma, g/2a$.

8. $\theta = (\varphi_0/10)(\cos 2\pi t - \cos 3\pi t)$ ($t$ in s); $\varphi_0/5, t = 1\,\mathrm{s}$.

9. $1.0025\,\mathrm{s}, 0.099\,75\,\mathrm{s}$; $0.5025\,\mathrm{mm}$.

10. $(0.401\sin 6.27t - 0.0399\sin 63.0t)\,\mathrm{mm}$, ($t$ in s).

11. $\phi = (q/4\pi\epsilon_0)[14/a + (4x^2 + y^2)/2a^3]$; $\omega_1^2 = q^2/4\pi\epsilon_0 a^3 m, \omega_2^2 = 4\omega_1^2$.

12. $\omega^2 = g/l, g/l, 3g/l$.

13. $A_{1,2} = (F/\sqrt{2m})(\omega_{1,2}^2 - \omega^2 + 2i\gamma_{1,2}\omega)$, with $\omega_1^2 = \omega_0^2, \omega_2^2 = \omega_0^2 + 2\omega_s^2$,
$\gamma_1 = \alpha/2m, \gamma_2 = (\alpha + 2\beta)/2m$; $\alpha > \sqrt{3}k/\omega_0$.

14. $q_{2r} = 0, q_{2r+1} = (-1)^r 4\sqrt{2}la/\pi^2(2r + 1)^2$.

## CHAPTER 12

1. $\omega^2 = g\cos\alpha/r\sin^2\alpha$; $\arcsin(1/\sqrt{3}) = 35.3°$.

2. $\Omega^2 = \omega^2 - g^2/l^2\omega^2$.

3. $H = (p_x - p_y)^2/6m + p_y^2/2m + \frac{1}{2}ky^2 - mgy$;
$y = mg(1 - \cos\omega t)/k$, $x = x_0 - y/4$, $\omega^2 = 4k/3m$.

4. $J^2 = p_\theta^2 + p_\varphi^2 / \sin^2 \theta.$

5. $H = \dfrac{p_\theta^2}{2ml^2} + \dfrac{(lp_x - p_\theta \cos \theta)^2}{2(M + m \sin^2 \theta)l^2} + mgl(1 - \cos \theta).$

6. $H = \dfrac{p_X^2 + p_Y^2}{2M} + \dfrac{p_\theta^2}{2(I_1^* + MR^2 \sin^2 \theta)} + \dfrac{(p_\varphi - p_\psi \cos \theta)^2}{2I_1^* \sin^2 \theta}$

   $+ \dfrac{p_\psi^2}{2I_3^*} + MgR \cos \theta;\ \omega_{3,\min}^2 = 4I_1^* MgR/I_3^2$, reduced by a factor $I_1^*/I_1$.

7. $(p_\varphi - p_\psi z)^2 - 2I_1(1 - z^2)(E - Mgrz - p_\psi^2/2I_3) = 0.$

8. $H = (\mathbf{p} - q\mathbf{A})^2/2m + q\phi.$

9. $H = \dfrac{p_\rho^2}{2m} + \dfrac{p_\varphi^2}{2m\rho^2} - \dfrac{qq'}{4\pi\epsilon_0\rho} + \dfrac{q^2 B^2 \rho^2}{2m}.$

10. $\sqrt{q'm/\pi\epsilon_0 q B^2};\ \sqrt{3}\omega_{\mathrm{L}}.$

11. $0 < b < a,\ a^2/4b.$

12. $M_1/M_2 > \frac{1}{2}(25 + \sqrt{621}) = 24.96.$

13. $\Omega^2 = -p^2 = \frac{1}{2}\omega^2\{1 \pm \sqrt{1 - 27M_1M_2/(M_1 + M_2)^2}\};\ 11.90\,\mathrm{yrs},$
    $147.4\,\mathrm{yrs}.$

14. $\dot{\varphi} = \dfrac{J}{m\rho^2};\ \dfrac{\partial U}{\partial \rho} = -\dfrac{J^2}{m\rho^3} - \dfrac{qJB_z}{m\rho},\ \dfrac{\partial U}{\partial z} = \dfrac{qJB_\rho}{m\rho}.$

## CHAPTER 13

1. (a) $\bar{f} = k^2/4\sigma$; equilibria $x_\pm = (k \pm \sqrt{k^2 - 4f\sigma})/2\sigma$, $x_+$ is
   asymptotically stable. (b) $\dot{x} \leq \bar{f} - f < 0 \implies t_0 \leq k/\sigma(f - \bar{f})$. In
   practice, $\bar{f}$ is not precisely known or constant.

2. Trajectories $\frac{1}{2}\dot{\theta}^2 - (g/l)\cos\theta = $ constant $(= g/l$ on separatrices).

4. $\dot{x} = y$, $\dot{y} = -\omega_0^2 x - \mu y$. Critical point $(0,0)$ which is (i) *asymptotically
   stable spiral*, (ii) (*improper*) *asymptotically stable node*, (iii) (*inflected*)
   *asymptotically stable node*.

5. $(0,0)$ and $(1,0)$ are *asymptotically stable nodes*; $(\frac{1}{2},0)$ is a *saddle*.
   [*Note* that the nodes are local minima of $U(x,y)$.]

6. $f(x) = -c\ln x + dx$ has a single minimum and $f \to +\infty$ when $x \to 0^+$
   and when $x \to +\infty$.

7. $(0,0)$ has $\lambda_{1,2} = 1, \frac{1}{2}$ with eigenvectors $(1,0),(0,1)$: *unstable node*;
   $(0,2)$ has $\lambda_{1,2} = -1, -\frac{1}{2}$ with $(1,3),(0,1)$: *asymptotically stable node*;
   $(1,0)$ has $\lambda_{1,2} = -1, -\frac{1}{4}$ with $(1,0),(1,-\frac{3}{4})$: *asymptotically stable
   node*; $(\frac{1}{2},\frac{1}{2})$ has $\lambda_{1,2} = (-5 \pm \sqrt{57})/16$ with $(-\frac{1}{2},\frac{1}{2} + \lambda_i)$: *saddle*.
   Nearly all initial conditions lead to extinction for one or other species
   in Fig. 13.9 $\implies$ no stable coexistence. Second set of parameters
   $\implies (0,0)$: *unstable node*; $(0,\frac{3}{2})$ and $(\frac{1}{2},0)$: *saddles*; $(\frac{1}{4},1)$:

*asymptotically stable node*; in this case there is asymptotically stable coexistence.

8. $M = \begin{bmatrix} -c_1 & a_2 \\ a_1 & -c_2 \end{bmatrix} \implies \lambda^2 + (c_1 + c_2)\lambda + (c_1 c_2 - a_1 a_2) = 0.$ Then $c_1 c_2 > a_1 a_2 \implies \lambda_1, \lambda_2$ real, negative: *asymptotically stable node,* coexistence in first quadrant; and $c_1 c_2 < a_1 a_2 \implies \lambda_1, \lambda_2$ real, opposite sign: *saddle* in third quadrant and trajectories run away $\to +\infty$.

9. Critical points at $(0,0)$, $\left[ \frac{\mu(R-1)}{(\mu R + a)}, \frac{a(R-1)}{R(a+\mu)} \right]$. $R < 1$: *asymptotically stable node, saddle.* $R > 1$: *saddle, asymptotically stable node.* Disease maintains itself only when $R > 1$.

10. At $(0,0)$, $\lambda_{1,2} = (\epsilon \pm \sqrt{\epsilon^2 - 4})/2$; *unstable spiral* $(\epsilon < 2)$ or *node* $(\epsilon \geq 2)$.

11. At $(0,0)$, $\lambda_{1,2} = \alpha\beta \pm i$, so we have an *asymptotically stable spiral* $(\beta < 0)$, *unstable spiral* $(\beta > 0)$ or *centre* $(\beta = 0)$. In polars, $\dot{\theta} = 1$ and $\dot{r} = \alpha r(\beta - r^2)$; for $\beta > 0$, $r \to \sqrt{\beta}$ as $t$ and $\theta \to \infty$; when $\beta < 0$, $r \to 0$.

12. (a) $J_1 \dot{J}_1 + J_2 \dot{J}_2 + J_3 \dot{J}_3 = 0$; then integrate to get $|\mathbf{J}| =$ constant. (b) $(\pm J, 0, 0), (0, 0, \pm J)$ are *centres* on the sphere; $(0, \pm J, 0)$ are *saddles.* (c) $\dot{J}_3 = 0 \implies J_3 = I_3 \Omega$. $J_{1,2}$ satisfy $\ddot{J}_i + [(I_1 - I_3)/I_1]^2 \Omega^2 J_i = 0$. (d) Critical point $(0,0,0)$, with $\lambda_{1,2,3} = -|\mu|/I_{1,2,3}$ and eigenvectors $(1,0,0), (0,1,0), (0,0,1)$. As $\omega \to 0$, $\boldsymbol{\omega} \to$ rotation about axis with the largest moment of inertia, $I_3$.

13. (a) $\mathrm{d}N/\mathrm{d}t \equiv 0$ and $\mathrm{d}I/\mathrm{d}S = \dot{I}/\dot{S} = -1 + b/aS$. (b) $\mathrm{d}I/\mathrm{d}S = 0$ when $S = b/a$ and $\mathrm{d}^2 I/\mathrm{d}S^2 = -b/aS^2 < 0$, so maximum. $I = 0$ at $S = S_-, S_+$. (c) $S_0 = b/a + \delta$, $I_0 = \epsilon$. Write $S = b/a + \xi$ and expand $\implies$ trajectory passes through $(b/a - \delta, \epsilon)$. Since $S_0$ can be made arbitrarily close to $S_-$, so $S_+$ is arbitrarily close to $b/a - \delta$ and these susceptibles escape infection.

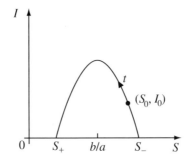

14. For a critical point $(X, Y, Z)$, the matrix $M = \begin{bmatrix} -\sigma & \sigma & 0 \\ \rho - Z & -1 & -X \\ Y & X & -\beta \end{bmatrix}$.

(a) $\lambda_1 = -\beta < 0$, $\lambda_{2,3} = -\frac{1}{2}(\sigma + 1) \pm \frac{1}{2}\sqrt{(\sigma + 1)^2 - 4(1 - \rho)\sigma}$, so that $\lambda_{2,3}$ are real and are both negative only when $0 < \rho < 1$.

(c) $\rho = 1 \implies$ cubic becomes $\lambda(\lambda + \beta)(\lambda + \sigma + 1) = 0$. If the cubic has the form $(\lambda + \mu)(\lambda^2 + \nu^2) = 0$ then $\mu = \sigma + \beta + 1$, $\nu^2 = \beta(\sigma + \rho)$, $\mu\nu^2 = 2\sigma\beta(\rho - 1) \implies$ result for $\rho_{\text{crit}}$. (d) Write $\rho = 1 + \epsilon$ (with $\sigma, \beta$

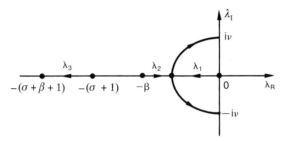

constant, $\epsilon$ small) and find changes in eigenvalues by putting $\lambda = \lambda_i + \xi(\epsilon)$ and performing a linear analysis of the cubic in (b).
(e) Evidently $RHS = 0$ is an ellipsoid and the distance of $(x, y, z)$ from $\bar{O} \equiv (0, 0, \rho + \sigma)$ decreases where $RHS < 0$, i.e., outside the ellipsoid and so *a fortiori* outside any sphere centred at $\bar{O}$ which contains it.

15. (b) For a critical point $(X_1, X_2, Y)$, the matrix
$$M = \begin{bmatrix} -\mu & Y & X_2 \\ Y - A & -\mu & X_1 \\ -X_2 & -X_1 & 0 \end{bmatrix}.$$
(c) $\nabla \cdot (\dot{x}) = -2\mu < 0$. (d) $X_1^2 + X_2^2 + \bar{Y}^2 = C$, where $C > 0$, is an ellipsoid (oblate). Trajectories move towards smaller $C$ values whenever $(X_1, X_2, \bar{Y})$ is below the paraboloid $\bar{Y} = \mu(X_1^2 + X_2^2)/\sqrt{2}$.

16. $\Delta x = \Delta x_0 \cos \omega t + (\Delta y_0 / \omega) \sin \omega t$, $\Delta y = -\omega \Delta x_0 \sin \omega t + \Delta y_0 \cos \omega t$. Each is bounded and so is $\sqrt{(\Delta x)^2 + (\Delta y)^2}$.

17. Trajectories in the $xy$-plane are parabolic arcs in $x \geq 0$. The higher the energy the longer the period of oscillation (between successive bounces). Between bounces, $\Delta x = \Delta x_0 + \Delta y_0 t$, $\Delta y = \Delta y_0$, so that $\sqrt{(\Delta x)^2 + (\Delta y)^2} \sim \kappa t$ for $t$ increasing, and bounces don't affect this result.

18. Evidently, we have $(1)$, $(\frac{1}{2}, \frac{1}{2})$, $(\frac{1}{4}, \frac{1}{2}, \frac{1}{4})$, etc. in successive rows, i.e., $(\frac{1}{2})^n \times$ (row of Pascal's triangle). When $n = 16$, we have

$(\frac{1}{2})^{16}[1,16,120,560,1820,4368,8008,11440, 12870,11440,8008,4368,$
$1820,560,120,16,1].$

## CHAPTER 14

1. Return trajectories have $-|k|/R \le E < 0$.
   $t_0 = \sqrt{2h^3 m/|k|}\left[\pi/2 - \arcsin(\sqrt{R/h}) + \sqrt{(R/h)(1 - R/h)}\right] \to$
   $\sqrt{h^3 m/2|k|}\,\pi$ as $R/h \to 0$, *i.e.*, Kepler's third law for a flat ellipse,
   major axis $h$ in this limit.

2. Explicitly, $[F,H] = 0$ in each case. In (c), the vector $\boldsymbol{A}$ may be
   considered using Cartesians or polars. (Then the orientation of the
   orbit may be specified using $\alpha = \arctan(A_y/A_x)$ and
   $|\boldsymbol{A}|^2 = 1 - (2J^2|E|/mk^2)$, consistent with (4.30).)

3. $H\ (\equiv E)$ and $p_\theta$ are constants of the motion. Steady $r = r_0$ when
   $p_\theta = mr_0^2\Omega_0$ with $\Omega_0 = \pm\sqrt{g/r_0}$. Small oscillations $r = r_0 + \Delta \implies$
   SHM for $\Delta$ with frequency $\sqrt{3g/2r_0}$.

4. $\omega = \sqrt{g\cos\alpha/r_0\sin^2\alpha}$, rotation rate at $r_0$. Small oscillations
   $r = r_0 + \Delta \implies$ SHM for $\Delta$ with frequency $\varpi \equiv \sqrt{3}(\omega\sin\alpha)$. Closure
   for rational $\sqrt{3}\sin\alpha$.

5. Effective potential $p_\theta^2/2mr^2 - |k|/r \implies$ minimum $-k^2 m/2p_\theta^2$ when
   $r = r_0 = p_\theta^2/m|k|$, circle; $(r, p_r)$ curves closed (around $r = r_0$) when
   $E < 0$, corresponding to $(r, \theta)$ ellipses; $E \ge 0$ curves stretch to
   $r \to \infty$, $(r, \theta)$ hyperbolae, parabola. We always have closure for this
   system — given $E$, Poincaré section is a single point on an $(r, p_r)$
   curve, a different point for each choice of $\theta$ section.

6. $H \equiv E$ here. Action $I = 2E\sqrt{l/g} \implies E = \frac{1}{2}\sqrt{g/l}\,I$, and frequency
   $\omega = \partial H/\partial I = \frac{1}{2}\sqrt{g/l}$.

7. Oscillation between $x = \pm[l + \sqrt{2E/k}]$. Action
   $I = (2l/\pi)\sqrt{2mE} + E/\Omega \implies \sqrt{E} = \sqrt{I\Omega + \beta^2} - \beta$.
   $\phi(x) = (\partial/\partial I)\int^x p\,dx$ with $\phi = \omega(I)t + \text{constant}$,
   $\omega = \Omega\sqrt{E}/(\sqrt{E} + \beta)$, $T = 2\pi/\omega$. Small $E \implies T \to \infty$ as $E^{-1/2}$;
   large $E \implies T \to 2\pi/\Omega$.

8. $I = \frac{1}{2}\sqrt{mk}[E/k - \sqrt{\lambda/k}] \implies E(I)$; $\phi(q) = (\partial/\partial I)\int^q p\,dq \equiv \omega t + \beta$,
   with $\omega = \partial E/\partial I = 2\sqrt{k/m}$. Evaluating $\phi(q)$ gives
   $q(t) = \left[\frac{E}{k} + \frac{\sqrt{E^2 - \lambda k}}{k}\sin(\omega t + \beta)\right]^{1/2}$. (*Note.* For the isotropic
   oscillator there are two radial oscillations for each complete elliptical
   trajectory.)

9. Action $I_1 = (\sqrt{-2mE}/\pi) \int_{r_1}^{r_2} \sqrt{(r_2 - r)(r - r_1)}(dr/r)$, where $r_1 + r_2 = -|k|/E$ and $r_1 r_2 = -I_2^2/2mE \implies$ result. So $E(I_1, I_2)$ and $\omega_1 = \partial E/\partial I_1 \equiv \omega_2$ (see (14.18)).

10. Consider the lines from a bounce point to the foci and the angles they make with the trajectory just before and just after the bounce. $\Lambda = [(x + ae)p_y - p_x y][(x - ae)p_y - p_x y]$. Change to $\lambda, \theta$ variables; $H$ from Chapter 3, Problem 24 using $p_\lambda = mc^2(\cosh^2 \lambda - \cos^2 \theta)\dot{\lambda}$, $p_\theta = mc^2(\cosh^2 \lambda - \cos^2 \theta)\dot{\theta}$. Turning value for $\lambda$ is when $p_\lambda = 0 \implies$ tangency condition.

11. (a) Motion within curve $(E/mg\mu r) = 1 - (1/\mu)\cos\theta$ (ellipse).
    (b) $\mu = 1$: bounding curve $E = 2mgr\sin^2(\theta/2)$; $\mu < 1$: curve as in (a) — hyperbola for $E \neq 0$, two straight lines for $E = 0$.

12. Hamilton's equations: $\dot{x} = p_x + \omega y$, $\dot{p}_x = \omega p_y - \omega^2 x + \partial U/\partial x$, $\dot{y} = p_y - \omega x$, $\dot{p}_y = -\omega p_x - \omega^2 y + \partial U/\partial y$ and $H = \frac{1}{2}(p_x + \omega y)^2 + \frac{1}{2}(p_y - \omega x)^2 - U$. System autonomous, $H = $ constant; $U \geq C$ for possible motion and for the Earth/Moon system the critical $C$ is that corresponding to the 'equilibrium' point between Earth and Moon.

13. Action $I = E/\omega$, so $E \propto \omega \propto l^{-1/2}$. Maximum sideways displacement $= l\theta_{max} \propto l^{1/4}$. Maximum acceleration $|\omega^2 l\theta_{max}| \propto l^{-3/4}$.

14. Action $I \propto L\sqrt{H}$ and $H \propto v^2 \implies v \propto 1/L$. Temperature $\propto 1/L^2 \implies$ pressure $\propto 1/L^5 \implies$ pressure $\propto$ (density)$^{5/3}$.

15. Action $I = (2m/3\pi)\sqrt{2gq_0^3 \sin\alpha}$ and $E = mgq_0 \sin\alpha \implies$ result. Frequency $\omega = 2\pi\sqrt{g\sin\alpha/8q_0}$. Evidently, $I = $ constant $\implies$ $E \propto (\sin\alpha)^{2/3}$, $q_0 \propto (\sin\alpha)^{-1/3}$, period $2\pi/\omega \propto (\sin\alpha)^{-2/3}$. So $E_2/E_1 = 0.69$, $q_{02}/q_{01} = 1.20$, $\omega_1/\omega_2 = 1.44$.

16. Evidently $E \propto k^2$ and $\tau \propto k^{-2}$. Since $\tau \propto \sqrt{a^3/|k|}$ (see (4.31)), $a \propto 1/|k| \implies k$ decreases, $a$ increases. Since eccentricity $e = \sqrt{1 + 2EI_2^2/mk^2}$ (see (4.30)), $e$ remains constant.

## APPENDIX A

1. (a) $1, 2, 3$; (b) $(-3, -3, 3), (-3, 7, 8), (-6, 4, 11)$; (c) $-15, 15$;
   (d) $(-3, 3, 0), (-1, 3, -3)$; (e) as (d).
2. $164.2°, 16.2°$.
4. $(3x^2 - yz, -xz, -xy)$; $6x$.
5. (a) $(1 - y^2/x^2, 2y/x, 0), (2x/y, 1 - x^2/y^2, 0)$; (b) circles passing through the origin with centres on $x$- or $y$-axes, respectively, intersecting at right angles.

8. $a(\nabla \cdot b) + (b \cdot \nabla)a - b(\nabla \cdot a) - (a \cdot \nabla)b$.

9. $2\omega k$.

11. $(\nabla^2 A)_\rho = \nabla^2(A_\rho) - \dfrac{1}{\rho^2}\left(A_\rho + 2\dfrac{\partial A_\varphi}{\partial \varphi}\right)$,

$(\nabla^2 A)_\varphi = \nabla^2(A_\varphi) - \dfrac{1}{\rho^2}\left(A_\varphi - 2\dfrac{\partial A_\rho}{\partial \varphi}\right)$, $(\nabla^2 A)_z = \nabla^2(A_z)$.

12. $A_r = A_\theta = 0$, $A_\varphi = \mu_0 \mu \sin\theta/4\pi r^2$;
$B_r = \mu_0 \mu \cos\theta/2\pi r^3$, $B_\theta = \mu_0 \mu \sin\theta/4\pi r^3$, $B_\varphi = 0$;
$A = (\mu_0/4\pi r^3)\mu \wedge r$; $B = (\mu_0/4\pi r^5)(3\mu \cdot rr - r^2\mu)$.

13. $B = (\mu_0 I/4\pi r^3)ds \wedge r$; $F = (\mu_0 II'/4\pi r^3)ds' \wedge (ds \wedge r)$;
$F + F' = (\mu_0 II'/4\pi r^3)r \wedge (ds \wedge ds') \neq 0$.

14. $A_r = r^{-1}\cos\theta$, $A_\theta = r^{-1}\ln r \sin\theta$, $A_\varphi = 0$; $r^{-2}(1 + 2\ln r)\cos\theta$;
$(0, 0, 2r^{-2}\sin\theta)$.

15. $\begin{bmatrix} \cos\theta & 0 & \sin\theta \\ 0 & 1 & 0 \\ -\sin\theta & 0 & \cos\theta \end{bmatrix}$.

17. $Q = \iiint \rho(r')(3r'r' - r'^2 1)d^3r'$.

## APPENDIX B

1. Distances $r_{1,2}$ are given by
$r_{1,2}^2 = (a\cos\psi \mp ae)^2 + (b\sin\psi)^2 = a^2(1 \mp e\cos\psi)^2$.

2. Tangent vector is $t = (-a\sin\psi, b\cos\psi)$, unit vectors from two foci are
$n_{1,2} = (1 \mp e\cos\psi)^{-1}(\cos\psi \mp e, (b/a)\sin\psi)$, scalar products are
$t \cdot n_{1,2} = \pm ae\sin\psi$.

## APPENDIX C

1. (a) $(0,0)$, $\lambda_{1,2} = -\frac{5}{2} \pm \frac{1}{2}\sqrt{17}$, eigenvectors $(1, 3 + \lambda_i)$, *asymptotically stable node*.

(b) $(0,0)$, $\lambda_{1,2} = 4, 1$, eigenvectors $(1, \lambda_i - 3)$, *unstable node*.

(c) $(-1,-1)$, $\lambda_{1,2} = -5 \pm \sqrt{5}$, eigenvectors $(-2, \lambda_i + 8)$, *asymptotically stable node*;
$(4,4)$, $\lambda_{1,2} = 5 \pm i\sqrt{55}$, *unstable spiral*.

(d) $(0,2)$, $\lambda_{1,2} = -1 \pm i$, *asymptotically stable spiral*;
$(1,0)$ $\lambda_{1,2} = 1, -2$, eigenvectors $(-1, \lambda_i + 2)$, *saddle* ($\therefore$ *unstable*).

(e) $(0,2)$, $\lambda_{1,2} = \pm i2\sqrt{6}$, *centre* ($\therefore$ *stable*);
$(0,-2)$, $\lambda_{1,2} = \pm i2\sqrt{6}$, *centre* ($\therefore$ *stable*);

$(1,0)$, $\lambda_{1,2} = -8, 3$, eigenvectors $(1,0), (0,1)$, resp., *saddle* ($\therefore$ unstable);

$(-1,0)$, $\lambda_{1,2} = 8, -3$, eigenvectors $(1,0), (0,1)$, resp., *saddle* ($\therefore$ unstable).

(f) $(0, n\pi)$, $n$ odd: $\lambda_{1,2} = \pm i$, *centres* ($\therefore$ stable);

$n$ even: $\lambda_{1,2} = \pm 1$, eigenvectors $(1,1), (1,-1)$, resp., *saddles* ($\therefore$ unstable).

(g) $(0,0)$, $\lambda_{1,2} = \pm\sqrt{\omega^2 - \alpha}$, so $\omega^2 > \alpha$, *saddle* ($\therefore$ unstable) and $\omega^2 < \alpha$, *centre* ($\therefore$ stable).

(*Note*: In each case, consider local sketches near the critical points and how these build into the global phase portrait.)

2. $\dot{x} = y$, $\dot{y} = x^3 - x \implies (0,0)$, *centre*, $(\pm 1, 0)$, *saddles*;

(a) $dy/dx = (x^3 - x)/y \implies \frac{1}{2}y^2 = \frac{1}{4}x^4 - \frac{1}{2}x^2 + c$, *i.e.*,

$y = \pm\sqrt{\frac{1}{2}x^4 - x^2 + 2c}$; symmetry about $x$ axis rules out spirals;

(b) separatrices given by $c = \frac{1}{4}$.

## APPENDIX D

1. (b) $\frac{d}{dx}F^{(2)}(x) = F'(x)F'(F(x)) \implies$ asymptotically stable when $|F'(X_1)F'(X_2)| < 1 \implies |-r^2 + 2r + 4| < 1 \implies 3 < r < 1 + \sqrt{6}$ (if $r \geq 0$).

(c) $a = 1 - 1/r, b = 2/r - 1, s = 2 - r$.

(d) Applying (c) to (a), (b), etc, yields asymptotically stable points/cycles:

| $s = 2 - r$ | $Y$ | $r$ | $X$ |
|---|---|---|---|
| $0 \to 1$ | $0$ | $2 \to 1$ | $1 - 1/r$ |
| $1 \to 3$ | $1 - 1/s$ | $1 \to -1$ | $0$ |
| $3 \to 1 + \sqrt{6}$ | $(Y_1(s), Y_2(s))$ | $-1 \to 1 - \sqrt{6}$ | $(X_1(r), X_2(r))$ |
| $3.57$ | $2^\infty$ accum. | $-1.57$ | $2^\infty$ accum. |
| $4$ | range limit | $-2$ | range limit |

2. $r = 1 \pm \sqrt{1 + 4a}$, $\beta = -2\alpha = r/a$.

3. (a) $X = 0$ asymptotically stable in circle $|r| < 1$, $X = 1 - 1/r$ asymptotically stable in circle $|2 - r| < 1$.

(b) asymptotically stable 2-cycle when $|-r^2 + 2r + 4| < 1$, or $|r - 1 - \sqrt{5}|.|r - 1 + \sqrt{5}| < 1$ (two disjoint ovals).

*Note*: If we continue in this way, adding the regions of the complex $r$ plane in which there are cycles of any finite length we arrive at the figure shown:

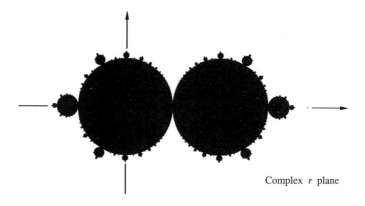

Complex $r$ plane

(c) In the complex $a$ plane, the circles become a heart-shaped region with a cusp at $a = -\frac{1}{4}$; the 2-cycles yield a circle $|a - 1| = \frac{1}{4}$. The entire set yields the Mandelbrot 'signature' (see Peitgen and Richter, *The Beauty of Fractals*, Springer, 1986):

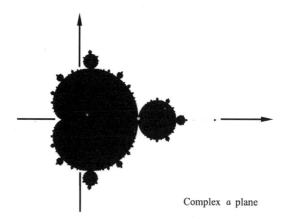

Complex $a$ plane

4. Fixed points at $X = 0$, $X = \frac{2}{3}$ have $|F'(X)| = 2 \implies$ instability. Uncertainty $\epsilon_n = \epsilon_0 2^n \geq 1$ when $n \geq \ln(1/\epsilon_0)/\ln 2$. *Note*: $\epsilon_n = \epsilon_0 2^n$ $\implies$ Lyapunov exponent $\lambda = \ln 2 > 0$.

5. $X = 0$ asymptotically stable if $|a| < 1$. $X = \pm\sqrt{a - 1}$ asymptotically stable when $|3 - 2a| < 1$, or $1 < a < 2$. At $a = 2$, there is period-doubling on each branch, followed by a period-doubling cascade.

7. (a) Both $|\lambda_1|, |\lambda_2| < 1 \implies |b| < 1$ and also $|2aX| < 1 - b$. $(X_-, Y_-)$ is always unstable; $(X_+, Y_+)$ is asymptotically stable if $a < \frac{3}{4}(1-b)^2$.
(b) Nontrivial 2-cycle $\implies X_i$ roots of
$a^2 X^2 - a(1-b)X + (1-b)^2 - a = 0$, with $Y = b(1 - aX^2)/(1-b)$; real roots if $a > \frac{3}{4}(1-b)^2$. Eigenvalues $\lambda$ of $M_1 M_2$ satisfy
$\lambda^2 - (4a^2 X_1 X_2 + 2b)\lambda + b^2 = 0 \implies$ for asymptotic stability $|b| < 1$
and $|4(1-b)^2 - 4a + 2b| < 1 + b^2 \implies a < (1-b)^2 + \frac{1}{4}(1+b)^2$ [using an argument similar to that in the Hint].

8. $x_n = X + \xi$, $y_n = Y + \eta \implies x_{n+1} = 1 - aX^2 + Y + \bar{\xi}$, $y_{n+1} = bX + \bar{\eta}$,
with $\bar{\xi} = -2aX\xi + \eta$, $\bar{\eta} = b\xi$. Circle $\xi^2 + \eta^2 = \epsilon^2$ becomes
$[\bar{\xi} + (2aX/b)\bar{\eta}]^2 + \bar{\eta}^2/b^2 = \epsilon^2$, ellipse with semi-axes $\epsilon/\sqrt{\mu_1}$, $\epsilon/\sqrt{\mu_2}$
where $\mu_1 \mu_2 = 1/b^2$, $\mu_1 + \mu_2 = 1 + (1 + 4a^2 X^2)/b^2$. Area of ellipse
$= \pi |b| \epsilon^2 \implies$ area reduction. $(1 - \mu_1)(1 - \mu_2) < 0 \implies$
$0 < \mu_1 < 1 < \mu_2$ (say) $\implies$ Lyapunov exponents $\lambda_1 > 0 > \lambda_2$.

9. (a) Fixed points $P_1 = (1 + a - b)^{-1}(1, b)$ if $a > b - 1$,
$P_2 = (1 - a - b)^{-1}(1, b)$ if $a > 1 - b$. If $|b| < 1$ and $|a| < 1 - b$ only $P_1$
exists, $\lambda - b/\lambda = -a \implies P_1$ asymptotically stable.
(b) For nontrivial 2-cycle $Q_1 = (X_1, Y_1) \rightleftarrows Q_2 = (X_2, Y_2)$, $X_1, X_2$ must
be of opposite sign. Hence $X_{1,2} = (1 - b \mp a)/[a^2 + (1-b)^2]$,
$Y_{1,2} = bX_{2,1}$.

11. (a) $\begin{pmatrix} x \\ y \end{pmatrix} = \begin{pmatrix} X + \xi \\ Y + \eta \end{pmatrix} \implies \begin{pmatrix} \xi_{n+1} \\ \eta_{n+1} \end{pmatrix} = \begin{pmatrix} 1 & 1 \\ 1 & 2 \end{pmatrix} \begin{pmatrix} \xi_n \\ \eta_n \end{pmatrix} \implies$ eigenvalues
$\mu_{1,2} = \frac{3}{2} \pm \frac{1}{2}\sqrt{5} \implies$ Lyapunov exponents $\lambda_{1,2} = \ln \mu_{1,2} = \pm \ln 2.618 =$
$\pm 0.9624$ — stretch and squeeze. See diagram:

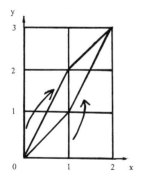

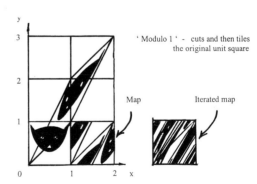

'Modulo 1' - cuts and then tiles the original unit square

Map            Iterated map

(b) eigenvalues $\mu_1 = 2$, $\mu_2 = \frac{1}{2} \implies \lambda_{1,2} = \pm \ln 2 = \pm 0.6931$. See diagram:

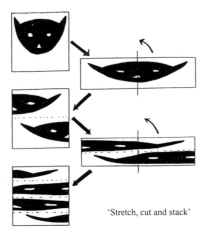

'Stretch, cut and stack'

12. (a) $\alpha, \beta$ distinct irrationals with $\beta - \alpha > \epsilon > 0$, then choose integer $N$
    such that $\epsilon > 1/N$; mesh integer multiples of $1/N$, at least one is
    between $\alpha$ and $\beta$. (b) Since $\sqrt{2}$ is irrational, then $e.g.$,
    $(1 - 1/\sqrt{2})(p_1/q_1) + (1/\sqrt{2})(p_2/q_2)$ is irrational and lies between the
    rationals $p_1/q_1$ and $p_2/q_2$.

13. We may take both $\theta$ $and$ $I$ to be $2\pi$-periodic here. See diagram:

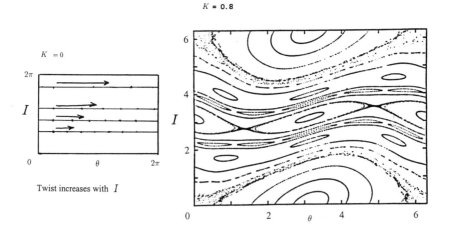

[from Reinhardt and Dana, *Proc. Roy. Soc.* **413**, 157–170, 1987]

# Bibliography

The following is a short list of selected books which may be found helpful for further reading on classical mechanics and related subjects.

Abraham, R. and Marsden, J.E.: *Foundations of Mechanics*, 2nd ed., Benjamin/Cummings, 1994.

Abraham, R.H. and Shaw, C.D.: *Dynamics — the Geometry of Behavior*, 2nd ed., Addison–Wesley, 1992 [originally published by Aerial Press].

Arnol'd, V.I.: *Mathematical Methods of Classical Mechanics*, 2nd ed., translated by K. Vogtmann and A. Weinstein, Springer–Verlag, 1995.

Arnol'd, V.I., Kozlov, V.V. and Neishtadt, A.I.: *Mathematical Aspects of Classical and Celestial Mechanics*, 2nd ed., Springer, 1996.

Baker, G.L. and Gollub, J.P.: *Chaotic Dynamics — an Introduction*, 2nd ed., Cambridge University Press, 1996.

Chandrasekhar, S.: *Newton's "Principia" for the Common Reader*, Clarendon, 1995.

Cvitanovic, P.: *Universality in Chaos — a reprint selection*, 2nd ed., Adam Hilger/IOP Publishing, 1989.

Fetter, A.L. and Walecka, J.D.: *Theoretical Mechanics of Particles and Continua*, McGraw–Hill, 1980.

Feynman, R.P.: *Lectures on Physics*, vol. I, Addison–Wesley, 1963.

Goldstein, H., Poole, C.P. and Safko, J.L.: *Classical Mechanics*, 3rd ed., Addison–Wesley, 2002.

Guckenheimer, J. and Holmes, P.: *Nonlinear Oscillations, Dynamical Systems, and Bifurcations of Vector Fields*, Springer–Verlag, 1983.

Gutzwiller, M.C.: *Chaos in Classical and Quantum Mechanics*, Springer–Verlag, 1990.

Hofbauer, J. and Sigmund, K.: *Evolutionary Games and Population Dynamics*, Cambridge University Press, 1998.

Hunter, S.C.: *Mechanics of Continuous Media*, 2nd ed., Ellis Horwood, 1983.

Jackson, E.A.: *Perspectives of Nonlinear Dynamics*, Cambridge University Press, 1989.

Jammer, M.: *Concepts of Force: A Study in the Foundations of Dynamics*, Dover, 1999.

Jammer, M.: *Concepts of Mass in Classical Mechanics*, Dover, 1997.

Landau, L.D. and Lifshitz, E.M.: *Mechanics* (vol. 1 of *Course of Theoretical Physics*), 3rd ed., Butterworth–Heinemann, 1982.

Lichtenberg, A.J. and Lieberman, M.A.: *Regular and Stochastic Motion*, Springer–Verlag, 1983.

Lorenz, E.N.: *The Essence of Chaos*, UCL Press/University of Washington Press, 1993.

May, R.M.: *Stability and Complexity in Model Ecosystems*, 2nd ed., Princeton University Press, 1974 and 2001.

Murray, C.D. and Dermott, S.F.: *Solar System Dynamics*, Cambridge University Press, 1999.

Ott, E.: *Chaos in Dynamical Systems*, 2nd ed., Cambridge University Press, 2002.

Park, D.: *Classical Dynamics and its Quantum Analogues*, 2nd ed., Springer, 1990.

Pars, L.A.: *A Treatise on Analytical Dynamics*, Heinemann, 1965 [reprinted by Oxbow Press].

Peitgen, H.-O., Jürgens, H. and Saupe, D.: *Chaos and Fractals*, Springer–Verlag, 1992.

Percival, I. and Richards, D.: *Introduction to Dynamics*, Cambridge University Press, 1982.

# Index